1,001 PROBLEMS
TO MASTER ALGEBRA

1,001 PROBLEMS TO MASTER ALGEBRA

Second Edition

LEARNINGEXPRESS®

NEW YORK

Copyright © 2015 LearningExpress.

All rights reserved under International and Pan American Copyright Conventions.
Published in the United States by LearningExpress, New York.

Cataloging-in-Publication Data is on file with the Library of Congress.

Printed in the United States of America

9 8 7 6 5 4 3 2 1

Second Edition

ISBN: 978-1-61103-027-3

For information on LearningExpress products, or bulk sales, please write to us at:
    80 Broad Street
    4th Floor
    New York, NY 10004

ABOUT THE AUTHOR ▶

Dr. Mark McKibben is currently a tenured associate professor of mathematics and computer science at Goucher College in Baltimore, Maryland. He earned his PhD in mathematics from Ohio University in 1999, where his area of study was nonlinear analysis and differential equations. His dedication to undergraduate mathematics education prompted him to write textbooks and more than 20 supplements for courses on algebra, statistics, trigonometry, pre-calculus, and calculus. He is an active research mathematician who has published more than 25 original research articles as well as a recent book entitled *Discovering Evolution Equations with Applications Volume 1: Deterministic Equations*, published by CRC Press/Chapman-Hall.

CONTENTS ▶

INTRODUCTION ▶

Many of the questions you ask in everyday life, such as "How many song downloads can I buy with a certain amount of money?" or "What percentage reduction in price would lower the cost of a particular shirt to $20?" are solved using algebra. Although you might not have realized it, you've been doing algebra for quite some time!

The set of rules and techniques that has come to be known as *algebra* revolves around finding values of some unknown quantity that, when used, make a given mathematical statement true. Such a value might represent the length of the side of a fence, the number of minutes a jogger needs to run in order to catch the nearest opponent, or the original cost of an item. Mastery of the rules and techniques embodied in the problem sets in this book will arm you with the tools necessary to attack applied problems accurately and with ease.

How to Use This Book

This book has been designed to provide you with a collection of problems to assist you in reviewing the basic techniques of algebra. It has been written with several audiences in mind. If you have taken an algebra course and need to refresh skills that have become a bit rusty—this book is for you. Instructors teaching an algebra course might find this repository of problems to be a useful supplement to their own problem sets. Teachers and tutors might use the problems in this book in help sessions. Or, if you are a student taking algebra for the first time, this book will provide you with some extra practice. Whatever your background or reason for picking up this book, we hope that you will find it to be a useful resource in your journey through algebra!

1 ▶ PRE-ALGEBRA FUNDAMENTALS

The basic arithmetic properties of whole numbers, integers, exponential expressions, fractions, and decimals are fundamental building blocks of algebra. In fact, the properties used to simplify algebraic expressions later in the text coincide with the rudimentary properties exhibited by these number systems. As such, it is time well spent to first gain familiarity with them and to then determine how to adapt them to a setting in which variables are involved. These properties are reviewed in the first five problem sets in this section. Translating verbal statements into mathematical ones and learning to deal with elementary algebraic expressions involving variables are the focus of the remaining four problem sets in this section.

Set 1 (Answers begin on page 153)

The arithmetic properties of the set of whole numbers are reviewed in this set.

1. $(15 + 32)(56 - 39) =$
   a. 142
   b. 799
   c. 4,465
   d. 30

2. What is the value of $65,715 \div 4$ rounded to the nearest thousand?
   a. 20,000
   b. 16,000
   c. 16,428
   d. 16,429

3. Estimate the value of $7,404 \div 74$.
   a. 1
   b. 10
   c. 100
   d. 1,000

4. $12(84 - 5) - (3 \times 54) =$
   a. 786
   b. 796
   c. 841
   d. 54,000

5. Which of the following expressions is equal to 60,802?
   a. $600 + 80 + 2$
   b. $6,000 + 800 + 2$
   c. $60,000 + 80 + 2$
   d. $60,000 + 800 + 2$

6. Which of the following whole numbers is divisible by both 7 and 8?
   a. 42
   b. 78
   c. 112
   d. 128

7. What is the estimated product when both 162 and 849 are rounded to the nearest hundred and then multiplied?
   a. 160,000
   b. 180,000
   c. 16,000
   d. 80,000

8. Which of the following choices is equivalent to $5 \times 5 \times 5$?
   a. 3×5
   b. 10×5
   c. 15
   d. 125

9. Which of the following choices is equivalent to 3^5?
   a. 8
   b. 15
   c. 243
   d. 125

10. The whole number p is greater than 0, a multiple of 6, and a factor of 180. How many possibilities are there for the value of p?
   a. 7
   b. 8
   c. 9
   d. 10
   e. 11

11. Which of the following is the prime factorization of 90?
   a. 9×10
   b. 90×1
   c. $2 \times 3 \times 3 \times 5$
   d. $2 \times 5 \times 9$
   e. $3 \times 3 \times 10$

12. Which of the following is the set of positive factors of 12 that are NOT multiples of 2?
   a. { }
   b. {1}
   c. {1,3}
   d. {1,2,3}
   e. {2,4,6,12}

13. Which of the following operations will result in an odd number?
   a. $36 + 48$
   b. 20×8
   c. $37 + 47$
   d. 7×12
   e. $13 + 12$

14. Which of the following equals 2^4?
   a. 10
   b. 15
   c. 32
   d. 16

15. Which of the following expressions is equal to 5?
   a. $(1 + 2)^2$
   b. $9 - 2^2$
   c. $11 - 10 \times 5$
   d. $45 \div 3 \times 3$

16. Which of the following is a prime number?
   a. 6
   b. 9
   c. 11
   d. 27

Set 2 (Answers begin on page 154)

The arithmetic properties of the integers are reviewed in this set.

17. $-25 \div |4 - 9| =$
   a. -30
   b. -20
   c. -5
   d. 5
   e. 13

18. $-4 \times -2 \times -6 \times 3 =$
   a. -144
   b. 144
   c. -9
   d. 9

19. $5 - (-17 + 7)^2 \times 3 =$
   a. -135
   b. 315
   c. -295
   d. -45
   e. 75

20. $(49 \div 7) - (48 \div (-4)) =$
   a. 19
   b. 5
   c. -5
   d. -19

21. In the equation $y = 6p - 23$, if p is a positive whole number, which of the following is the least value of p for which y is positive?
   a. 1
   b. 2
   c. 3
   d. 4
   e. 5

22. $-(5 \cdot 3) + (12 \div (-4)) =$
   a. -12
   b. -18
   c. 12
   d. 18

23. $-2\,(-2)^2 - 2^2 =$
   a. 4
   b. -4
   c. -12
   d. 12

24. $(3^2 + 6) \div (-24 \div 8) =$
   a. -5
   b. 5
   c. 4
   d. -4

25. $(-2[1 -2(4 - 7)])^2 =$
   a. -36
   b. 36
   c. 28
   d. 196

26. $3(5 - 3)^2 -3(5^2 - 3^2)=$
   a. 9
   b. -36
   c. 15
   d. 0

27. $-(-2 -(-11 - (-3^2 - 5) - 2)) =$
   a. 3
   b. -3
   c. 4
   d. -4

28. If $g > 0$ and $h < 0$, which of the following quantities is always positive?
   a. gh
   b. $g + h$
   c. $g - h$
   d. $|h| - |g|$

29. If $g < 0$ and $h < 0$, which of the following quantities cannot be negative?
   a. $h - g$
   b. $g + h$
   c. $-g -h$
   d. $2g + 3h$

30. If $g < 0$ and $h < 0$, which of the following quantities is the largest?
   a. $-g + h$
   b. $g - h$
   c. $g + h$
   d. $-g -h$

31. If $g < 0$ and $h < 0$, which of the following quantities is the smallest?
   a. $-g + h$
   b. $g - h$
   c. $g + h$
   d. $-g -h$

32. If $g < -2$, which of the following quantities is the largest?
   a. g
   b. $-g$
   c. $-g^2$
   d. $(-g)^2$

Set 3 (Answers begin on page 155)

The arithmetic properties of the set of fractions are reviewed in this set.

33. $\frac{5}{9} - \frac{1}{4} =$
   a. $\frac{11}{36}$
   b. $\frac{4}{5}$
   c. $\frac{3}{4}$
   d. $\frac{5}{18}$

34. $\frac{2}{15} + \frac{1}{5} + \frac{1}{6} + \frac{3}{10} =$
   a. $\frac{7}{36}$
   b. $\frac{4}{5}$
   c. $\frac{1}{750}$
   d. none of these

35. What fraction of the following figure is shaded?

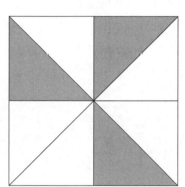

   a. $\frac{1}{2}$
   b. $\frac{1}{4}$
   c. $\frac{2}{3}$
   d. $\frac{3}{8}$

36. $\frac{17}{20} - \frac{5}{6} =$
   a. $\frac{1}{5}$
   b. $\frac{12}{14}$
   c. $\frac{1}{60}$
   d. none of these

37. $\frac{18}{5} \div \frac{9}{20} =$
   a. $\frac{63}{100}$
   b. $\frac{300}{45}$
   c. 8
   d. 10

38. Which of the following fractions is the largest?
   a. $\frac{5}{8}$
   b. $\frac{2}{3}$
   c. $\frac{8}{11}$
   d. $\frac{4}{10}$

39. Which of the following fractions is between $\frac{1}{4}$ and $\frac{2}{3}$?
   a. $\frac{5}{8}$
   b. $\frac{5}{6}$
   c. $\frac{8}{11}$
   d. $\frac{7}{10}$

40. Irma has read $\frac{3}{5}$ of the novel assigned for her English class. The novel is 360 pages long. How many pages has she read?
   a. 216
   b. 72
   c. 300
   d. 98

41. $\frac{5}{8} \times \frac{4}{7} =$

    a. $\frac{5}{14}$

    b. $\frac{20}{8}$

    c. $\frac{25}{32}$

    d. $\frac{9}{16}$

42. What is the reciprocal of $\frac{21}{42}$?

    a. $\frac{1}{2}$

    b. $-\frac{21}{42}$

    c. $-\frac{1}{2}$

    d. 2

43. What is the additive inverse of $1\frac{3}{8}$?

    a. $\frac{8}{11}$

    b. $-\frac{3}{8}$

    c. $\frac{11}{8}$

    d. $-\frac{11}{8}$

44. Danny addressed 14 out of 42 envelopes. What fraction of the envelopes still needs to be addressed?

    a. $\frac{23}{42}$

    b. $\frac{13}{21}$

    c. $\frac{2}{3}$

    d. $\frac{4}{7}$

45. $1 + \dfrac{\left(\frac{-5}{3}\right)(-2)}{\left(\frac{-10}{7}\right)} \div \left(\frac{7}{5} \cdot \frac{10}{3}\right) =$

    a. $\frac{3}{2}$

    b. $\frac{1}{2}$

    c. $-\frac{1}{49}$

    d. $\frac{99}{49}$

46. In Judy's math class, there are m men in a class of n students. Which expression gives the ratio of men to women in the class?

    a. $\frac{m}{n}$

    b. $\frac{n}{m}$

    c. $\frac{m}{m-n}$

    d. $\frac{n}{n-m}$

    e. $\frac{m}{n-m}$

47. Which of the following fractions is closest in value to $\frac{1}{2}$?

    a. $\frac{2}{3}$

    b. $\frac{3}{10}$

    c. $\frac{5}{6}$

    d. $\frac{3}{5}$

48. $7\left(\frac{5}{6}\right) - 3\left(\frac{1}{2}\right)^2 =$

    a. $\frac{17}{24}$

    b. $\frac{17}{6}$

    c. $\frac{61}{12}$

    d. $\frac{5}{4}$

Set 4 (Answers begin on page 157)

The basic exponent rules in the context of signed arithemetic are reviewed in this set.

49. $-5^3 =$
   a. -15
   b. 15
   c. 125
   d. -125

50. $(-11)^2 =$
   a. 121
   b. -121
   c. -22
   d. 22

51. What is the value of the expression $5(4^0)$?
   a. 0
   b. 1
   c. 5
   d. 20

52. $(2^2)^{-3} =$
   a. $\frac{1}{64}$
   b. $-\frac{1}{32}$
   c. -12
   d. 2^{-5}

53. $\frac{(1-3)^2}{-8} =$
   a. 1
   b. -1
   c. $-\frac{1}{2}$
   d. $\frac{1}{2}$

54. $-5(-1-5^{-2}) =$
   a. -45
   b. $\frac{26}{5}$
   c. $\frac{24}{5}$
   d. $-\frac{24}{5}$

55. $-\left[\left(-\frac{3}{2}\right)^{-2} - \left(\frac{2}{3}\right)^2\right] =$
   a. $-\frac{5}{3}$
   b. $-\frac{8}{9}$
   c. 0
   d. $-\frac{18}{4}$

56. $-\left(-\frac{1}{2}\right)^{-3} - \frac{\left(-\frac{1}{3}\right)^2}{9^{-2}} =$
   a. $-\frac{1}{54}$
   b. -1
   c. $\frac{7}{8}$
   d. $-\frac{9}{8}$

57. $-\left(\frac{2}{5}\right)^0 \cdot (-3^2 + 2^{-3})^{-1} =$
   a. 0
   b. -17
   c. $-\frac{71}{9}$
   d. $\frac{8}{71}$

58. $4^{-2}\left(1 - 2(-1)^{-3}\right)^{-2} =$
   a. 32
   b. $-\frac{1}{144}$
   c. 12^{-2}
   d. $\left(\frac{1}{28}\right)^2$

59. $-2^{-2} + \dfrac{(-1^3 + (-1)^3)^{-2}}{-2^2} =$

   a. $-\dfrac{1}{10}$

   b. 3

   c. -5

   d. $-\dfrac{5}{16}$

60. Which of the following quantities has the greatest value?

   a. $\left(-\dfrac{1}{4}\right)^{-1}$

   b. $-\dfrac{3}{8\left(-\frac{1}{4}\right)}$

   c. $4\left(-\dfrac{1}{4}\right) + 3$

   d. $-\left(-\dfrac{1}{4}\right)^{0}$

61. If p is a fraction strictly between 0 and 1, which of the following has the largest value?

   a. p

   b. p^2

   c. p^3

   d. p^{-1}

62. If p is a fraction strictly between 0 and 1, which of the following has the smallest value?

   a. p

   b. p^2

   c. p^3

   d. p^{-1}

63. If p is a fraction strictly between -1 and 0, which of the following has the largest value?

   a. p

   b. p^2

   c. p^3

   d. p^{-1}

64. If p is a fraction strictly between 1 and 2, which of the following has the smallest value?

   a. p

   b. p^2

   c. p^{-2}

   d. p^{-1}

Set 5 (Answers begin on page 158)

Arithmetic involving decimals and percentages is the focus of this set.

65. On an exam, Bart is asked to choose two ways to determine $n\%$ of 40. He is given these four choices:

   I. $n \div 100 \times 40$

   II. $(n \times 0.01) \times 40$

   III. $(n \times 100) \div 40$

   IV. $(n \div 0.01) \times 40$

Which two ways are correct?

   a. I and II

   b. I and IV

   c. II and III

   d. II and IV

   e. III and IV

66. What is the result of increasing 48 by 55%?

   a. 26.4

   b. 30.9

   c. 69.6

   d. 74.4

67. Which of the following expressions show how to determine the sale price of a $250 car stereo that is being offered at a 25% discount?

    I. $0.25 \times \$250$
    II. $0.75 \times \$250$
    III. $(1 + 0.25) \times \$250$
    IV. $(1 - 0.25) \times \$250$

  a. I and III
  b. I and IV
  c. II and III
  d. II and IV
  e. III and IV

68. If $\frac{1}{3} < x < \frac{1}{2}$, which of the following is NOT true?

  a. $\frac{1}{x} > 1$
  b. $\frac{1}{x} > 0$
  c. $2 > \frac{1}{x} > 3$
  d. $x < 2$

69. Rounding 117.3285 to the nearest hundredth results in which of the following decimals?
  a. 100
  b. 117.3
  c. 117.33
  d. 117.329

70. What percentage of 300 results in 400?
  a. 200%
  b. $133\frac{1}{3}\%$
  c. 500%
  d. 1,200%

71. Which of the following inequalities is true?
  a. $0.52 < 0.0052$
  b. $0.52 < 0.052$
  c. $0.00052 > 0.052$
  d. $0.052 > 0.0052$

72. Which of the following is 400% of 30?
  a. 1.2
  b. 12
  c. 120
  d. 1,200

73. If $\frac{5}{16} < x < \frac{9}{20}$, which of the following is a possible value for x?
  a. $\frac{1}{4}$
  b. $\frac{3}{5}$
  c. $\frac{3}{8}$
  d. $\frac{4}{7}$

74. 22.5% is equivalent to which of the following decimals?
  a. 2.25
  b. 0.225
  c. 0.025
  d. 0.0225

75. Which of the following is NOT less than $\frac{2}{5}$?
  a. $\frac{1}{3}$
  b. 0.04
  c. $\frac{3}{8}$
  d. $\frac{3}{7}$
  e. 0.0404

76. Which of the following decimals is between −0.01 and 1.01?
- **a.** −0.015
- **b.** −0.005
- **c.** 1.5
- **d.** 1.15

77. Which of the following decimals is equivalent to $\frac{5}{8} - \frac{2}{5}$?
- **a.** −0.25
- **b.** 0.225
- **c.** 0.25
- **d.** 0.275

78. $(3.09 \times 10^{12}) \div 3 =$
- **a.** 1.03×10^4
- **b.** 3.09×10^4
- **c.** 1.03×10^{12}
- **d.** 1.03×3.33^{12}

79. 0.00000321 is equivalent to which of the following?
- **a.** 3.21×10^{-6}
- **b.** 3.21×10^{-5}
- **c.** 3.21×10^6
- **d.** 3.21×10^5

80. What percentage of $\frac{8}{9}$ results in $\frac{1}{3}$?
- **a.** $\frac{1}{3}\%$
- **b.** 29.6%
- **c.** 37.5%
- **d.** $40\frac{1}{3}\%$

Set 6 (Answers begin on page 159)

This set contains problems that focus on evaluating algebraic expressions at numerical values.

81. What is the value of the expression $-2x^2 + 3x - 7$ when $x = -3$?
- **a.** −34
- **b.** −27
- **c.** −16
- **d.** −10
- **e.** 2

82. What is the value of the expression $\frac{7a}{a^2 + a}$ when $a = -2$?
- **a.** −14
- **b.** −7
- **c.** $-\frac{7}{4}$
- **d.** $\frac{7}{4}$
- **e.** 7

83. What is the value of the expression $2ax - z$ when $a = 3$, $x = 6$, and $z = -8$?
- **a.** 28
- **b.** 44
- **c.** 288
- **d.** 20

84. If $y = -x^3 + 3x - 3$, what is the value of y when $x = -3$?
- **a.** −35
- **b.** −21
- **c.** 15
- **d.** 18
- **e.** 33

85. What is the value of the expression $bx + z \div y$ when $b = -5$, $x = 6$, $y = \frac{1}{2}$ and $z = -8$?

a. 46
b. -46
c. -76
d. 76

84. What is the value of the expression $\frac{m^2}{3} - 4m + 10$ when $m = 6$?

a. -12
b. -2
c. 6
d. 12
e. 22

87. What is the value of the expression $4(x^{-y})(2x^y)(3y^x)$ when $x = 2$ and $y = -2$?

a. 6
b. 8
c. 12
d. 24
e. 384

88. What is the value of the expression $7x + \frac{12}{x} - z$ when $x = 6$ and $z = -8$?

a. 52
b. 36
c. 58
d. 46

89. What is the value of the expression $(3xy + x) \cdot \left(\frac{x}{y}\right)$ when $x = 2$ and $y = 5$?

a. 16
b. 12.8
c. 32.4
d. 80

90. What is the value of the expression $\left(\frac{7}{5a^2} + \frac{3}{10a}\right)^a$ when $a = -2$?

a. $-\frac{1}{10}$
b. $-\frac{2}{25}$
c. $-\frac{2}{5}$
d. 25

91. What is the value of the expression $\frac{6x^2}{2y^2} + \frac{4x}{3y}$ when $x = 2$ and $y = 3$?

a. $\frac{4}{9}$
b. $\frac{4}{3}$
c. $\frac{20}{9}$
d. $\frac{21}{9}$
e. $\frac{13}{3}$

92. What is the value of the expression $ab + \frac{a}{b} + a^2 - b^2$ when $a = 1$ and $b = -1$?

a. -4
b. -3
c. -2
d. -1
e. 0

93. What is the value of the expression $\left((xy)^y\right)^x$ if $x = 2$ and $y = -x$?

a. -4
b. $\frac{1}{256}$
c. $\frac{1}{16}$
d. 4
e. 16

94. What is the value of the expression

$y\left[\left(\frac{x}{2}-3\right)-4a\right]$ when $a=3$, $x=6$, and $y=\frac{1}{2}$?

a. 6
b. −6
c. 12
d. −12

95. What is the value of the expression $z^2-4a^2\,y$ when $a=3$, $z=-8$, and $y=\frac{1}{2}$?

a. 4
b. −28
c. −82
d. 46

96. What is the value of the expression $3x^2b(5a-3b)$ when $a=3$, $b=-5$, and $x=6$?

a. −16,200
b. −1,800
c. 0
d. 1,800

Set 7 (Answers begin on page 160)

The problems in this set focus on simplifying algebraic expressions using the exponent rules.

97. Simplify the expression $\frac{(3x^2)^3}{x^2x^4}$

a. 9
b. 27
c. $\frac{9}{x}$
d. $\frac{27}{x}$

98. Simplify the expression $(4w^9)^3$.

a. $4w^{12}$
b. $4w^{27}$
c. $12w^{27}$
d. $64w^{27}$

99. Simplify the expression $6(e^{-2})^{-2}$.

a. $6e^{-4}$
b. $6e^4$
c. $36e^{-4}$
d. $36e^4$

100. What is the simplified result of the operation $(-45a^4b^9c^5) \div (9ab^3c^3)$?

a. $-5a^3b^6c^2$
b. $-5a^4b^3c^3$
c. $-5a^4b^36c^2$
d. $-5a^5b^{12}c^8$

101. Simplify the expression $(2x^3)^3 \cdot (4x^3)^2$.

a. $64x^{15}$
b. $64x^{54}$
c. $48x^{15}$
d. $48x^{54}$

102. Simplify the expression $\left(\frac{(ab)^3}{b}\right)^4$.

a. a^7
b. a^{12}
c. a^7b^6
d. $a^{12}b^8$
e. $a^{12}b^{11}$

103. Simplify the expression $\frac{\left(\frac{x}{y}\right)^2\left(\frac{y}{x}\right)^{-2}}{xy}$.

a. $\frac{1}{xy}$
b. $\frac{x^3}{y^5}$
c. $\frac{x^3}{y^3}$
d. x^3y^3
e. x^5y^5

104. Simplify the expression $\left(\frac{2a}{b}\right)\left(\frac{a^{-1}}{(2b)^{-1}}\right)$.

a. 1
b. 2
c. 2a
d. 4
e. $\frac{a^2}{b^2}$

105. Simplify the expression $3x^2y(2x^3y^2)$.
   a. $6x^6y^2$
   b. $6x^5y^2$
   c. $6x^5y^3$
   d. $6x^6y^3$

106. Simplify the expression $\left(\frac{a}{b}\right)^2\left(\frac{b}{a}\right)^{-2}\left(\frac{1}{a}\right)^{-1}$.
   a. a
   b. $\frac{1}{a}$
   c. $\frac{a^3}{b^4}$
   d. $\frac{a^4}{b^4}$
   e. $\frac{a^5}{b^4}$

107. Simplify the expression $(3xy^5)^2 - 11x^2y^2(4y^4)^2$.
   a. $-82x^2y^{10}$
   b. $6x^2y^7 - 88x^2y^8$
   c. $-167x^2y^{10}$
   d. $9x^2y^7 - 176x^2y^8$

108. Simplify the expression $\frac{2(3x^2y)^2\,(xy)^3}{3(xy)^2}$.
   a. $6x^5y^3$
   b. $4x^5y^3$
   c. $4x^{\frac{7}{2}}y^{\frac{5}{2}}$
   d. $6x^{\frac{7}{2}}y^{\frac{5}{2}}$

109. Simplify the expression $\frac{(4b)^2x^{-2}}{(2ab^2x)^2}$.
   a. $\frac{16}{a^2b^2x^4}$
   b. $\frac{4}{a^2b^2}$
   c. $\frac{4}{a^2b^2x^4}$
   d. $\frac{2}{a^2b^2}$

110. The product of $6x^2$ and $4xy^2$ is divided by $3x^3y$. What is the simplified expresson?
   a. $8y$
   b. $\frac{4y}{x}$
   c. $4y$
   d. $\frac{8y}{x}$

111. If $3x^2$ is multiplied by the quantity $2x^3y$ raised to the fourth power, to what would this expression simplify?
   a. $48x^{14}y^4$
   b. $1{,}296x^{16}y^4$
   c. $6x^9y^4$
   d. $6x^{14}y^4$

112. Express the product of $-9p^3r$ and the quantity $5p - 6r$ in simplified form.
   a. $-4p^4r - 15p^3r^2$
   b. $-45p^4r + 54p^3r^2$
   c. $-45p^4r - 6r$
   d. $-45p^3r + 54p^3r^2$

Set 8 (Answers begin on page 160)

The problems in this set focus on simplifying arithmetic combinations of algebraic expressions by using exponent rules and combining like terms.

113. Simplify the expression $5ab^4 - ab^4$.
   a. $-5ab^4$
   b. $-5a^2b^8$
   c. $4ab^4$
   d. The expression cannot be simplified further.

114. Simplify the expression $5c^2 + 3c - 2c^2 + 4 - 7c$.
   a. $3c^2 - 4c + 4$
   b. $-3c^4 - 4c^2 + 4$
   c. $-10c^2 - 21c + 4$
   d. The expression cannot be simplified further.

115. Simplify the expression $-5(x-(-3y)) + 4(2y + x)$.
   a. $x + 7y$
   b. $x - 7y$
   c. $-x - 7y$
   d. $-x + 7y$

116. Simplify the expression
$3x^2 + 4ax - 8a^2 + 7x^2 - 2ax + 7a^2$.
a. $21x^2 - 8a^2x - 56a^2$
b. $10x^2 + 2ax - a^2$
c. $10x^4 + 2a^2x^2 - a^4$
d. The expression cannot be simplified further.

117. Simplify the expression $9m^3n + 8mn^3 + 2m^3n^3$.
a. $19m^7n^7$
b. $19(m^3n + nm^3 + m^3n^3)$
c. $17(mn)^3 + 2m^3n^3$
d. The expression cannot be simplified further.

118. Simplify the expression $-7g^6 + 9h + 2h - 8g^6$.
a. $-4g^6h$
b. $-2g^6 - 4h$
c. $-5g^6 + h$
d. $-15g^6 + 11h$

119. Simplify the expression $(2x^2)(4y^2) + 6x^2y^2$.
a. $12x^2y^2$
b. $14x^2y^2$
c. $2x^2 + 4y^2 + 6x^2y^2$
d. $8x^2 + y^2 + 6x^2y^2$
e. $8x^4y^4 + 6x^2y^2$

120. Simplify the expression $(5a^2 \cdot 3ab) + 2a^3b$.
a. $15a^2b + 2a^3b$
b. $17a^6b^2$
c. $17a^3b$
d. The expression cannot be simplified further.

121. Simplify the expression $2x^{-3} - \frac{3x^{-1}}{x^4} - (x^3)^{-1}$.
a. $-2x^{-3}$
b. $x^{-3} - 3x^{-5}$
c. $-x^{-3} - x^{-2}$
d. The expression cannot be simplified further.

122. Simplify the expression $(ab^2)^3 + 2b^2 - (4a)^3b^6$.
a. $2b^2 - 63a^3b^6$
b. $2b^2 - 11a^3b^6$
c. $a^3b^5 + 2b^2 - 12a^3b^6$
d. The expression cannot be simplified further.

123. Simplify the expression $\frac{(-3x^{-1})^{-2}}{x^{-2}} + \frac{8}{9}(x^2)^2$.
a. $\frac{7}{9}x^4$
b. x^4
c. $\frac{62}{9}x^4$
d. The expression cannot be simplified further.

124. Simplify the expression $-(-a^{-2}bc^{-3})^{-2} + 5\left(\frac{b}{a^2c^3}\right)^{-2}$.
a. $\frac{a^4c^6}{24b^2}$
b. $\frac{4a^4c^6}{b^2}$
c. $\frac{6a^4c^6}{b^2}$
d. $\frac{a^4c^6}{4b^2}$

125. Simplify the expression
$3(z + 1)^2w^3 - \frac{2w(z + 1)}{((z + 1)w^2)^{-1}}$.
a. $3(z + 1)^2w^3 - w^3$
b. $3(z + 1)^2w^3 - \frac{1}{w}$
c. $(z + 1)^2w^3$
d. The expression cannot be simplified further.

126. Simplify the expression
$\left(-2(4x + 1)^5\, y^{-5} - \frac{2y(4x + 1)^2}{((4x + 1)y^{-2})^{-3}}\right)^{-2}$.
a. $\frac{y^{10}}{16(4x + 1)^{10}}$
b. $8(4x + 1)^{-10}\, y^{10}$
c. $8(4x + 1)^{-3}\, y^{-7}$
d. none of the above

127. Simplify the expression
$4z\big((xy^{-2})^{-3} + (x^{-3}y^6)^{-1}\big) - \left[\frac{1}{z}\left(\frac{2y^6}{x^3}\right)\right]^{-1}$.
a. $\frac{8zy^6}{x^3} - \frac{zx^3}{2y^6}$
b. $\frac{3zx^3}{2y^6}$
c. $\frac{2zx^3}{3y^6}$
d. $\frac{8zy^6}{x^3} + \frac{zx^3}{2y^6}$

128. Simplify the expression $(0.2x^{-2})^{-1} + \frac{2}{5}x^2 - \frac{5x^4}{(2x)^2}$.

  a. $\frac{83}{20}x^2$

  b. $\frac{2}{9}x^2$

  c. $\frac{-21}{4}x^2$

  d. none of the above

Set 9 (Answers begin on page 161)

This problem set focuses on interpreting verbal mathematical statements as symbolic algebraic expressions.

129. *Two less than four times the square of a number* can be represented as which of the following?

  a. $2 - 4x^2$

  b. $4x^2 - 2$

  c. $(4x)^2 - 2$

  d. both **b** and **c**

130. If the volume V in a water tank is increased by 25%, which of the following expressions represents the new volume of water?

  a. $V + \frac{1}{4}V$

  b. $1.25V$

  c. $V + 0.25V$

  d. All three choices are correct.

131. Jonathon is paying a math tutor a $30 one-time fee plus $40 per hour for time spent tutoring. Which of the following equations indicates how to compute x, the total amount Jonathon will be charged for h hours?

  a. $x = \$30h + \$40$

  b. $x = \$30 + \$40h$

  c. $x = (\$30 + \$40)h$

  d. $x = \$30h - \$40$

  e. $(\$30 - \$40)h$

132. Which of the following expressions represents *nine less than three times the sum of a number and 5?*

  a. $3(x + 5) - 9$

  b. $(3x + 5) - 9$

  c. $9 - 3(x + 5)$

  d. $9 - (3x + 5)$

133. A hotel charges $0.35 for the first minute of a phone call and $0.15 for each additional minute of the call. Which of the following equations represents the cost y of a phone call lasting x minutes?

  a. $y = 0.15(x - 1) + 0.35$

  b. $x = 0.15(y - 1) + 0.35$

  c. $y = 0.15x + 0.35$

  d. $x = 0.15y + 0.35$

134. Which of the following expressions represents *half the difference between a number and five?*

  a. $x - 5$

  b. $\frac{1}{2}(x - 5)$

  c. $\frac{1}{2}x - 5$

  d. $5 - \frac{1}{2}x$

135. Which of the following expressions describes the sum of three numbers multiplied by the sum of their reciprocals?

  a. $(a + b + c) + (\frac{1}{a})(\frac{1}{b})(\frac{1}{c})$

  b. $a(\frac{1}{a}) + b(\frac{1}{b}) + c(\frac{1}{c})$

  c. $(a + b + c) (\frac{1}{a} + \frac{1}{b} + \frac{1}{c})$

  d. $(a)(b)(c) + (\frac{1}{a})(\frac{1}{b})(\frac{1}{c})$

136. Which of the following statements represents the equation $3x + 15 = 32$?

  a. 15 less than 3 times a number is 32.

  b. 32 times 2 is equal to 15 more than a number.

  c. 15 more than 3 times a number is 32.

  d. 3 more than 15 times a number is 32.

137. Suppose that a desk costs D dollars, a chair costs E dollars, and a file cabinet costs F dollars. If an office needs to purchase x desks, y chairs, and z file cabinets, which of the following equations can be used to calculate the total cost T?
   a. $xF + yE + zD = T$
   b. $xE + yD + zF = T$
   c. $xD + yE + zF = T$
   d. $xF + yD + zD = T$

138. The value of d is increased by 50%, and then the resulting quantity is decreased by 50%. How does the resulting quantity compare to d?
   a. It is 25% smaller than d.
   b. It is 25% larger than d.
   c. It is 50% smaller than d.
   d. It is 50% larger than d.
   e. It is the same as d.

139. There are m months in a year, w weeks in a month, and d days in a week. Which of the following expressions represents the number of days in a year?
   a. mwd
   b. $m + w + d$
   c. $\frac{mw}{d}$
   d. $d + \frac{w}{d}$

140. If 40% of j is equal to 50% of k, then j is
   a. 10% larger than k.
   b. 15% larger than k.
   c. 20% larger than k.
   d. 25% larger than k.
   e. 80% larger than k.

141. If q is decreased by p percent, then the resulting quantity is represented by which of the following expressions?
   a. $q - p$
   b. $q - \frac{p}{100}$
   c. $-\frac{pq}{100}$
   d. $q - \frac{pq}{100}$
   e. $pq - \frac{pq}{100}$

142. Two brothers decide to divide the entire cost of taking their father out to dinner evenly between the two of them. If the three meals cost a, b, and c dollars, and a 15% tip will be added in for the waiter, which of the following expressions represents how much each brother will spend?
   a. $0.15(a + b + c) \div 2$
   b. $\frac{1.15(a + b + c)}{2}$
   c. $\frac{(a + b + c) + 0.15(a + b + c)}{2}$
   d. both **b** and **c**

143. If the enrollment E at a shaolin kung fu school is increased by 75%, which of the following expressions represents the new enrollment?
   a. $0.75E$
   b. $E - 0.75E$
   c. $\frac{3}{4}E$
   d. $E + \frac{3}{4}E$

144. Mary gets a 15% discount on all orders that she places at the copy store. She places four orders that cost W dollars, X dollars, Y dollars, and Z dollars before the discount is applied. Which of the following expressions represents how much it will cost her after the discount is deducted from the total for these four orders?
   a. $0.85(W + X + Y + Z)$
   b. $0.15(W + X + Y + Z)$
   c. $(W + X + Y + Z) + 0.15(W + X + Y + Z)$
   d. $(W + X + Y + Z) - 15(W + X + Y + Z)$

S E C T I O N

2 ▶ LINEAR EQUATIONS AND INEQUALITIES

Equations and inequalities, and systems thereof, that are composed of expressions in which the unknown quantity is a variable that is raised only to the first power throughout, are said to be *linear*. Elementary arithmetic properties (e.g., the associative and distributive properties of addition and multiplication), properties of inequalities, and the order of operations are used to solve them.

A graph of a line can be obtained using its slope and a point on the line; the same is true for linear inequalities, with the additional step of shading the region on the appropriate side of the line that depicts the set of ordered pairs satisfying the inequality. Systems are handled similarly, although there are more possibilities regarding the final graphical representation of the solution. These topics are explored in the following 13 problem sets.

Set 10 (Answers begin on page 163)

This set is devoted to problems focused on solving elementary linear equations.

145. What value of z satisfies the equation $z - 7 = -9$?
a. -2
b. -1
c. 2
d. 16

146. What value of k satisfies the equation $\frac{k}{8} = 8$?
a. $\frac{1}{8}$
b. $\sqrt{8}$
c. 8
d. 16
e. 64

147. What value of k satisfies the equation $-7k - 11 = 10$?
a. -3
b. -1
c. 2
d. 21

148. What value of a satisfies the equation $9a + 5 = -22$?
a. -27
b. -9
c. -3
d. -2
e. $-\frac{17}{9}$

149. What value of p satisfies the equation $\frac{p}{6} + 13 = p - 2$?
a. 6
b. 12
c. 15
d. 18

150. What value of p satisfies the equation $2.5p + 6 = 18.5$?
a. 5
b. 10
c. 15
d. 20

151. What value of x satisfies the equation $\frac{3x}{10} = \frac{15}{25}$?
a. 2
b. 2.5
c. 3
d. 3.5

152. What value of x satisfies the equation $2.3(4 - 3.1x) = 1 - 6.13x$?
a. 8.5
b. $\frac{41}{5}$
c. 8.1
d. -8.5

153. If $11c - 7 = 8$, what is the value of $33c - 21$?
a. $\frac{15}{11}$
b. $\frac{8}{3}$
c. 16
d. 24
e. 45

154. What value of x satisfies the equation $\frac{x}{2} + \frac{1}{6}x = 4$?
a. $\frac{1}{24}$
b. $\frac{1}{6}$
c. 3
d. 6

155. What value of b satisfies the equation $-3b - \frac{2}{3} = \frac{b}{3}$?
a. $-\frac{2}{3}$
b. $-\frac{1}{5}$
c. $\frac{2}{3}$
d. $-\frac{20}{9}$

156. What value of c satisfies the equation
$\frac{3c}{4} - 9 = 3$?
- **a.** 4
- **b.** 12
- **c.** 16
- **d.** 20

157. What value of a satisfies the equation
$-\frac{2}{3}a = -54$?
- **a.** −81
- **b.** 81
- **c.** −36
- **d.** 36

158. What value of x satisfies the equation
$1.3 + 5x - 0.1 = -1.2 - 3x$?
- **a.** 0.3
- **b.** 3
- **c.** 3.3
- **d.** −0.3

159. What value of v satisfies the equation
$4(4v + 3) = 6v - 28$?
- **a.** 3.3
- **b.** −3.3
- **c.** −0.25
- **d.** −4

160. What value of k satisfies the equation
$13k + 3(3 - k) = -3(4 + 3k) - 2k$?
- **a.** 1
- **b.** −1
- **c.** 0
- **d.** −2

Set 11 (Answers begin on page 164)

This problem set is focused on solving intermediate linear equations.

161. What value of v satisfies the equation
$-2(3v + 5) = 14$?
- **a.** −4
- **b.** −2
- **c.** 1
- **d.** 3

162. What value of x satisfies the equation
$\frac{5}{2}(x - 2) + 3x = 3(x + 2) - 10$?
- **a.** $\frac{1}{5}$
- **b.** $\frac{2}{5}$
- **c.** $-\frac{1}{5}$
- **d.** $-\frac{2}{5}$

163. Twice a number increased by 11 is equal to 32 less than three times the number. Find the number.
- **a.** −21
- **b.** $\frac{21}{5}$
- **c.** 43
- **d.** $\frac{43}{5}$

164. What value of a satisfies the equation
$\frac{4a + 4}{7} = -\frac{2 - 3a}{4}$?
- **a.** $-\frac{30}{37}$
- **b.** $\frac{12}{5}$
- **c.** 4
- **d.** 6
- **e.** 16

165. The sum of two consecutive even integers is 126. What are the integers?
   a. 62, 64
   b. 62, 63
   c. 64, 66
   d. 2, 63

166. What value of x satisfies the equation $0.8(x + 20) - 4.5 = 0.7(5 + x) - 0.9x$?
   a. 8
   b. −8
   c. 80
   d. −80

167. If $4x + 5 = 15$, then $10x + 5 =$
   a. 2.5
   b. 15
   c. 22.5
   d. 25
   e. 30

168. Ten times 40% of a number is equal to four less than six times the number. Find the number.
   a. 12
   b. 8
   c. 4
   d. 2

169. $\frac{7}{8}$ of nine times a number is equal to ten times the number minus 17. Find the number.
   a. 18.6
   b. 80
   c. 1.86
   d. 8

170. Solve the following equation for b: $a = \frac{7b - 4}{4}$
   a. $\frac{a}{7}$
   b. $\frac{4a}{7}$
   c. $\frac{a + 1}{7}$
   d. $\frac{4a + 4}{7}$
   e. $\frac{7a - 4}{7}$

171. What value of x satisfies the equation $\frac{2x + 8}{5} = \frac{5x - 6}{6}$?
   a. $\frac{14}{3}$
   b. 6
   c. $\frac{76}{20}$
   d. $\frac{-14}{3}$

172. When ten is subtracted from the opposite of a number, the resulting difference is 5. What is the number?
   a. 15
   b. −15
   c. 12
   d. −52

173. What value of x satisfies the equation $9x + \frac{8}{3} = \frac{8}{3}x + 9$?
   a. $\frac{3}{8}$
   b. 1
   c. $\frac{8}{3}$
   d. 9

174. Convert 50° Fahrenheit into degrees Celsius using the formula $F = \frac{9}{5}C + 32$.
   a. 45° Celsius
   b. 2° Celsius
   c. 10° Celsius
   d. 122° Celsius

175. Determine a number such that a 22.5% decrease in its value is the number 93.
   a. 27
   b. 114
   c. 115
   d. 120

176. Negative four is multiplied by the quantity $x + 8$. If $6x$ is then added to this, the result is $2x + 32$. What is the value of x?

a. There can be no such number x.

b. 1

c. 0

d. 16

Set 12 (Answers begin on page 166)

In this problem set, we consider more advanced linear equations and word problems that can be solved using linear equations.

177. What value of x satisfies the equation

$$\frac{\frac{1}{2}x - 4}{3} = \frac{x + 8}{5} ?$$

a. −8

b. 8

c. −88

d. 88

178. What value of x satisfies the equation $5x - 2[x - 3(7 - x)] = 3 - 2(x - 8)$?

a. 23

b. −23

c. $\frac{23}{5}$

d. $\frac{-23}{5}$

179. Assuming that $a \neq c$, solve the following equation for x: $ax + b = cx + d$

a. $\frac{d + b}{a + c}$

b. $\frac{d - b}{c - a}$

c. $\frac{b - d}{a - c}$

d. $\frac{d - b}{a - c}$

180. The sum of four consecutive, odd whole numbers is 48. What is the value of the smallest number?

a. 9

b. 11

c. 13

d. 15

e. 17

181. If $PV = nRT$, which of the following represents an equivalent equation solved for T?

a. $T = \frac{PV}{nR}$

b. $PVnR = T$

c. $\frac{PVR}{n} = T$

d. $T = \frac{1}{PV} \times nR$

182. Solve the following equation for A:
$$B = \frac{C + A}{D - A}$$

a. $A = \frac{BD - C}{1 + B}$

b. $A = \frac{D - C}{1 + B}$

c. $A = \frac{B - C}{C + B}$

d. $A = \frac{B + D}{C + B}$

183. If 30% of r is equal to 75% of s, what is 50% of s if $r = 30$?

a. 4.5

b. 6

c. 9

d. 12

e. 15

184. If $fg + 2f - g = 2 - (f + g)$, what is the value of g in terms of f?

a. −1

b. $\frac{1}{f}$

c. $\frac{4}{f}$

d. $2 - 2f$

e. $\frac{2 - 3f}{f}$

185. The length of a room is three feet more than twice the width of the room. The perimeter of the room is 66 feet. What is the length of the room?
   a. 10 feet
   b. 23 feet
   c. 24 feet
   d. 25 feet

186. Solve the following equation for y:
$$\frac{4-2x}{3} - 1 = \frac{1-y}{2}$$
   a. $y = \frac{1-4x}{3}$
   b. $y = \frac{1+4x}{3}$
   c. $y = \frac{4x-1}{3}$
   d. $y = -\frac{1+4x}{3}$

187. The average of five consecutive odd integers is -21. What is the least of these integers?
   a. -17
   b. -19
   c. -21
   d. -23
   e. -25

188. If $-6b + 2a - 25 = 5$ and $\frac{a}{b} + 6 = 4$, what is the value of $\left(\frac{b}{a}\right)^2$?
   a. $\frac{1}{4}$
   b. 1
   c. 4
   d. -4

189. If three more than one-fourth of a number is three less than the number, what is the value of the number?
   a. $\frac{3}{4}$
   b. 4
   c. 6
   d. 8
   e. 12

190. Solve the following equation for x:
$$\frac{5x-2}{2-x} = y$$
   a. $x = \frac{2-2y}{5+y}$
   b. $x = \frac{2+2y}{5-y}$
   c. $x = \frac{2+2y}{5+y}$
   d. $x = -\frac{2+2y}{5+y}$

191. A grain elevator operator wants to mix two batches of corn with a resultant mix of 54 pounds per bushel. If he uses 20 bushels of 56 pounds per bushel corn, which of the following equations gives the amount of 50 pounds per bushel corn needed?
   a. $56x + 50x = 2x \times 54$
   b. $20 \times 56 + 50x = (20 + x) \times 54$
   c. $20 \times 56 + 50x = 2x + 54$
   d. $56x + 50x = (x + 20) \times 54$

192. What value of x satisfies the equation $-5[x - (3 - 4x - 5) - 5x] - 2^2 = 4[2 - (x-3)]$?
   a. 11.5
   b. 10.5
   c. 9.5
   d. 8.5

Set 13 (Answers begin on page 168)

Solving basic linear inequalities is the focus of this problem set.

193. What is the solution set for $3x + 2 < 11$?
   a. $\{x : x > 3\}$
   b. $\{x : x > -3\}$
   c. $\{x : x < 3\}$
   d. $\{x : x < -3\}$

194. What is the solution set for $-7x \geq -35$?

    a. $\{x : x \leq 5\}$

    b. $\{x : x \geq 5\}$

    c. $\{x : x < 5\}$

    d. the set of all real numbers

195. What is the solution set for $1 - 2x > -5$?

    a. $\{x : x < 3\}$

    b. $\{x : x > 3\}$

    c. $\{x : x < -3\}$

    d. $\{x : x > -3\}$

196. What inequality is represented by the following graph?

    a. $x < -4$

    b. $x > -4$

    c. $x \leq -4$

    d. $x \geq -4$

197. What is the solution set for $4x + 4 > 24$?

    a. $\{x : x < 5\}$

    b. $\{x : x > 5\}$

    c. $\{x : x < 7\}$

    d. $\{x : x > 7\}$

198. What is the solution set for $-8x + 11 < 83$?

    a. $\{x : x > -9\}$

    b. $\{x : x < -9\}$

    c. $\{x : x > 9\}$

    d. $\{x : x < 9\}$

199. What is the solution set for $-4(x - 1) \leq 2(x + 1)$?

    a. $\{x : x \geq -\frac{1}{3}\}$

    b. $\{x : x \leq -\frac{1}{3}\}$

    c. $\{x : x \geq \frac{1}{3}\}$

    d. $\{x : x \leq \frac{1}{3}\}$

    e. $\{x : x \leq 3\}$

200. What is the solution set for $x + 5 \geq 3x + 9$?

    a. $\{x : x \geq \frac{7}{2}\}$

    b. $\{x : x \geq -2\}$

    c. $\{x : x \leq -2\}$

    d. $\{x : x \leq 2\}$

201. What is the solution set for $-6(x + 1) \geq 60$?

    a. $\{x : x \geq -9\}$

    b. $\{x : x \leq -9\}$

    c. $\{x : x \geq -11\}$

    d. $\{x : x \leq -11\}$

202. Which of the following statements accurately describes the inequality $2x - 4 < 7(x - 2)$?

    a. The sum of seven and the quantity two less than a number is greater than four less than two times the number.

    b. The product of seven and the quantity two less than a number is greater than four less than two times the number.

    c. The product of seven and the quantity two less than a number is less than four less than two times the number.

    d. The product of seven and the quantity two less than a number is greater than four less than two more than the number.

203. What is the solution set for $\frac{-x}{0.3} \leq 20$?

    a. $\{x : x \geq -6\}$

    b. $\{x : x \leq -6\}$

    c. $\{x : x \geq -60\}$

    d. $\{x : x \leq -60\}$

204. What is the solution set for $-8(x + 3) \leq 2(-2x + 10)$?

    a. $\{x : x \leq -10\}$

    b. $\{x : x \geq -10\}$

    c. $\{x : x \leq -11\}$

    d. $\{x : x \geq -11\}$

205. What is the solution set for
$3(x-16)-2 < 9(x-2)-7x$?
   a. $\{x : x < -32\}$
   b. $\{x : x < 32\}$
   c. $\{x : x > 32\}$
   d. $\{x : x > -32\}$

206. What is the solution set for
$-5[9+(x-4)] \geq 2(13-x)$?
   a. $\{x : x \leq 17\}$
   b. $\{x : x \leq -17\}$
   c. $\{x : x \geq -17\}$
   d. $\{x : x \geq 17\}$

207. What is the solution set for the compound
inequality $-4 < 3x-1 \leq 11$?
   a. $\{x : -1 < x \leq 4\}$
   b. $\{x : 1 > x \geq -4\}$
   c. $\{x : -4 \leq x < 1\}$
   d. The solution set is the empty set.

208. What is the solution set for the compound
inequality $10 \leq 3(4-2x)-2 < 70$?
   a. $\{x : 0 \leq x < 10\}$
   b. $\{x : -10 < x \leq 0\}$
   c. $\{x : 0 < x \leq 10\}$
   d. $\{x : -10 \leq x < 0\}$

Set 14 (Answers begin on page 169)

This problem set focuses on solving linear equations
and inequalities that involve the absolute value of
certain linear expressions.

209. What values of x satisfy the equation
$|-x|-8=0$?
   a. 8 only
   b. −8 only
   c. both −8 and 8
   d. There are no solutions to this equation.

210. How many different values of x satisfy the
equation $2|x|+4=0$?
   a. 0
   b. 1
   c. 2
   d. more than 2

211. How many different values of x satisfy the
equation $-3|x|+2=5|x|-14$?
   a. 0
   b. 1
   c. 2
   d. more than 2

212. What values of x satisfy the equation
$\left|3x-\frac{2}{3}\right|-\frac{1}{9}=0$?
   a. $\frac{5}{27}$ and $\frac{7}{27}$
   b. $\frac{5}{27}$ and $-\frac{5}{27}$
   c. $\frac{7}{27}$ and $-\frac{7}{27}$
   d. $-\frac{5}{27}$ and $-\frac{7}{27}$

213. What values of x satisfy the equation
$|3x+5|=8$?
   a. $\frac{13}{3}$ and 1
   b. $-\frac{13}{3}$ and 1
   c. $-\frac{13}{3}$ and −1
   d. $\frac{13}{3}$ and −1

214. How many different values of x satisfy the
equation $-6(4-|2x+3|)=-24$?
   a. 0
   b. 1
   c. 2
   d. more than 2

215. How many different values of x satisfy the
equation $1-(1-(2-|1-3x|))=5$?
   a. 0
   b. 1
   c. 2
   d. more than 2

216. How many different values of x satisfy the equation $|2x + 1| = |4x - 5|$?
   a. 0
   b. 1
   c. 2
   d. 3

217. What is the solution set for $|x| > 3$?
   a. $(3,\infty)$
   b. $(-3,\infty)$
   c. $(-3,3)$
   d. $(-\infty,-3) \cup (3,\infty)$

218. What is the solution set for $|-2x| > 0$?
   a. $(-\infty,0) \cup (0,\infty)$
   b. the set of all real numbers
   c. $(-\infty,0)$
   d. The solution set is the empty set.

219. What is the solution set for $-|-x-1| \geq 0$?
   a. $(-\infty,1) \cup (1,\infty)$
   b. $(-\infty,-1) \cup (-1,\infty)$
   c. The only solution is $x = -1$.
   d. The solution set is the empty set.

220. What is the solution set for $|8x + 3| \geq 3$?
   a. $[0,\infty)$
   b. $(-\infty,-\frac{3}{4}]$
   c. $[0,\infty), \cup (-\infty,-\frac{3}{4}]$
   d. none of these choices

221. What is the solution set for $|2x - 3| < 5$?
   a. $(-\infty,4)$
   b. $(4,\infty)$
   c. $(-4,1)$
   d. $(-1,4)$

222. What is the solution set for $2 - (1 - (2 - |1 - 2x|)) > -6$?
   a. $(-4,5)$
   b. $(-\infty,-4) \cup (5,\infty)$
   c. $(-5, 4)$
   d. $(-\infty,-5) \cup (4,\infty)$

223. What is the solution set for $-7|1 - 4x| + 20 \leq -2|1 - 4x| - 15$?
   a. $(-\infty,-\frac{3}{2}]$
   b. $[2,\infty)$
   c. $(-\infty,-\frac{3}{2}] \cup [2,\infty)$
   d. $[-\frac{3}{2},2]$

224. What is the solution set for $|1 - (-2^2 + x) - 2x| \geq |3x - 5|$?
   a. $(\frac{5}{3},\infty)$
   b. $(-\infty,\frac{5}{3})$
   c. The solution set is the empty set.
   d. the set of all real numbers

Set 15 (Answers begin on page 171)

The basics of the Cartesian coordinate system are explored in this problem set.

225. What are the signs of the coordinates of points in the shaded quadrant?

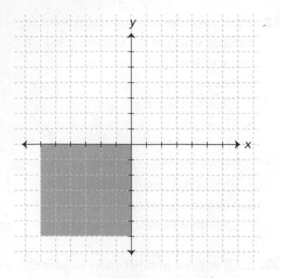

  a. *x* value is negative, *y* value is positive
  b. *x* value is positive, *y* value is negative
  c. *x* value is negative, *y* value is negative
  d. *x* value is positive, *y* value is positive
  e. none of these choices

226. What coordinates are identified by point *J* shown in the following Cartesian plane?

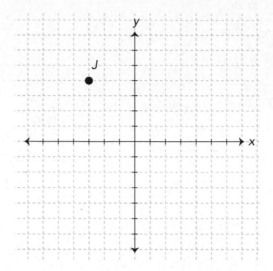

  a. (−4,−3)
  b. (−4,3)
  c. (−3,−4)
  d. (3,−4)
  e. (−3,4)

227. Consider the following graph and assume that *ABCD* is a square. What are the coordinates of point *B*?

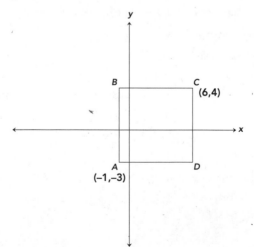

  a. (−1,−4)
  b. (−1,4)
  c. (−1,6)
  d. (−3,1)
  e. (−3,4)

228. Consider the following graph and assume that *ABCD* is a square. What are the coordinates of point *D*?

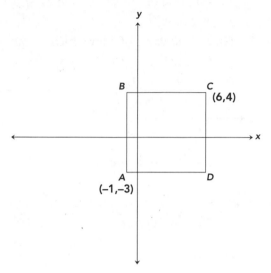

a. (6,–4)
b. (–6,4)
c. (–6,–4)
d. (–4,6)
e. (6,–3)

229. The point (2,–5) lies in which quadrant?
a. Quadrant I
b. Quadrant II
c. Quadrant III
d. Quadrant IV

230. For all nonzero real numbers *x* and *y*, points whose coordinates are given by $(x^2,(-y)^2)$ lie in which quadrant?
a. Quadrant I
b. Quadrant II
c. Quadrant III
d. Quadrant IV

231. For all real numbers $x < -2$, points whose coordinates are given by $(|-x-2|,-|-x-1|)$ must lie in which quadrant?
a. Quadrant I
b. Quadrant II
c. Quadrant III
d. Quadrant IV

232. If *x* is a positive real number and *y* is any real number, which of the following is an accurate characterization of the point (x, y)?
a. The point (x, y) can be in Quadrant I, in Quadrant II, or on the *x*-axis.
b. The point (x, y) can be in Quadrant I, in Quadrant IV, or on the *x*-axis.
c. The point (x, y) can be in Quadrant I, in Quadrant II, on the *x*-axis, or on the *y*-axis.
d. The point (x, y) can be in Quadrant I, in Quadrant IV, on the *x*-axis, or on the *y*-axis.

233. If *x* is any real number and *y* is a nonnegative real number, which of the following is an accurate characterization of the point (x, y)?
a. The point (x, y) can be in Quadrant I, in Quadrant II, or on the *x*-axis.
b. The point (x, y) can be in Quadrant I, in Quadrant IV, or on the *x*-axis.
c. The point (x, y) can be in Quadrant I, in Quadrant II, on the *x*-axis, or on the *y*-axis.
d. The point (x, y) can be in Quadrant I, in Quadrant IV, on the *x*-axis, or on the *y*-axis.

234. Assume $a < 0$. Which of the following points lies in Quadrant IV?
a. $(-a,a)$
b. $(a,-a)$
c. (a,a)
d. $(-a,-a)$

235. Assume $a < 0$. Which of the following points lies in Quadrant III?

  a. $(-a^2, a^2)$
  b. $(a, -a^2)$
  c. (a^2, a^2)
  d. $((-a)^2, -a^2)$

236. Assume $a > 0$. Which of the following points lies in Quadrant II?

  a. $(-a, a)$
  b. $(a, -a)$
  c. (a, a)
  d. $(-a, -a)$

237. For all negative integers x and y, points whose coordinates are given by $(-x^3, xy^2)$ lie in which quadrant?

  a. Quadrant I
  b. Quadrant II
  c. Quadrant III
  d. Quadrant IV

238. For all negative integers x and y, points whose coordinates are given by $\left(\frac{-x^2}{(-y)^3}, \frac{1}{xy} \right)$ lie in which quadrant?

  a. Quadrant I
  b. Quadrant II
  c. Quadrant III
  d. Quadrant IV

239. If x is any real number, which of the following is an accurate characterization of points of the form $(-x, -2)$?

  a. For some values of x, the point $(-x, -2)$ will lie in Quadrant III.
  b. The point $(-x, -2)$ is never on the x-axis
  c. Both **a** and **b** are true.
  d. Neither **a** nor **b** is true.

240. If y is a nonpositive real number, which of the following is an accurate characterization of points of the form $(1, -y)$?

  a. For some values of y, the point $(1, -y)$ will lie in Quadrant IV.
  b. There is no value of y for which the point $(1, -y)$ is on the x axis.
  c. Both **a** and **b** are true.
  d. Neither **a** nor **b** is true.

Set 16 (Answers begin on page 173)

The problems in this set deal with determining the equations of lines using information provided about the line.

241. What is the slope of the line whose equation is $3y - x = 9$?

  a. $\frac{1}{3}$
  b. -3
  c. 3
  d. 9

242. What is the slope of the line whose equation is $y = -3$?

  a. -3
  b. 0
  c. 3
  d. There is no slope.

243. What is the y-intercept of the line whose equation is $8y = 16x - 4$?

  a. $(0, -\frac{1}{2})$
  b. $(0, 2)$
  c. $(0, 8)$
  d. $(0, 16)$

244. Which of the following lines contains the point (3,1)?

a. $y = 2x + 1$

b. $y = 2x + 2$

c. $y = \frac{2}{3}x - 2$

d. $y = \frac{2}{3}x - 1$

e. none of the above

245. A line is known to have a slope of -3 and a y-intercept of $(0, 2)$. Which of the following equations describes this line?

a. $y = 2x - 3$

b. $y = -3x + 2$

c. $y = -2x + 3$

d. $y = 3x - 2$

246. Which of the following equations was used to construct this input/output table?

x	y
1	7
2	10
3	13
4	16
5	19

a. $y = 3x + 4$

b. $y = 4x - 1$

c. $y = 5x - 2$

d. $y = 7x$

247. Transform the equation $3x + y = 5$ into slope-intercept form.

a. $y = 3x + 5$

b. $y = -3x + 5$

c. $x = \frac{1}{3}y + 5$

d. $x = -\frac{1}{3}y + 5$

248. What is the equation of the line that passes through the points $(2, 3)$ and $(-2, 5)$?

a. $y = x + 1$

b. $y = -\frac{1}{2}x + 4$

c. $y = -\frac{1}{2}x$

d. $y = -\frac{3}{2}x$

e. $y = -\frac{3}{2}x + 2$

249. Transform the equation $y = -\frac{2}{15}x - \frac{3}{5}$ into standard form.

a. $-2x + 15y = -9$

b. $2x + 15y = 9$

c. $2x + 15y = -9$

d. $2x - 15y = -9$

250. What is the slope of the line whose equation is $-3y = 12x - 3$?

a. -4

b. -3

c. 1

d. 4

e. 12

251. Which of the following lines has a negative slope?

a. $6 = y - x$

b. $y = 4x - 5$

c. $-5x + y = 1$

d. $6y + x = 7$

252. Determine the missing value of z that completes the following table, assuming that all of the points are collinear.

x	y
−4	15
−2	11
2	z
5	−3
7	−7

a. −11
b. 0
c. 3
d. 8

253. A line is known to pass through the points $(0, −1)$ and $(2, 3)$. What is the equation of this line?

a. $y = \frac{1}{2}x − 1$
b. $y = \frac{1}{2}x + 1$
c. $y = 2x − 1$
d. $y = 2x + 1$

254. Which of these statements is true?
a. A vertical line need not have a y-intercept.
b. A horizontal line need not have a y-intercept.
c. A line with positive slope need not cross the x-axis.
d. A line with negative slope need not cross the x-axis.

255. A line has a y-intercept of $(0, −6)$ and an x-intercept of $(9, 0)$. Which of the following points must also lie on this line?
a. $(−6, −10)$
b. $(1, 3)$
c. $(0, 9)$
d. $(3, −8)$
e. $(6, 13)$

256. Determine the value of y if the points $(−3, −1)$, $(0, y)$, and $(3, −9)$ are assumed to be collinear.
a. 1
b. −1
c. −3
d. −5

Set 17 (Answers begin on page 174)

The problems in this set deal with graphing straight lines.

257. Which of the following is the graph of $y = -3$?

a.

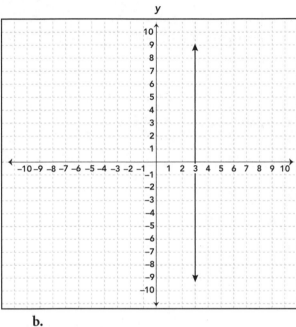

c.

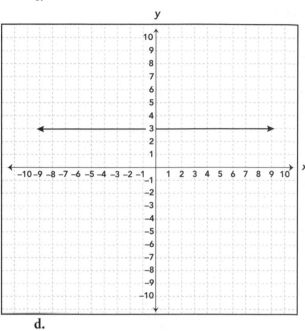

b.

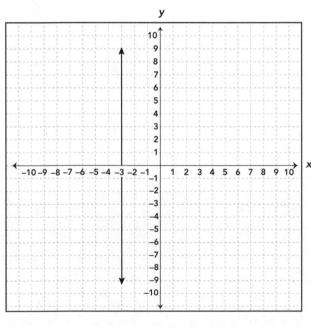

d.

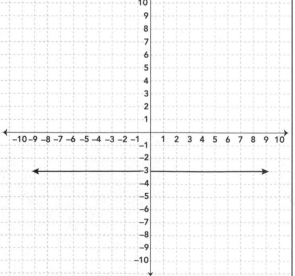

258. What is the slope of the line segment shown in the following graph?

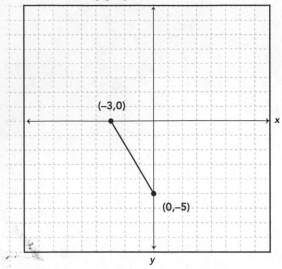

a. $\frac{5}{3}$

b. $-\frac{5}{3}$

c. $-\frac{3}{5}$

d. $\frac{3}{5}$

259. Which of the following is an accurate characterization of the slope of the y-axis?

a. It has a slope of zero.

b. Its slope is undefined.

c. It has a positive slope.

d. It has a negative slope.

260. What is the slope of the line segment in the following graph?

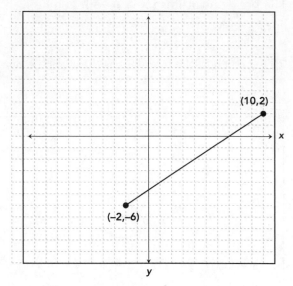

a. -2

b. 2

c. $-\frac{2}{3}$

d. $\frac{2}{3}$

261. Which of the following is the graph of
$y = 2x + 3$?

a.

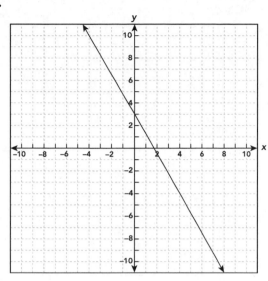

c.

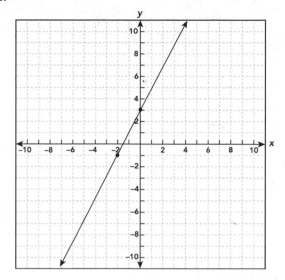

b.

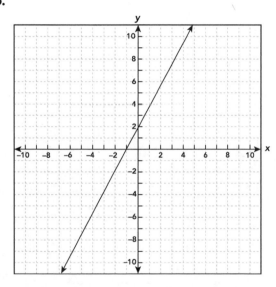

d.

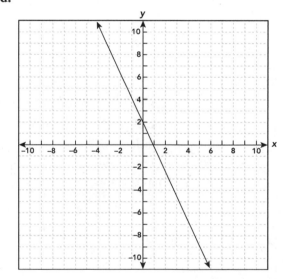

262. Which of the following is the graph of
$y = -2x + 9$?

a.

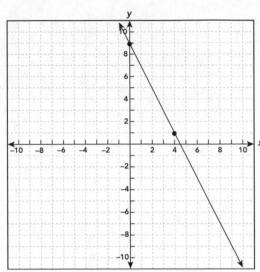

c.

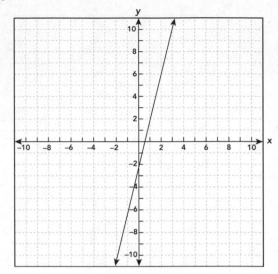

b.

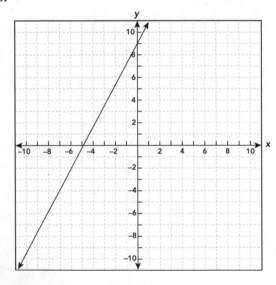

d.

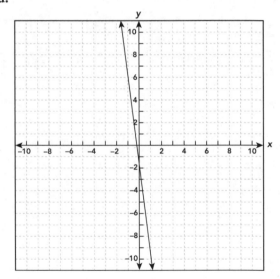

263. Which of the following is the graph of $y = -\frac{5}{2}x - 5$?

a.

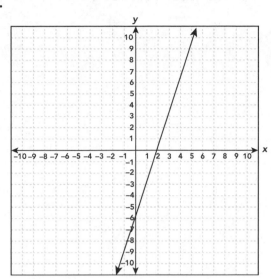

c.

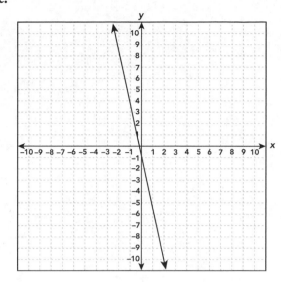

b.

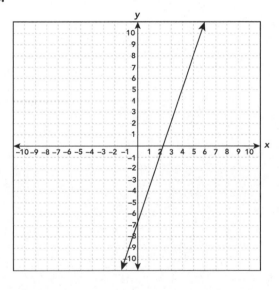

d.

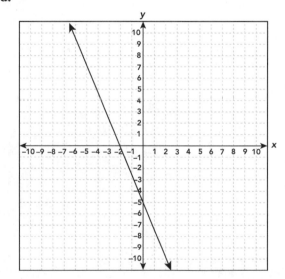

264. What is the equation of the line shown in the following graph?

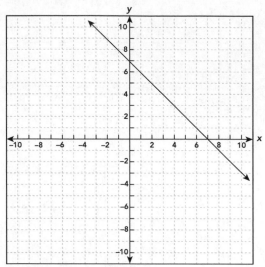

a. $y = x + 7$
b. $y = x - 7$
c. $y = -x - 7$
d. $y = -x + 7$

265. What is the equation of the line in the following graph?

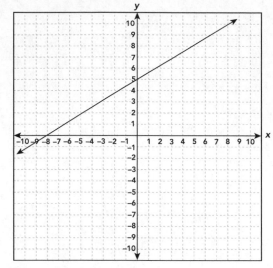

a. $y = -\frac{2}{3}x + 5$
b. $y = \frac{2}{3}x + 5$
c. $y = -\frac{3}{2}x + 5$
d. $y = \frac{3}{2}x + 5$

266. Which of the following is the graph of
$\frac{2}{3}y - \frac{1}{2}x = 0$?

a.

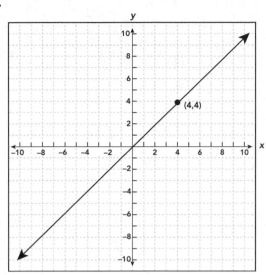

c.

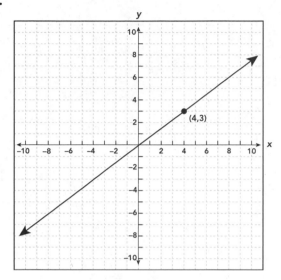

b.

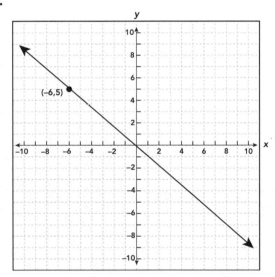

d.

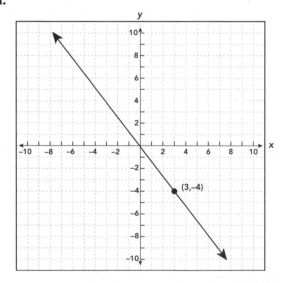

267. Which of the following lines has a positive slope?

a.

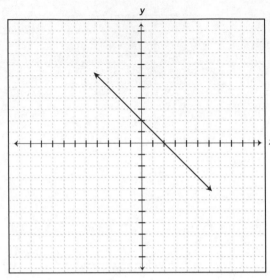

c.

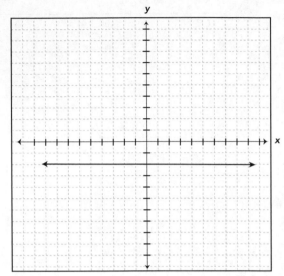

b.

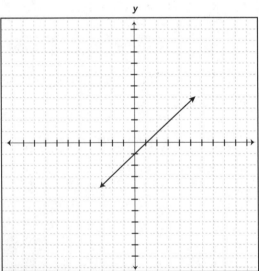

d.

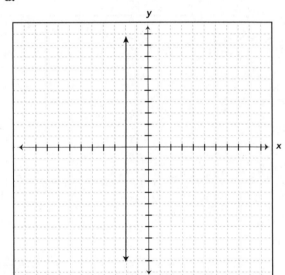

268. Which of the following lines has an undefined slope?

a.

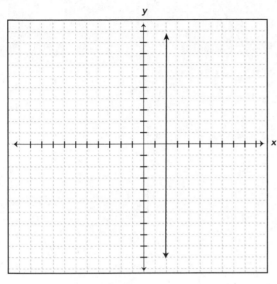

c.

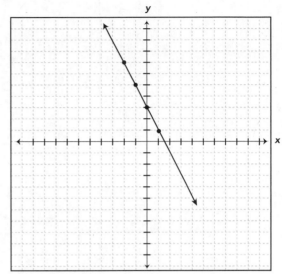

b.

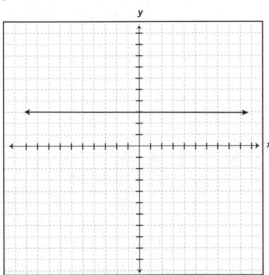

d.

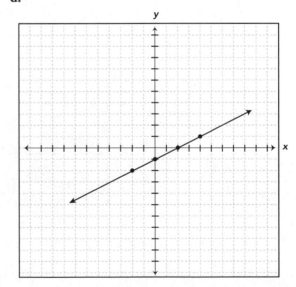

269. The equation $0.1x - 0.7y = 1.4$ is shown in which of the following graphs?

a.

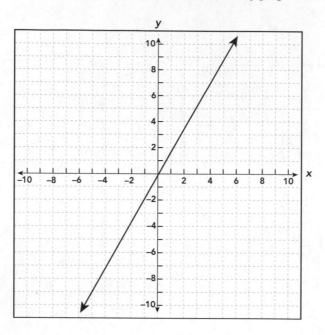

b.

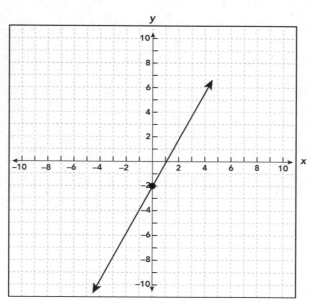

c.

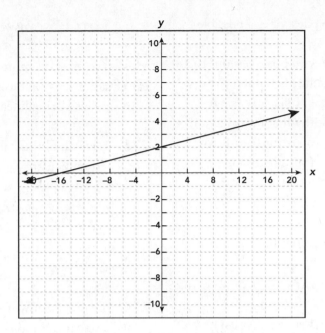

d.

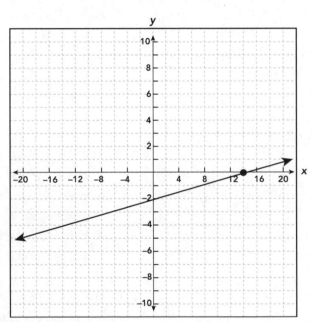

270. Which of the following describes a possible scenario?

  a. The graph of a line with positive slope can cross into both Quadrants II and IV.

  b. The graph of a line with negative slope cannot cross into both Quadrants I and II.

  c. The graph of $y = c$, where $c \neq 0$, can cross into only two of the four quadrants.

  d. The graph of a vertical line cannot cross into both Quadrants II and III.

271. Which of the following describes a possible scenario?

  a. The graph of $x = c$, where $c \neq 0$, cannot cross the x-axis.

  b. The graph of $y = c$, where $c \neq 0$, must have a y-intercept.

  c. A line with an undefined slope can cross into both Quadrant I and Quadrant II.

  d. A line whose graph rises from left to right has a negative slope.

272. Which of the following describes a possible scenario?

  a. A line whose equation is of the form $y = -x + c$ can cross into three of the four quadrants.

  b. A line with positive slope need not cross the x-axis.

  c. A line with negative slope need not cross the y-axis.

  d. A horizontal line has an undefined slope.

Set 18 (Answers begin on page 175)

This set focuses on more advanced properties of linear equations, as well as more advanced word problems modeled using linear equations.

273. To which of the following lines is $y = \frac{2}{3}x - 5$ perpendicular?

  a. $y = \frac{2}{3}x + 5$

  b. $y = 5 - \frac{2}{3}x$

  c. $y = -\frac{2}{3}x - 5$

  d. $y = \frac{3}{2}x - 5$

  e. $y = -\frac{3}{2}x + 5$

274. The graphs of which of the following pairs of linear equations would be parallel to each other?

  a. $y = 2x + 4$, $y = x + 4$

  b. $y = 3x + 3$, $y = -\frac{1}{3} - 3$

  c. $y = 4x + 1$, $y = -4x + 1$

  d. $y = 5x + 5$, $y = \frac{1}{5}x + 5$

  e. $y = 6x + 6$, $y = 6x - 6$

275. The line $y = -2x + 8$ is

  a. parallel to the line $y = \frac{1}{2}x + 8$

  b. parallel to the line $\frac{1}{2}y = -x + 3$

  c. perpendicular to the line $2y = -\frac{1}{2}x + 8$

  d. perpendicular to the line $\frac{1}{2}y = -2x - 8$

  e. perpendicular to the line $y = 2x - 8$.

276. Which of the following is the equation of the line perpendicular to $y = \frac{3}{4}x - 2$ and passing through the point $(-6, 4)$?

  a. $y = \frac{3}{4}x + 4$

  b. $y = \frac{3}{4}x - 4$

  c. $y = \frac{4}{3}x - 4$

  d. $y = -\frac{4}{3}x - 4$

  e. $y = -\frac{4}{3}x + 4$

277. Which of the following is the equation of the line parallel to $y = 3x + 8$ and passing through the point (4,4)?

a. $y = 3x + 4$

b. $y = 3x - 8$

c. $y = \frac{1}{3}x + 8$

d. $y = -\frac{1}{3}x + 8$

278. Which of the following is the equation of the line that has y-intercept (0,12) and is parallel to the line passing through the points (4,2) and (−5,6)?

a. $y = \frac{4}{9}x + 12$

b. $y = -\frac{4}{9}x + 12$

c. $y = -\frac{9}{4}x + 2$

d. $y = \frac{9}{4}x + 12$

279. Which of the following is the equation of the line perpendicular to $y = -\frac{13}{18}x + 5$ and passing through the origin?

a. $y = -\frac{18}{13}x$

b. $y = \frac{13}{18}x$

c. $y = \frac{18}{13}x$

d. $y = -\frac{13}{18}x$

280. Which of the following lines must be perpendicular to a line with an undefined slope?

a. $x = -2$

b. $y = -2$

c. both **a** and **b**

d. neither **a** nor **b**

281. Which of the following lines must be parallel to a line with zero slope?

a. $x = -2$

b. $y = -2$

c. both **a** and **b**

d. neither **a** nor **b**

282. A 60-foot piece of rope is cut into three pieces. The second piece must be 1 foot shorter in length than twice the first piece, and the third piece must be 10 feet longer than three times the length of the second piece. How long should the longest piece be?

a. 37 feet

b. 40 feet

c. 43 feet

d. 46 feet

283. At Zides Sport Shop, a canister of Ace tennis balls costs $3.50 and a canister of Longline tennis balls costs $2.75. The high school tennis coach bought canisters of both brands of balls, spending exactly $40.25 before the sales tax. If he bought one more canister of Longline balls than he did Ace balls, how many canisters of each did he purchase?

a. 7 canisters of Ace balls and 6 canisters of Longline balls

b. 6 canisters of Ace balls and 7 canisters of Longline balls

c. 5 canisters of Ace balls and 6 canisters of Longline balls

d. 7 canisters of Ace balls and 8 canisters of Longline balls

284. One essential step to ensure the success of a microgravity bean seed germination project is that 10 gallons of a 70% concentrated nutrient solution be administered to the bean seeds. If the payload specialist has some 90% nitrogen and some 30% nitrogen, how many gallons (accurate to 2 decimal places) of each should she mix in order to obtain the desired solution?

   a. 2.50 gallons of the 30% nitrogen solution with 7.50 gallons of the 90% nitrogen solution

   b. 7.50 gallons of the 30% nitrogen solution with 2.50 gallons of the 90% nitrogen solution

   c. 6.67 gallons of the 30% nitrogen solution with 3.33 gallons of the 90% nitrogen solution

   d. 3.33 gallons of the 30% nitrogen solution with 6.67 gallons of the 90% nitrogen solution

285. How long would it take a girl riding a bicycle at 17 mph to overtake her instructor riding at 7 mph along the same path, assuming that her instructor had a 3-hour head start?

   a. 2 hours 6 minutes

   b. 2 hours 15 minutes

   c. 3 hours

   d. 3 hours 12 minutes

286. Kari invested some money at 10% interest and $1,500 more than that amount at 11% interest. Her total yearly interest was $795. How much did she invest at each rate?

   a. $2,000 at 10% interest and $3,500 at 11% interest

   b. $2,500 at 10% interest and $4,000 at 11% interest

   c. $4,000 at 10% interest and $5,500 at 11% interest

   d. $3,000 at 10% interest and $4,500 at 11% interest

287. A small piggy bank is full of just nickels and dimes. If the bank contains 65 coins with a total value of 5 dollars, how many nickels and how many dimes are in the bank?

   a. 32 nickels and 33 dimes

   b. 30 nickels and 35 dimes

   c. 28 nickels and 37 dimes

   d. 25 nickels and 40 dimes

288. Lori is twice as old as her sister, Lisa. In 5 years, Lisa will be the same age as her sister was 10 years ago. What are their current ages?

   a. Lisa is 12 years old and Lori is 24 years old.

   b. Lisa is 15 years old and Lori is 30 years old.

   c. Lisa is 20 years old and Lori is 40 years old.

   d. Lisa is 23 years old and Lori is 46 years old.

Set 19 (Answers begin on page 178)

The problems in this set deal with graphing linear inequalities in the Cartesian plane.

289. Which inequality is illustrated by the following graph?

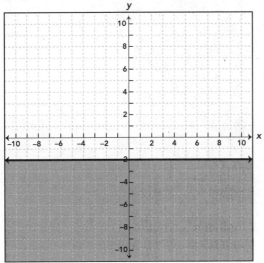

a. $y < -2$
b. $y \leq -2$
c. $y > -2$
d. $y \geq -2$

290. Which inequality is illustrated by the following graph?

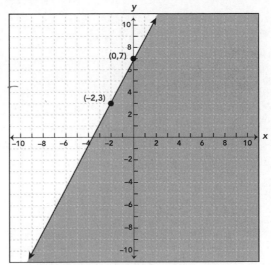

a. $y \leq 2x + 7$
b. $y \geq 2x + 7$
c. $y > -2x + 7$
d. $y \geq -2x + 7$

291. Which inequality is illustrated by the following graph?

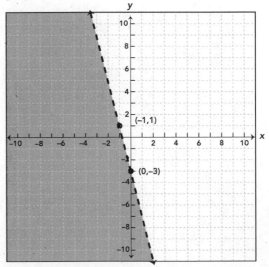

a. $y < 4x - 3$
b. $y < -4x - 3$
c. $y \leq -4x - 3$
d. $y > 4x - 3$

292. Which inequality is illustrated by the following graph?

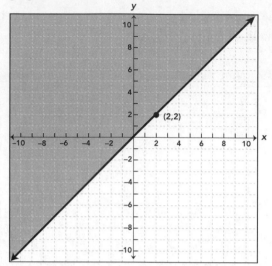

   a. $y > 8$
   b. $y < 8$
   c. $x < 8$
   d. $x > 8$
   e. $x \geq 8$

293. Which inequality is illustrated by the following graph?

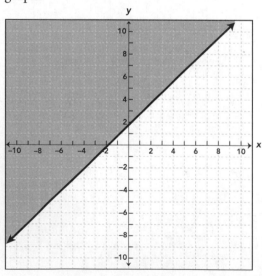

   a. $x + y \leq 2$
   b. $x - y \leq 2$
   c. $x - y \leq -2$
   d. $x + y \leq -2$

294. Which inequality is illustrated by the following graph?

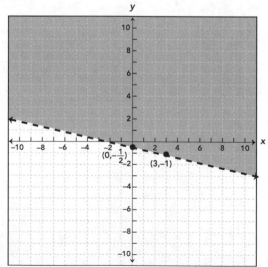

   a. $y - x > 0$
   b. $x - y > 0$
   c. $y - x \geq 0$
   d. $x - y \geq 0$

295. Which inequality is illustrated by the following graph?

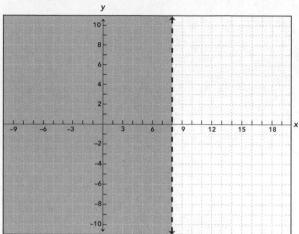

   a. $\frac{1}{3}x + 2y > -1$
   b. $x + 2y \geq -3$
   c. $x + 6y > -1$
   d. $\frac{1}{3}x + 2 < -1$

296. Which inequality is illustrated by the following graph?

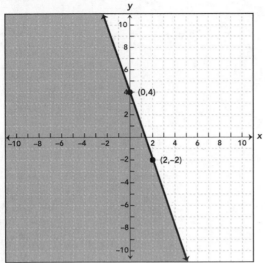

a. $2y + 6x \leq 8$

b. $2y - 6x \leq 8$

c. $2y + 6x \geq 8$

d. $2y - 6x \geq 8$

298. Which inequality is illustrated by the following graph?

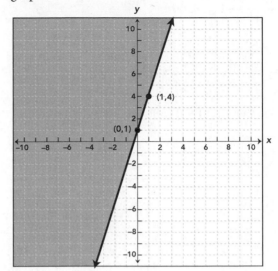

a. $y \geq 3x + 1$

b. $y > 3x + 1$

c. $y < 3x + 1$

d. $y \leq 3x + 1$

297. Which inequality is illustrated by the following graph?

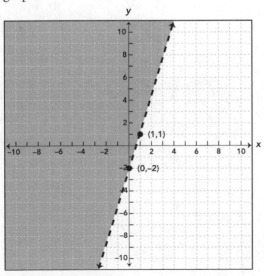

a. $3x + y + 2 > 0$

b. $3x - y + 2 < 0$

c. $3x - y - 2 < 0$

d. $3x + y - 2 > 0$

299. Which of the inequalities is illustrated by the following graph?

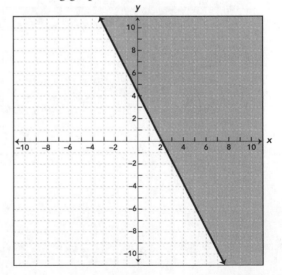

a. $3x + y \geq 7x + y - 8$

b. $3x - y \geq 7x + y + 8$

c. $3x + y \leq 7x + y - 8$

d. $3x - y \leq 7x + y - 8$

300. For which of the following inequalities is the
point (3,–2) a solution?

   a. $2y - x \geq 1$

   b. $x + y > 5$

   c. $3y < -3x$

   d. $9x - 1 > y$

301. Which of the following graphs illustrates the
inequality $y > 4$?

a.

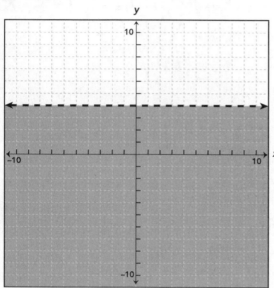

c.

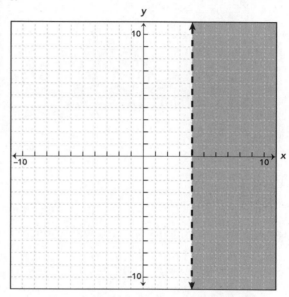

b.

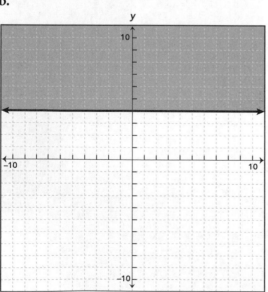

d.

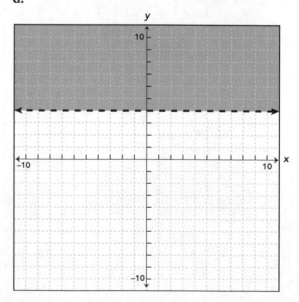

302. Which of the following graphs illustrates the inequality $x > 4$?

a.

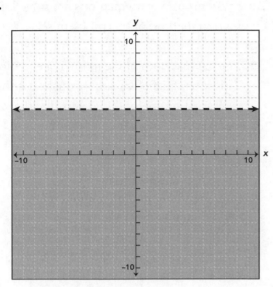

c.

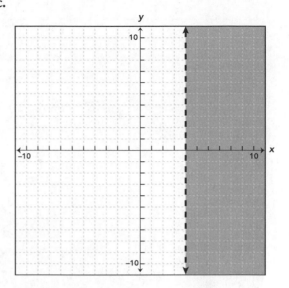

b.

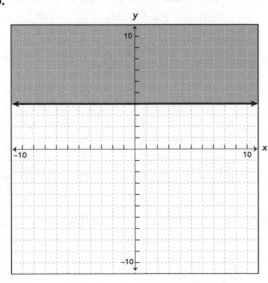

d.

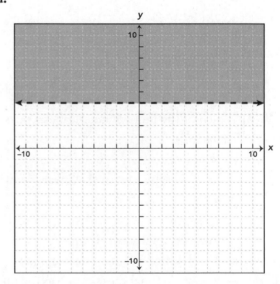

303. Which of the inequalities is illustrated by the following graph?

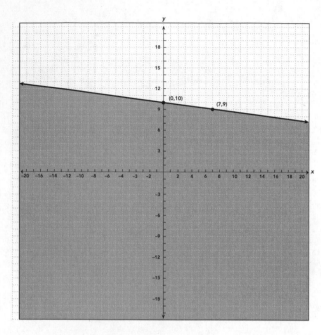

a. $28y \leq -2x - 14(y + 10)$
b. $-28y \geq 2x - 14(y + 10)$
c. $28y \leq 2x - 14(y + 10)$
d. $-28y \geq 2x - 14(y - 10)$

304. Which of the following describes a possible scenario?
a. Points of the form $(x, 2x)$, where $x > 0$, are in the solution set of the linear inequality $y < x$.
b. The solution set of a linear inequality can intersect all four quadrants.
c. The solution set of a linear inequality $y - 2x < -1$ includes points on the line $y = 2x - 1$.
d. Points of the form $(8, y)$ satisfy the linear inequality $x < 8$.

Set 20 (Answers begin on page 180)

2×2 systems of linear equations are solved using the elimination method in this problem set.

305. Given that both of the following equations must be satisfied simultaneously, use the elimination method to determine the value of b.

$$5a + 3b = -2$$
$$5a - 3b = -38$$

a. -6
b. -4
c. 6
d. 12
e. 13

306. Use the elimination method to determine the solution of the following system of linear equations:

$$-x + 3y = 11$$
$$x - 5y = -3$$

a. $x = 17, y = 4$
b. $x = 1, y = 4$
c. $x = 1, y = -4$
d. $x = -23, y = -4$

307. Given that both of the following equations must be satisfied simultaneously, use the elimination method to determine the value of x.

$$3(x + 4) - 2y = 5$$
$$2y - 4x = 8$$

a. -2
b. -1
c. 1
d. 13
e. 15

308. Given that both of the following equations must be satisfied simultaneously, use the elimination method to determine the value of x.

$$2x + y = 6$$
$$\tfrac{y}{2} + 4x = 12$$

a. -2
b. 0
c. 1
d. 3
e. 6

309. Given that both of the following equations must be satisfied simultaneously, use the elimination method to determine the value of $\tfrac{x}{y}$.

$$4x + 6 = -3y$$
$$-2x + 3 = y + 9$$

a. -6
b. -1
c. 0
d. 1
e. 6

310. Given that both of the following equations must be satisfied simultaneously, use the elimination method to determine the value of b.

$$-7a + \tfrac{b}{4} = 25$$
$$b + a = 13$$

a. -3
b. 4
c. 12
d. 13
e. 16

311. Given that both of the following equations must be satisfied simultaneously, use the elimination method to determine the value of n.

$$2(m + n) + m = 9$$
$$3m - 3n = 24$$

a. -5
b. -3
c. 3
d. 5
e. 8

312. Given that both of the following equations must be satisfied simultaneously, use the elimination method to determine the value of a.

$$7(2a + 3b) = 56$$
$$b + 2a = -4$$

a. -5
b. -4
c. -2
d. 4
e. 6

313. Given that both of the following equations must be satisfied simultaneously, use the elimination method to determine the value of y.

$$\tfrac{1}{2}x + 6y = 7$$
$$-4x - 15y = 10$$

a. -10
b. $-\tfrac{1}{2}$
c. 2
d. 5
e. 6

314. Given that both of the following equations must be satisfied simultaneously, use the elimination method to determine the value of $a + b$.

$$4a + 6b = 24$$
$$6a - 12b = -6$$

a. 2
b. 3
c. 4
d. 5
e. 6

315. Given that both of the following equations must be satisfied simultaneously, use the elimination method to determine the value of $a + b$.

$$\tfrac{1}{2}(a + 3) - b = -6$$
$$3a - 2b = -5$$

a. 5
b. 15
c. 20
d. 25
e. 45

316. Given that both of the following equations must be satisfied simultaneously, use the elimination method to determine the value of $\frac{c}{d}$.

$$\tfrac{c-d}{5} - 2 = 0$$
$$c - 6d = 0$$

a. 2
b. 6
c. 8
d. 12
e. 14

317. Given that both of the following equations must be satisfied simultaneously, use the elimination method to determine the value of xy.

$$-5x + 2y = -51$$
$$-x - y = -6$$

a. -27
b. -18
c. -12
d. -6
e. -3

318. Given that both of the following equations must be satisfied simultaneously, use the elimination method to determine the value of $(y - x)^2$.

$$9(x - 1) = 2 - 4y$$
$$2y + 7x = 3$$

a. 1
b. 4
c. 16
d. 25
e. 36

319. Given that both of the following equations must be satisfied simultaneously, use the elimination method to determine the value of $(p + q)^2$.

$$8q + 15p = 26$$
$$-5p + 2q = 24$$

a. 4
b. 5
c. 25
d. 49
e. 81

320. Use the elimination method to determine the solution of the following system of linear equations:

$$4x - 3y = 10$$
$$5x + 2y = 1$$

a. $x = 4, y = -3$
b. $x = 1, y = -2$
c. $x = -1, y = -\frac{1}{3}$
d. $x = 2, y = -\frac{2}{3}$

Set 21 (Answers begin on page 182)

2×2 systems of linear equations are solved using the substitution method and graphical techniques in this problem set.

321. Given that both of the following equations must be satisfied simultaneously, use the substitution method to solve the following system:

$$x = -5y$$
$$2x + 2y = 16$$

a. $x = 10, y = -2$
b. $x = -2, y = 10$
c. $x = 20, y = -4$
d. $x = -5, y = 1$

322. Given that both of the following equations must be satisfied simultaneously, use the substitution method to determine the value of x.

$$2x + y = 6$$
$$\frac{y}{2} + 4x = 12$$

a. -2
b. 0
c. 1
d. 3

323. Given that both of the following equations must be satisfied simultaneously, use the substitution method to determine the value of $\sqrt{\frac{a}{b}}$.

$$\frac{a}{2} = b + 1$$
$$3(a - b) = -21$$

a. $\frac{16}{9}$
b. $\frac{2}{3}$
c. $\frac{3}{4}$
d. $\frac{4}{3}$
e. $\frac{3}{2}$

324. Given that both of the following equations must be satisfied simultaneously, use the substitution method to determine the value of b.

$$-7a + \frac{b}{4} = 25$$
$$b + a = 13$$

a. -3
b. 4
c. 12
d. 13
e. 16

325. Given that both of the following equations must be satisfied simultaneously, use the substitution method to determine the value of a.

$$7(2a + 3b) = 56$$
$$b + 2a = -4$$

a. -5
b. -4
c. -2
d. 4
e. 6

326. Given that both of the following equations must be satisfied simultaneously, use the substitution method to determine the value of $\frac{c}{d}$.

$$\frac{c-d}{5} - 2 = 0$$
$$c - 6d = 0$$

  a. 2
  b. 6
  c. 8
  d. 12
  e. 14

327. Given that both of the following equations must be satisfied simultaneously, use the substitution method to determine the value of xy.

$$-5x + 2y = -51$$
$$-x - y = -6$$

  a. −27
  b. −18
  c. −12
  d. −6
  e. −3

328. Given that both of the following equations must be satisfied simultaneously, use the substitution method to determine the value of ab.

$$10b - 9a = 6$$
$$b - a = 1$$

  a. −12
  b. −7
  c. 1
  d. 7
  e. 12

329. Given that both of the following equations must be satisfied simultaneously, use the substitution method to determine the value of $x - y$.

$$\frac{x+y}{3} = 8$$
$$2x - y = 9$$

  a. −24
  b. −2
  c. 0
  d. 1
  e. 2

330. Which of the following linear systems contains two parallel lines?
  a. $x = 5$, $y = 5$
  b. $y = -x$, $y = x - 1$
  c. $x - y = 7$, $2 - y = -x$
  d. $y = 3x + 4$, $2x + 4 = y$

331. How many solutions are there of the system of equations shown in the following graph?

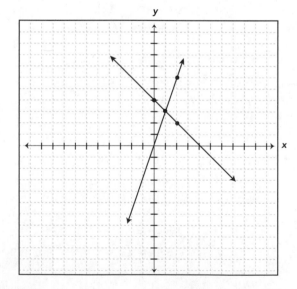

  a. 0
  b. 1
  c. 2
  d. infinitely many

332. Determine the number of solutions of the following system of equations:

$$y = 3x + 2$$
$$y - 3x = -2$$

a. 1
b. 0
c. infinitely many
d. none of the above

333. Use the substitution method to determine the value of $\frac{2x}{y}$ given that both of the following equations must be satisfied simultaneously:

$$3x - y = 2$$
$$2y - 3x = 8$$

a. $\frac{4}{3}$
b. $\frac{4}{5}$
c. 5
d. 8
e. 12

334. Determine the number of solutions of the linear system that has the following graphical depiction:

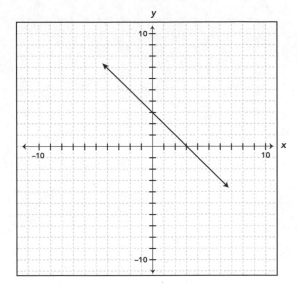

a. 1
b. 0
c. infinitely many
d. none of the above

335. Determine the number of solutions of the linear system that has the following graphical depiction:

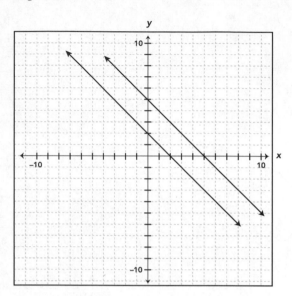

a. 1
b. 0
c. infinitely many
d. none of the above

336. Determine the number of solutions of the following system of equations:

$$y = 3x + 2$$
$$-3y + 9x = -6$$

a. 1
b. 0
c. infinitely many
d. none of the above

Set 22 (Answers begin on page 184)

The problems in this set consist of graphing systems of linear inequalities.

337. The graphs of the lines $y = 4$ and $y = x + 2$ form the boundaries of the shaded region. The solution set of which of the following systems of linear inequalities is given by the shaded region?

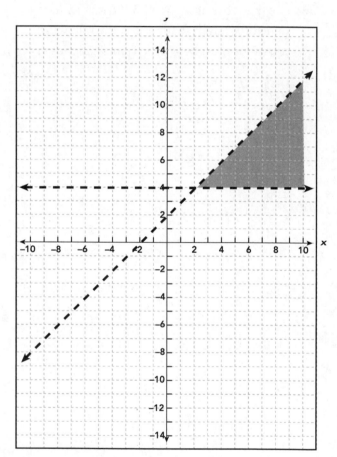

a. $y \geq 4, y \leq x + 2$
b. $y < 4, y > x + 2$
c. $y > 4, y < x + 2$
d. $y \leq 4, y \leq x + 2$

338. The graphs of the lines $y = 5$ and $x = 2$ form the boundaries of the shaded region. The solution set of which of the following systems of linear inequalities is given by the shaded region?

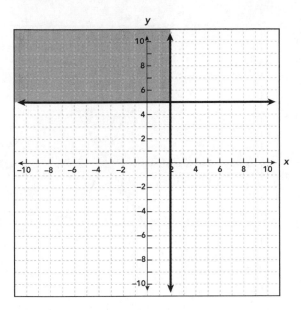

a. $y \geq 5, x \leq 2$
b. $y \leq 5, x \leq 2$
c. $y > 5, x < 2$
d. $y < 5, x < 2$

339. The graphs of the lines $y = -x + 4$ and $y = x + 2$ form the boundaries of the shaded region. The solution set of which of the following systems of linear inequalities is given by the shaded region?

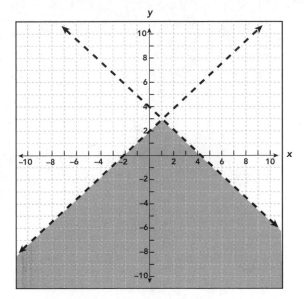

a. $y < x + 2, y < -x + 4$
b. $y \leq x + 2, y \leq -x + 4$
c. $y > x + 2, y < -x + 4$
d. $y \geq x + 2, y < -x + 4$

340. The graphs of the lines $y = \frac{1}{4}x$ and $y = -4x$ form the boundaries of the shaded region. The solution set of which of the following systems of linear inequalities is given by the shaded region?

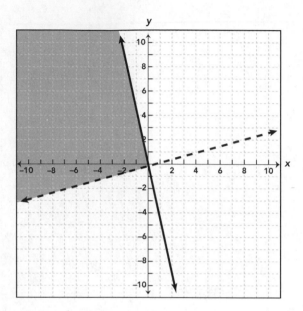

a. $y > \frac{1}{4}x, y \leq -4x$

b. $y < \frac{1}{4}x, y \leq -4x$

c. $y > \frac{1}{4}x, y \geq -4x$

d. $y < \frac{1}{4}x, y \geq -4x$

341. The graphs of the lines $2y - 3x = -6$ and $y = 5 - \frac{5}{2}x$ form the boundaries of the shaded region. The solution set of which of the following systems of linear inequalities is given by the shaded region?

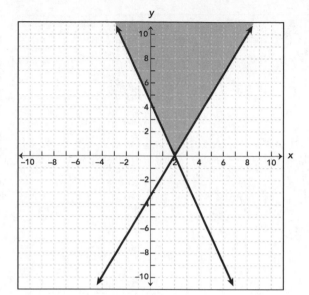

a. $2y - 3x > -6, y > 5 - \frac{5}{2}x$

b. $2y - 3x \leq -6, y \geq 5 - \frac{5}{2}x$

c. $2y - 3x > -6, y < 5 - \frac{5}{2}x$

d. $2y - 3x \geq -6, y \geq 5 - \frac{5}{2}x$

342. Which of the following graphs depicts the solution set for the following system of linear inequalities?

$$y > 2$$
$$y \leq 2x + 1$$

a.

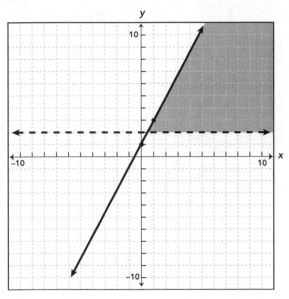

c.

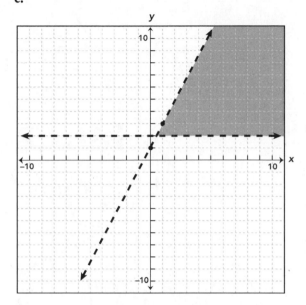

b.

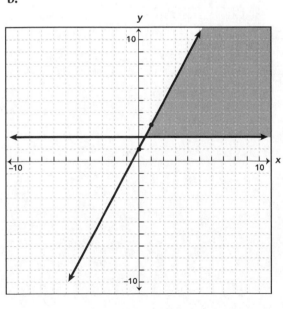

d.

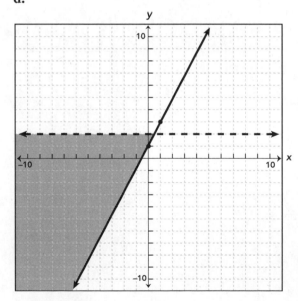

343. The graphs of the lines $5y = 8(x + 5)$ and $12(5 - x) = 5y$ form the boundaries of the shaded region. The solution set of which of the following systems of linear inequalities is given by the shaded region?

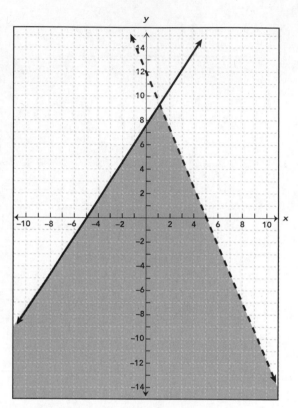

a. $5y < 8(x + 5), 12(5 - x) \geq 5y$
b. $5y \leq 8(x + 5), 12(5 - x) > 5y$
c. $5y < 8(x + 5), 12(5 - x) \leq 5y$
d. $5y > 8(x + 5), 12(5 - x) \leq 5y$

344. The graphs of the lines $y = 3x$ and $y = -5$ form the boundaries of the shaded region. The solution set of which of the following systems of linear inequalities is given by the shaded region?

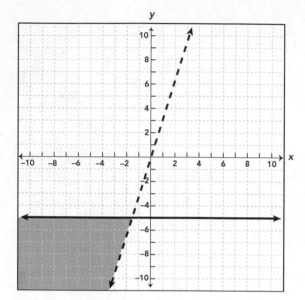

a. $y < 3x, y \geq -5$
b. $y < 3x, y \leq -5$
c. $y \geq 3x, y < -5$
d. $y > 3x, y \leq -5$

345. The graphs of the lines $9(y - 4) = 4x$ and $-9y = 2(x + 9)$ form the boundaries of the shaded region. The solution set of which of the following systems of linear inequalities is given by the shaded region?

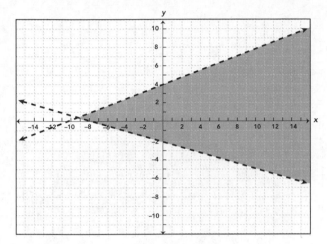

    a. $9(y - 4) \leq 4x, -9y \leq 2(x + 9)$
    b. $9(y - 4) < 4x, -9y < 2(x + 9)$
    c. $9(y - 4) \geq 4x, -9y \geq 2(x + 9)$
    d. $9(y - 4) > 4x, -9y > 2(x + 9)$

346. The graphs of the lines $y - x = 6$ and $11y = -2(x + 11)$ form the boundaries of the shaded region. The solution set of which of the following systems of linear inequalities is given by the shaded region?

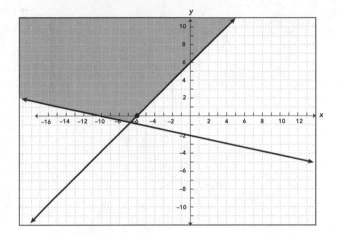

    a. $y - x \leq 6, 11y \leq -2(x + 11)$
    b. $y - x > 6, 11y > -2(x + 11)$
    c. $y - x \geq 6, 11y \geq -2(x + 11)$
    d. $y - x < 6, 11y < -2(x + 11)$

347. The graphs of the lines $5x - 2(y + 10) = 0$ and $2x + y = -3$ form the boundaries of the shaded region. The solution set of which of the following systems of linear inequalities is given by the shaded region?

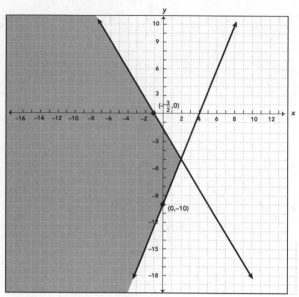

  a. $5x - 2(y + 10) > 0, 2x + y > -3$
  b. $5x - 2(y + 10) < 0, 2x + y < -3$
  c. $5x - 2(y + 10) \le 0, 2x + y \le -3$
  d. $5x - 2(y + 10) \ge 0, 2x + y \ge -3$

348. The graphs of the lines $7(y - 5) = -5x$ and $-3 = \frac{1}{4}(2x - 3y)$ form the boundaries of the shaded region below. The solution set of which of the following systems of linear inequalities is given by the shaded region?

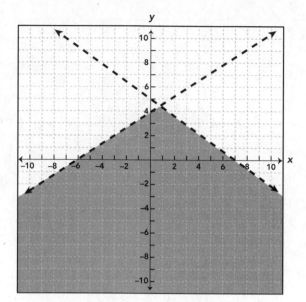

  a. $7(y-5) > -5x, -3 < \frac{1}{4}(2x - 3y)$
  b. $7(y-5) < -5x, -3 < \frac{1}{4}(2x - 3y)$
  c. $7(y-5) < -5x, -3 > \frac{1}{4}(2x - 3y)$
  d. $7(y-5) \ge -5x, -3 \le \frac{1}{4}(2x - 3y)$

349. For which of the following systems of linear inequalities is the solution set the entire Cartesian plane?
   a. $y > x + 3, y < x - 1$
   b. $2y - 6x \leq 4, y \geq 2 + 3x$
   c. $y \leq x, y \geq x$
   d. none of the above

350. For which of the following systems of linear inequalities is the solution set the empty set?
   a. $y > x - 3, y < x - 1$
   b. $y > x + 3, y < x - 1$
   c. $y \leq x, y \leq 2x$
   d. $y \leq 3x + 4, y \leq 3x + 6$

351. For which of the following systems of linear inequalities does the solution set consist precisely of the points in Quadrant III, not including either axis?
   a. $x \geq 0, y < 0$
   b. $x > 0, y > 0$
   c. $x \leq 0, y > 0$
   d. $x < 0, y < 0$

352. For which of the following systems of linear inequalities does the solution set consist of the points on a single line?
   a. $2y - 6x < 4, y \geq 2 - 3x$
   b. $2y - 6x \leq 4, y \geq 2 + 3x$
   c. $2y - 6x < 4, y > 2 + 3x$
   d. $2y - 6x \leq 4, y > 2 + 3x$

SECTION

3 ▶ POLYNOMIAL EXPRESSIONS

Algebraic expressions consisting of sums of constant multiples of positive integer powers of the variable are called *polynomials*. Simplyfying polynomials and understanding their graphical properties rely heavily on the use of factoring. These topics are the focus of the following seven problem sets.

Set 23 (Answers begin on page 187)

The problems in this set focus on the basic definition of and addition/subtraction of polynomials.

353. Compute $(x^2 - 3x + 2) + (x^3 - 2x^2 + 11)$.
   a. $x^3 + x^2 + 3x + 13$
   b. $x^3 - x^2 + 3x + 13$
   c. $x^3 + x^2 - 3x + 13$
   d. $x^3 - x^2 - 3x + 13$

354. Compute $(3x^2 - 5x + 4) - (-\frac{2}{3}x + 5)$.
   a. $3x^2 - \frac{13}{3}x - 1$
   b. $3x^2 - \frac{7}{3}x - 1$
   c. $3x^2 - \frac{13}{3}x - 9$
   d. $3x^2 - \frac{7}{3}x - 9$

355. Compute $(\frac{1}{3}x^2 - \frac{1}{5}x - \frac{2}{3}) - (\frac{2}{3}x^2 - \frac{7}{10}x + \frac{1}{2})$.
   a. $-\frac{1}{3}x^2 - \frac{9}{10}x - \frac{1}{6}$
   b. $-\frac{1}{3}x^2 + \frac{1}{2}x - \frac{7}{6}$
   c. $-\frac{1}{3}x^2 - \frac{9}{10}x + \frac{1}{6}$
   d. $\frac{1}{3}x^2 - \frac{9}{10}x + \frac{1}{6}$

356. Compute $(9a^2b + 2ab - 5a^2) - (-2ab - 3a^2 + 4a^2b)$.
   a. $5a^2b + 8a^2$
   b. $13a^2b + 4ab - 8a^2$
   c. $5a^2b + 4ab - 2a^2$
   d. $13a^2b - 4ab - 8a^2$

357. Compute $(\frac{1}{6}x^2 + \frac{2}{3}x + 1) + (2x - \frac{2}{3}x^2 + 4) - (\frac{7}{2} + 3x + \frac{1}{2}x^2)$.
   a. $-x^2 - \frac{1}{3}x + \frac{3}{2}$
   b. $-\frac{4}{3}x^2 + \frac{5}{3}x + \frac{3}{2}$
   c. $-x^2 + \frac{5}{3}x + \frac{3}{2}$
   d. $-x^2 + \frac{1}{3}x + \frac{3}{2}$

358. Compute $(2 - 3x^3) - [(3x^3 + 1) - (1 - 2x^3)]$.
   a. $2 + 8x^3$
   b. $-2 + 8x^3$
   c. $-2 - 8x^3$
   d. $2 - 8x^3$

359. What is the degree of the polynomial $-5x^8 + 9x^4 - 7x^3 - x^2$?
   a. -5
   b. 8
   c. 9
   d. 2

360. What is the degree of the polynomial $-\frac{3}{2}x + 5x^4 - 2x^2 + 12$?
   a. 5
   b. $-\frac{3}{2}$
   c. 4
   d. 1

361. What is the degree of the constant polynomial 4?
   a. 0
   b. 1
   c. A constant polynomial does not have a degree.
   d. none of the above

362. Which of the following is not a polynomial?
   a. 2
   b. $2 - 3x - x^2$
   c. $x - 3x^{-2}$
   d. $1 - [1 - x^2 - (2 - x)]$

363. Which of the following is not a polynomial?
   a. $-2^{-2}x - 3^{-1}$
   b. $(2x^0)^{-3} + 5^{-2}x^2 - 3^{-1}x$
   c. $(-2x)^{-1} - 2$
   d. All of the choices are polynomials.

364. Which of the following statements is always true?

    a. The difference of two polynomials is a polynomial.

    b. The sum of three polynomials is a polynomial.

    c. A trinomial minus a binomial is a polynomial.

    d. All of the above statements are true.

365. Which of the following statements is NOT true?

    a. The quotient of two polynomials is a polynomial.

    b. The product of a constant and a polynomial is a polynomial.

    c. The degree of the polynomial, in simplified form, is the highest power to which the variable is raised in the expression.

    d. The degree of a constant polynomial is zero.

366. Write the expression $-(-2x^0)^{-3} + 4^{-2}x^2 - 3^{-1}x - 2$ in simplified form.

    a. $\frac{1}{8}x^2 - \frac{1}{3}x - \frac{15}{8}$

    b. $\frac{1}{16}x^2 - \frac{1}{3}x - \frac{15}{8}$

    c. $-\frac{1}{8}x^2 - \frac{1}{3}x - 8$

    d. $-\frac{1}{8}x^2 + \frac{1}{3}x - \frac{15}{8}$

367. Compute $-(2 - (1 - 2x^2 - (2x^2 - 1))) - (3x^2 - (1 - 2x^2))$.

    a. $-9x^2 - 1$

    b. $9x^2 - 1$

    c. $9x^2 + 1$

    d. $-9x^2 + 1$

368. Compute $-2^2(2^{-3} - 2^{-2}x^2) + 3^3(3^{-2} - 3^{-3}x^3)$.

    a. $x^3 + x^2 + \frac{5}{2}$

    b. $-x^3 + x^2 + \frac{5}{2}$

    c. $-x^3 - x^2 + \frac{5}{2}$

    d. $x^3 - x^2 - \frac{5}{2}$

Set 24 (Answers begin on page 188)

The problems in this set focus on the multiplication of polynomials.

369. Compute $(3x^3)(7x^2)$.

    a. $21x^5$

    b. $21x^6$

    c. $10x^5$

    d. $10x^6$

370. $2x(5x^2 + 3y)$ is equivalent to which of the following expressions?

    a. $5x^3 + 6xy$

    b. $10x^2 + 6xy$

    c. $10x^3 + 6xy$

    d. $10x^3 + 6y$

371. Which of the following expressions is equivalent to $x^3 + 6x$?

    a. $x(x^2 + 6)$

    b. $x(x + 6)$

    c. $x(x^2 + 6x)$

    d. $x^2(x + 6)$

372. Compute $2x^2(3x + 4xy - 2xy^3)$.

    a. $6x^3 + 8x^2y - 4x^3y^3$

    b. $6x^3 + 8x^3y - 4x^3y^3$

    c. $6x^3 + 8x^3y - 4x^2y^3$

    d. $6x^2 + 8x^2y - 4x^3y^3$

373. Compute $7x^5(x^8 + 2x^4 - 7x - 9)$.

    a. $7x^{13} + 9x^9 - 14x^6 - 16x^5$

    b. $7x^{40} + 14x^{20} - 49x^5 - 63$

    c. $7x^{13} + 2x^4 - 7x - 9$

    d. $7x^{13} + 14x^9 - 49x^6 - 63x^5$

374. Compute $4x^2z(3xz^3 - 4z^2 + 7x^5)$.
   a. $12x^3z^4 - 8x^2z^3 + 28x^7z$
   b. $12x^2z^3 - 16x^2z^2 + 28x^{10}z$
   c. $12x^3z^4 - 16x^2z^3 + 28x^7z$
   d. $12x^3z^4 - 4z^2 + 7x^5$

375. What is the product of $(x - 3)(x + 7)$?
   a. $x^2 - 21$
   b. $x^2 - 3x - 21$
   c. $x^2 + 4x - 21$
   d. $x^2 + 7x - 21$
   e. $x^2 - 21x - 21$

376. What is the product of $(x - 6)(x - 6)$?
   a. $x^2 + 36$
   b. $x^2 - 36$
   c. $x^2 - 12x - 36$
   d. $x^2 - 12x + 36$
   e. $x^2 - 36x + 36$

377. What is the product of $(x - 1)(x + 1)$?
   a. $x^2 - 1$
   b. $x^2 + 1$
   c. $x^2 - x - 1$
   d. $x^2 - x + 1$
   e. $x^2 - 2x - 1$

378. What is the value of $(x + c)^2$?
   a. $x^2 + c^2$
   b. $x^2 + cx + c^2$
   c. $x^2 + c^2x^2 + c^2$
   d. $x^2 + cx^2 + c^2x + c^2$
   e. $x^2 + 2cx + c^2$

379. What is the product of $(2x + 6)(3x - 9)$?
   a. $5x^2 - 54$
   b. $6x^2 - 54$
   c. $6x^2 + 18x - 15$
   d. $6x^2 - 18x - 15$
   e. $6x^2 + 36x - 54$

380. Compute $-3x(x + 6)(x - 9)$.
   a. $-3x^3 + 6x - 54$
   b. $-x^3 + 3x^2 + 24x$
   c. $-3x^3 - 3x^2 - 54$
   d. $-3x^2 + 6x - 72$
   e. $-3x^3 + 9x^2 + 162x$

381. Compute $(x - 4)(3x^2 + 7x - 2)$.
   a. $3x^3 + 5x^2 - 30x - 8$
   b. $3x^3 + 5x^2 - 30x + 8$
   c. $3x^3 - 5x^2 - 30x + 8$
   d. $3x^3 - 5x^2 - 30x - 8$

382. Compute $(x - 6)(x - 3)(x - 1)$.
   a. $x^3 - 18$
   b. $x^3 - 9x - 18$
   c. $x^3 - 8x^2 + 27x - 18$
   d. $x^3 - 10x^2 - 9x - 18$
   e. $x^3 - 10x^2 + 27x - 18$

383. Which of the following equations is equivalent to $(5x + 1)(2y + 2) = 10xy + 12$?
   a. $10x + 2y + 2 = 10$
   b. $10x + y = 10$
   c. $5x + y = 5$
   d. $5x - y = 5$

384. Compute $(2x^3 - 2x^2 + 1)(6x^3 + 7x^2 - 5x - 9)$.
   a. $12x^6 + 2x^5 - 24x^4 - 2x^3 + 25x^2 - 5x - 9$
   b. $12x^6 - 2x^5 - 24x^4 + 2x^3 + 25x^2 - 5x - 9$
   c. $12x^6 - 2x^5 - 24x^4 - 2x^3 - 25x^2 + 5x - 9$
   d. $12x^6 + 2x^5 - 24x^4 + 2x^3 - 25x^2 + 5x - 9$

Set 25 (Answers begin on page 189)

The method of factoring out the greatest common factor (GCF) from a polynomial is the focus of this problem set.

385. Factor out the GCF: $15x - 10$
  a. $-5(3x - 2)$
  b. $5(3x - 2)$
  c. $-5(3x + 2)$
  d. This polynomial cannot be factored further.

386. Factor out the GCF: $9x^5 + 24x^2 - 6x$
  a. $3(3x^5 + 8x^2 - 2x)$
  b. $3x(3x^4 + 8x - 2)$
  c. $x(9x^4 + 24x - 6)$
  d. This polynomial cannot be factored further.

387. Factor out the GCF: $36x^4 - 90x^3 - 18x$
  a. $9x(4x^3 - 10x^2 - 2)$
  b. $18(2x^4 - 5x^3 - x)$
  c. $18x(2x^3 - 5x^2 - 1)$
  d. This polynomial cannot be factored further.

388. Factor out the GCF: $x^3 - x$
  a. $x(x^2 - 1)$
  b. $-x(x^2 + 1)$
  c. $-x(x^2 - 1)$
  d. This polynomial cannot be factored further.

389. Factor out the GCF: $5x^2 + 49$
  a. $5(x^2 + 49)$
  b. $5(x^2 + 44)$
  c. $5x(x + 49)$
  d. This polynomial cannot be factored further.

390. Factor out the GCF: $36 - 81x^2$
  a. $9(4 - 9x^2)$
  b. $9(4 - x^2)$
  c. $9(x^2 - 4)$
  d. This polynomial cannot be factored further.

391. Factor out the GCF: $125x^3 - 405x^2$
  a. $-5x^2(25x - 81)$
  b. $5x(25x^2 - 81)$
  c. $5x^2(25x - 81)$
  d. This polynomial cannot be factored further.

392. Factor out the GCF: $7^3x^3 - 7^2x^2 + 7x - 49$
  a. $-7(49x^3 + 7x^2 + x - 7)$
  b. $7(14x^3 - 7x^2 - x + 7)$
  c. $7(49x^3 - 7x^2 + x - 7)$
  d. This polynomial cannot be factored further.

393. Factor out the GCF: $5x(2x + 3) - 7(2x + 3)$
  a. $(2x + 3)(7 - 5x)$
  b. $(2x + 3)(5x - 7)$
  c. $(2x + 3)(5x + 7)$
  d. This polynomial cannot be factored further.

394. Factor out the GCF: $5x(6x - 5) + 7(5 - 6x)$
  a. $(5x - 7)(5 - 6x)$
  b. $(5x + 7)(6x - 5)$
  c. $(5x - 7)(6x - 5)$
  d. This polynomial cannot be factored further.

395. Factor out the GCF:
  $6(4x + 1) - 3y(1 + 4x) + 7z(4x + 1)$
  a. $(6 - 3y + 7z)(4x + 1)$
  b. $(6 - 3y - 7z)(4x + 1)$
  c. $(-6 + 3y - 7z)(1 + 4x)$
  d. This polynomial cannot be factored further.

396. Factor out the GCF: $5x(\frac{2}{3}x + 7) - (\frac{2}{3}x + 7)$
  a. $5x(\frac{2}{3}x + 7)$
  b. $(5x - 1)(\frac{2}{3}x + 7)$
  c. $(5x + 1)(\frac{2}{3}x + 7)$
  d. This polynomial cannot be factored further.

397. Factor out the GCF: $3x(x+5)^2 - 8y(x+5)^3 + 7z(x+5)^2$

   a. $(x+5)(3x - 8yx - 40y + 7z)$

   b. $(x+5)^2(-3x + 8yx + 40y - 7z)$

   c. $(x+5)^2(3x - 8yx - 40y + 7z)$

   d. This polynomial cannot be factored further.

398. Factor out the GCF: $8x^4y^2(x-9)^2 - 16x^3y^5$ $(x-9)^3 + 12\,x^5y^3(9-x)$

   a. $4x^3y^2(x-9)[2x^2 - 18x - 4y^3x^2 + 72xy^3 - 324y^3 - 3x^2y]$

   b. $4x^3y^2(x-9)[2x^2 - 18x + 4y^3x^2 - 72y^3x +324y^3 - 3x^2y]$

   c. $4x^3y^2(x-9)[2x^2 + 18x - 4y^3x^2 + 72y^3x - 324y^3 + 3x^2y]$

   d. This polynomial cannot be factored further.

399. Factor out the GCF: $8x^4y^2z(2w-1)^3 - 16x^2y^4z^3(2w-1)^3 + 12x^4y^4z(2w-1)^4$

   a. $4xyz(2w-1)^2[2x^3 - 4y^3z^3 + 6x^3y^3w - 3x^3y^3]$

   b. $4x^2y^2z(2w-1)^2[2x^2 - 4y^2z^2 + 6x^2y^2w - 3x^2y^2]$

   c. $4x^2y^2z(2w-1)^3[2x^2 - 4y^2z^2 + 6x^2y^2w - 3x^2y^2]$

   d. This polynomial cannot be factored further.

400. Factor out the GCF: $-22a^3bc^2(d-2)^3(1-e)^2 + 55a^2b^2c^2(d-2)^2(1-e) - 44a^2bc^4(d-2)(1-e)$

   a. $11a^2bc^2(d-2)(1-e)[2a(d-2)^2(1-e) + 5b(d-2) + 4c^2]$

   b. $11a^2bc^2(d-2)(1-e)[-2a(d-2)^2(1-e) + 5b(d-2) - 4c^2]$

   c. $11a^2bc^2(d-2)(1-e)[-2a(d-2)^2(1-e) + 5b(d-2) + 4c^2]$

   d. This polynomial cannot be factored further.

Set 26 (Answers begin on page 190)

The problems in this set focus on factoring polynomials that can be viewed as the difference of squares or as perfect trinomials squared.

401. Factor completely: $x^2 - 36$

   a. $(x-6)^2$

   b. $(x-6)(x+6)$

   c. $(x+6)^2$

   d. This polynomial cannot be factored further.

402. Factor completely: $144 - y^2$

   a. $(12-y)(12+y)$

   b. $(11-y)(11+y)$

   c. $(y-12)(y+12)$

   d. This polynomial cannot be factored further.

403. Factor completely: $4x^2 + 1$

   a. $(2x+1)^2$

   b. $(2x+1)(2x-1)$

   c. $(2x-1)^2$

   d. This polynomial cannot be factored further.

404. Factor completely: $9x^2 - 25$

   a. $(3x-5)$

   b. $(3x-5)(3x+5)$

   c. $(5x-3)(5x+3)$

   d. This polynomial cannot be factored further.

405. Factor completely: $121x^4 - 49z^2$

   a. $(11x^2 - 7z)(11x^2 + 7z)$

   b. $(12x^2 - 7z)(12x^2 + 7z)$

   c. $(7z - 11x^2)(7z + 11x^2)$

   d. This polynomial cannot be factored further.

406. Factor completely: $6x^2 - 24$

   a. $(6x-2)(x+2)$

   b. $6(x-2)^2$

   c. $6(x-2)(x+2)$

   d. This polynomial cannot be factored further.

407. Factor completely: $32x^5 - 162x$
  a. $2x(4x^2 + 9)$
  b. $2x(2x - 3)^2(2x + 3)^2$
  c. $2x(2x - 3)(2x + 3)(4x^2 + 9)$
  d. This polynomial cannot be factored further.

408. Factor completely: $28x(5 - x) - 7x^3(5 - x)$
  a. $7x(x - 2)(x + 2)(5 - x)$
  b. $7x(2 - x)(2 + x)(5 - x)$
  c. $7x(5 - x)(x^2 + 4)$
  d. This polynomial cannot be factored further.

409. Factor completely: $x^2(3x - 5) + 9(5 - 3x)$
  a. $(x - 3)(x + 3)(3x - 5)$
  b. $(x^2 + 9)(3x - 5)$
  c. $(x^2 + 9)(5 - 3x)$
  d. This polynomial cannot be factored further.

410. Factor completely: $x(x^2 + 7x) - 9x^3(x^2 + 7x)$
  a. $x^2(1 - 3x)(1 + 3x)(x + 7)$
  b. $x^2(x + 7)(1 + 9x^2)$
  c. $x^2(3x - 1)(3x + 1)(x + 7)$
  d. This polynomial cannot be factored further.

411. Factor completely: $1 + 2x + x^2$
  a. $(x - 1)^2$
  b. $(x + 1)^2$
  c. $(x + 1)(x + 2)$
  d. This polynomial cannot be factored further.

412. Factor completely: $4x^2 - 12x + 9$
  a. $(2x - 3)(2x + 3)$
  b. $(2x + 3)^2$
  c. $(2x - 3)^2$
  d. This polynomial cannot be factored further.

413. Factor completely: $75x^4 + 30x^3 + 3x^2$
  a. $3x^2(5x + 1)(5x - 1)$
  b. $3x^2(5x - 1)^2$
  c. $3x^2(5x + 1)^2$
  d. This polynomial cannot be factored further.

414. Factor completely: $9x^2(3 + 10x) - 24x(10x + 3) + 16(3 + 10x)$
  a. $(3 + 10x)(3x - 4)^2$
  b. $(3 + 10x)(3x - 4)(3x + 4)$
  c. $(3 + 10x)(3x + 4)^2$
  d. This polynomial cannot be factored further.

415. Factor completely: $1 - 6x^2 + 9x^4$
  a. $(1 + 3x^2)^2$
  b. $(1 - 3x^2)^2$
  c. $(1 - 3x^2)(1 + 3x^2)$
  d. This polynomial cannot be factored further.

416. Factor completely: $8x^7 - 24x^4 + 18x$
  a. $2x(2x^3 + 3)^2$
  b. $2x(2x^3 - 3)^2$
  c. $2x(2x^3 - 3)(2x^3 + 3)$
  d. This polynomial cannot be factored further.

Set 27 (Answers begin on page 190)

Factoring polynomials using the trinomial method is the focus of this problem set.

417. Factor completely: $x^2 + 2x - 8$
  a. $(x + 4)(x - 2)$
  b. $(x - 4)(x + 2)$
  c. $(x + 1)(x - 8)$
  d. This polynomial cannot be factored further.

418. Factor completely: $x^2 - 9x + 20$
  a. $(x - 4)(x - 5)$
  b. $(x + 2)(x - 10)$
  c. $-(x + 4)(x + 5)$
  d. This polynomial cannot be factored further.

419. Factor completely: $6x^2 + 11x - 2$
  a. $(2x + 2)(3x - 1)$
  b. $(3x + 2)(2x - 1)$
  c. $(x + 2)(6x - 1)$
  d. This polynomial cannot be factored further.

420. Factor completely: $12x^2 - 37x - 10$
  a. $(4x - 10)(3x + 1)$
  b. $(3x - 10)(4x + 1)$
  c. $(3x - 2)(4x + 5)$
  d. This polynomial cannot be factored further.

421. Factor completely: $7x^2 - 12x + 5$
  a. $(7x - 1)(x - 5)$
  b. $(7x + 1)(x + 5)$
  c. $(7x - 5)(x - 1)$
  d. This polynomial cannot be factored further.

422. Factor completely: $9 - 7x - 2x^2$
  a. $(9 + 2x)(1 - x)$
  b. $(3 + 2x)(3 - x)$
  c. $(3 + x)(3 - 2x)$
  d. This polynomial cannot be factored further.

423. Factor completely: $2x^3 + 6x^2 + 4x$
  a. $2(x + 2)(x^2 + 1)$
  b. $2(x^2 + 2)(x + 1)$
  c. $2x(x + 2)(x + 1)$
  d. This polynomial cannot be factored further.

424. Factor completely: $-4x^5 + 24x^4 - 20x^3$
  a. $4x^3(5 - x)(1 - x)$
  b. $4x^3(x - 5)(x - 1)$
  c. $-4x^3(x - 5)(x - 1)$
  d. This polynomial cannot be factored further.

425. Factor completely: $-27x^4 + 27x^3 - 6x^2$
  a. $-3x^2(3x + 1)(3x + 2)$
  b. $-3x^2(3x - 1)(3x - 2)$
  c. $3x^2(3x - 1)(3x - 2)$
  d. This polynomial cannot be factored further.

426. Factor completely: $x^2(x + 1) - 5x(x + 1) + 6(x + 1)$
  a. $(x + 1)(x + 3)(x - 2)$
  b. $(x + 1)(x - 3)(x - 2)$
  c. $(x - 1)(x - 3)(x + 2)$
  d. This polynomial cannot be factored further.

427. Factor completely:
  $2x^2(x^2 - 4) - x(x^2 - 4) + (4 - x^2)$
  a. $(x - 2)(x + 2)(2x + 1)(x - 1)$
  b. $(x - 2)(x + 2)(2x - 1)(x + 1)$
  c. $(x - 2)(x + 2)(2x - 1)(x - 1)$
  d. This polynomial cannot be factored further.

428. Factor completely: $27(x - 3) + 6x(x - 3) - x^2(x - 3)$
  a. $-(x - 3)(x + 3)(x + 9)$
  b. $-(x - 3)(x + 3)(x - 9)$
  c. $(x - 3)(x + 3)(x - 9)$
  d. This polynomial cannot be factored further.

429. Factor completely:
  $(x^2 + 4x + 3)x^2 + (x^2 + 4x + 3)3x + 2(x^2 + 4x + 3)$
  a. $(x + 1)^2(x - 2)(x - 3)$
  b. $(x - 1)^2(x + 2)(x + 3)$
  c. $(x + 1)^2(x + 2)(x + 3)$
  d. This polynomial cannot be factored further.

430. Factor completely: $18(x^2 + 6x + 8) - 2x^2(x^2 + 6x + 8)$
  a. $2(x + 2)(x + 4)(3 - x)(3 + x)$
  b. $2(x + 2)(x - 4)(x - 3)(x + 3)$
  c. $2(x - 2)(x + 4)(3 - x)(3 + x)$
  d. This polynomial cannot be factored further.

431. Factor completely:
  $2x^2(16 + x^4) + 3x(16 + x^4) + (16 + x^4)$
  a. $(16 + x^4)(2x + 1)(x + 1)$
  b. $(4 + x^2)(4 - x^2)(2x + 1)(x + 1)$
  c. $(4 + x^2)(2 - x^2)(2 + x)(2x + 1)(x + 1)$
  d. This polynomial cannot be factored further.

432. Factor completely:

$6x^2(1 - x^4) + 13x(1 - x^4) + 6(1 - x^4)$

a. $(1 - x)^2(1 + x)^2(2x + 3)(3x + 2)$

b. $(1 - x)^2(1 + x)^2(2x + 3)(3x + 2)$

c. $(1 - x)(1 + x)(1 + x^2)(2x + 3)(3x + 2)$

d. This polynomial cannot be factored further.

Set 28 (Answers begin on page 191)

This problem set focuses on finding roots of polynomials using factoring techniques and the Zero Factor Property.

433. Which of the following is a complete list of zeros for the polynomial $x^2 - 36$?

a. 6

b. −6

c. −6 and 6

d. 4 and 9

434. Which of the following is a complete list of zeroes for the polynomial $9x^2 - 25$?

a. $-\frac{5}{3}$ and $\frac{5}{3}$

b. −3 and 3

c. $-\frac{3}{5}$ and $\frac{3}{5}$

d. −3 and 5

435. Which of the following is a complete list of zeros for the polynomial $5x^2 + 49$?

a. 0

b. −1 and 0

c. −2 and 0

d. There are no zeros for this polynomial.

436. Which of the following is a complete list of zeros for the polynomial $6x^2 - 24$?

a. −2 and 4

b. −2 and 2

c. 2 and −4

d. There are no zeros for this polynomial.

437. Which of the following is a complete list of zeros for the polynomial $5x(2x + 3) - 7(2x + 3)$?

a. $\frac{2}{3}$ and $-\frac{5}{7}$

b. $\frac{2}{3}$ and $\frac{7}{5}$

c. $-\frac{2}{3}$ and $\frac{7}{5}$

d. $-\frac{3}{2}$ and $\frac{7}{5}$

438. Which of the following is a complete list of zeros for the polynomial $5x(\frac{2}{3}x + 7) - (\frac{2}{3}x + 7)$?

a. 0 and $-\frac{21}{2}$

b. $\frac{1}{5}$ and $\frac{21}{2}$

c. $-\frac{1}{5}$ and $-\frac{21}{2}$

d. $\frac{1}{5}$ and $-\frac{21}{2}$

439. Which of the following is a complete list of zeros for the polynomial $28x(5 - x) - 7x^3(5 - x)$?

a. 0, −2, 2, and 5

b. −2, 2, and 5

c. 0, −2, 2, and −5

d. −2, 2, and −5

440. Which of the following is a complete list of zeros for the polynomial $75x^4 + 30x^3 + 3x^2$?

a. $\frac{1}{5}$ and $-\frac{1}{5}$

b. 0 and $\frac{1}{5}$

c. 0 and $-\frac{1}{5}$

d. 0, $\frac{1}{5}$ and $-\frac{1}{5}$

441. Which of the following is a complete list of zeros for the polynomial $x^2 - 9x + 20$?

a. 4 and 5

b. −4 and 5

c. 4 and −5

d. −4 and −5

442. Which of the following is a complete list of zeros for the polynomial $12x^2 - 37x - 10$?

   a. $\frac{3}{10}$ and -4

   b. $\frac{3}{10}$ and $-\frac{1}{4}$

   c. $\frac{10}{3}$ and $-\frac{1}{4}$

   d. $-\frac{3}{10}$ and -4

443. Which of the following is a complete list of zeros for the polynomial $9 - 7x - 2x^2$?

   a. $-\frac{2}{9}$ and -1

   b. $-\frac{2}{9}$ and 1

   c. $-\frac{9}{2}$ and -1

   d. $-\frac{9}{2}$ and 1

444. Which of the following is a complete list of zeros for the polynomial $2x^3 + 6x^2 + 4x$?

   a. $0, 1,$ and 2

   b. $-2, -1,$ and 0

   c. -1 and 2

   d. 1 and 2

445. Which of the following is a complete list of zeros for the polynomial $-4x^5 + 24x^4 - 20x^3$?

   a. $0, 2,$ and 4

   b. $0, -2$ and 4

   c. $0, 1,$ and 5

   d. $1, 2,$ and 5

446. Which of the following is a complete list of zeros for the polynomial $2x^2(x^2 - 4) - x(x^2 - 4) + (4 - x^2)$?

   a. $1, 2, -2$ and $-\frac{1}{2}$

   b. $-2, -1,$ and 2

   c. -2 and 2

   d. $-2, 1,$ and 2

447. Which of the following is a complete list of zeros for the polynomial $2x^2(16 + x^4) + 3x(16 + x^4) + (16 + x^4)$?

   a. $-1, -\frac{1}{2}, 2,$ and -2

   b. -1 and $\frac{1}{2}$

   c. 1 and $-\frac{1}{2}$

   d. -1 and $-\frac{1}{2}$

448. Which of the following is a complete list of zeros for the polynomial $18(x^2 + 6x + 8) - 2x^2(x^2 + 6x + 8)$?

   a. $2, 3,$ and 4

   b. $-4, -2, -3,$ and 3

   c. $-2, -3,$ and 4

   d. $-2, 3,$ and 4

Set 29 (Answers begin on page 193)

This problem set focuses on solving polynomial inequalities.

449. Which of the following is the solution set for $x^2 - 36 > 0$?

   a. $(6, \infty)$

   b. $(-\infty, -6) \cup (6, \infty)$

   c. $(-\infty, -6] \cup [6, \infty)$

   d. the set of all real numbers

450. Which of the following is the solution set for $9x^2 - 25 \leq 0$?

   a. $\left(-\frac{5}{3}, \frac{5}{3}\right)$

   b. $\left[-\frac{5}{3}, \frac{5}{3}\right]$

   c. $\left(-\infty, -\frac{5}{3}\right) \cup \left(\frac{5}{3}, \infty\right)$

   d. the empty set

451. Which of the following is the solution set for
$5x^2 + 49 < 0$?
   a. $(-\infty, -\frac{49}{5})$
   b. $(-\infty, 0)$
   c. the empty set
   d. the set of all real numbers

452. Which of the following is the solution set for
$6x^2 - 24 \geq 0$?
   a. $(-\infty, -2] \cup [2, \infty)$
   b. $(-\infty, -2) \cup (2, \infty)$
   c. $(2, \infty)$
   d. $[-2, 2]$

453. Which of the following is the solution set for
$5x(2x + 3) - 7(2x + 3) > 0$?
   a. $(-\infty, -\frac{3}{2}) \cup (\frac{7}{5}, \infty)$
   b. $(-\infty, -\frac{3}{2}] \cup [\frac{7}{5}, \infty)$
   c. $(-\frac{3}{2}, \frac{7}{5})$
   d. $[-\frac{3}{2}, \frac{7}{5}]$

454. Which of the following is the solution set for
$5x(\frac{2}{3}x + 7) - (\frac{2}{3}x + 7) \leq 0$?
   a. $(-\frac{21}{2}, \frac{1}{5})$
   b. $[\frac{1}{5}, \infty)$
   c. $[-\frac{21}{2}, \frac{1}{5}]$
   d. $(-\infty, -\frac{21}{2}) \cup (\frac{1}{5}, \infty)$

455. Which of the following is the solution set for
$28x(5 - x) - 7x^3(5 - x) \geq 0$?
   a. $(-\infty, -2] \cup [0, 2] \cup [5, \infty)$
   b. $[-2, 0] \cup [2, 5]$
   c. $(-\infty, -2] \cup [-2, 0] \cup [5, \infty)$
   d. the set of all real numbers

456. Which of the following is the solution set for
$75x^4 + 30x^3 + 3x^2 \leq 0$?
   a. $(-\infty, 0)$
   b. $[-\frac{1}{5}, 0]$
   c. $\{-\frac{1}{5}, 0\}$
   d. the empty set

457. Which of the following is the solution set for
$x^2 - 9x + 20 < 0$?
   a. $(-\infty, 4) \cup (5, \infty)$
   b. $(-\infty, 4] \cup [5, \infty)$
   c. $(-\infty, 5]$
   d. $(4, 5)$

458. Which of the following is the solution set for
$12x^2 - 37x - 10 < 0$?
   a. $(-\infty, -\frac{1}{4}] \cup [\frac{10}{3}, \infty)$
   b. $(-\infty, -\frac{1}{4}) \cup (\frac{10}{3}, \infty)$
   c. $[-\frac{1}{4}, \frac{10}{3}]$
   d. $(-\frac{1}{4}, \frac{10}{3})$

459. Which of the following is the solution set for
$9 - 7x - 2x^2 > 0$?
   a. $[-\frac{9}{2}, 1]$
   b. $(-\infty, -\frac{9}{2}) \cup (1, \infty)$
   c. the set of all real numbers *except* $-\frac{9}{2}$ and 1
   d. $(-\frac{9}{2}, 1)$

460. Which of the following is the solution set for
$2x^3 + 6x^2 + 4x \geq 0$?
   a. $[-2, 0]$
   b. $[-2, -1] \cup [0, \infty)$
   c. $(-\infty, -2] \cup [-1, \infty)$
   d. $(-\infty, -1] \cup [0, \infty)$

461. Which of the following is the solution set for
$-4x^5 + 24x^4 - 20x^3 \geq 0$?

   a. $(-\infty,0]\cup[1,5]$

   b. $(-\infty,1]$

   c. $[0,1]\cup[5,\infty)$

   d. the set of all real numbers

462. Which of the following is the solution set for
$2x^2(x^2 - 4) - x(x^2 - 4) + (4 - x^2) < 0$?

   a. $(-2,-\frac{1}{2})\cup(1,2)$

   b. $(-\frac{1}{2},1)\cup(1,2)$

   c. $(-\infty,-\frac{1}{2})\cup(2,\infty)$

   d. $(-\infty,-2)\cup(1,2)$

463. Which of the following is the solution set for
$2x^2(16 + x^4) + 3x(16 + x^4) + (16 + x^4) \leq 0$?

   a. $(-\infty,-1]\cup[-\frac{1}{2},\infty)$

   b. $(-\infty,-1)\cup(-\frac{1}{2},\infty)$

   c. $[-1,-\frac{1}{2}]$

   d. $(-1,-\frac{1}{2})$

464. Which of the following is the solution set for
$18(x^2 + 6x + 8) - 2x^2(x^2 + 6x + 8) > 0$?

   a. $(-\infty,-4)\cup(-2,\infty)$

   b. $(-4,-3)\cup(-2,3)$

   c. $[-4,-3]\cup[-2,3]$

   d. the set of all real numbers *except* $-4, -3, -2,$ and 3

4 ▶ RATIONAL EXPRESSIONS

Quotients of polynomials are called *rational expression*s. The arithmetic of rational expressions closely resembles that of fractions. Simplifying and understanding the graphical properties of both polynomials and rational expressions relies heavily on the use of factoring. This is the focus of the following six problem sets.

Set 30 (Answers begin on page 197)

This problem set focuses on basic properties and simplification of rational expressions.

465. Simplify: $\frac{2z^2 - z - 15}{z^2 + 2z - 15}$

   a. $\frac{2z - 5}{z - 5}$

   b. $\frac{2z + 5}{z - 5}$

   c. $\frac{2z - 5}{z + 5}$

   d. $\frac{2z + 5}{z + 5}$

466. Simplify: $\frac{25(-x)^4}{x(5x^2)^2}$

   a. $-\frac{1}{x}$

   b. $\frac{5}{2x}$

   c. $-\frac{5}{2x}$

   d. $\frac{1}{x}$

467. Simplify: $\frac{z^3 - 16z}{8z - 32}$

   a. $\frac{z(z + 4)}{8}$

   b. $\frac{z(z - 4)}{8}$

   c. $\frac{-z(z + 4)}{8}$

   d. $\frac{-z(z - 4)}{8}$

468. Simplify: $\frac{y^2 - 64}{8 - y}$

   a. $-y + 8$

   b. $-(y + 8)$

   c. $-(y - 8)$

   d. $y + 8$

469. Simplify: $\frac{x^2 + 8x}{x^3 - 64x}$

   a. $\frac{1}{x - 8}$

   b. $\frac{x}{x - 8}$

   c. $\frac{x + 8}{x - 8}$

   d. $x - 8$

   e. $x + 8$

470. Simplify: $\frac{2x^2 + 4x}{4x^3 - 16x^2 - 48x}$

   a. $\frac{x + 2}{x - 6}$

   b. $\frac{x}{(x + 2)(x + 6)}$

   c. $\frac{1}{2x - 12}$

   d. $\frac{x + 2}{4x(x - 6)}$

   e. $\frac{2x(x + 2)}{x - 6}$

471. Which of the following makes the fraction $\frac{x^2 + 11x + 30}{4x^3 + 44x^2 + 120x}$ undefined?

   a. −6

   b. −4

   c. −3

   d. −2

   e. −1

472. The domain of the expression $\frac{2x}{x^3 - 4x}$ is

   a. $(-\infty, -2) \cup (2, \infty)$

   b. $(-\infty, 2) \cup (2, \infty)$

   c. $(-\infty, -2) \cup (-2, 0) \cup (0, 2) \cup (2, \infty)$

   d. $(-\infty, -2) \cup (-2, 2) \cup (2, \infty)$

473. Simplify: $\frac{x^2 - 16}{x^3 + x^2 - 20x}$

   a. $\frac{4}{x + 5}$

   b. $\frac{x + 4}{x}$

   c. $\frac{x + 4}{x + 5}$

   d. $\frac{x + 4}{x^2 + 5x}$

   e. $-\frac{16}{x^3 - 20x}$

474. Which of the following could be equal to $\frac{x}{4x}$?

   a. $-\frac{1}{4}$

   b. $\frac{0}{4}$

   c. 0.20

   d. $\frac{4}{12}$

   e. $\frac{5}{20}$

475. Which of the following values makes the expression $\frac{x-16}{x^2-16}$ undefined?

 a. -16

 b. -4

 c. -1

 d. 1

 e. 16

476. Which of the following lists of values makes the expression $\frac{x^2+7x+12}{x^3+3x^2-4x}$ undefined?

 a. $-4, 1$

 b. $-4, 0, 1$

 c. $-4, -1, 0$

 d. $-1, 0, 4$

 e. $0, 1, 4$

477. Simplify: $\frac{5x^2(x-1)-3x(x-1)-2(x-1)}{10x^2(x-1)+9x(x-1)+2(x-1)}$

 a. $\frac{x+1}{2x+1}$

 b. $\frac{x-1}{2x+1}$

 c. $\frac{x-1}{2x-1}$

 d. $\frac{x+1}{2x-1}$

478. Simplify: $\frac{6x^3-12x}{24x^2}$

 a. $\frac{x^2+2}{4x}$

 b. $\frac{x^2-2}{4x}$

 c. $-\frac{x^2-2}{4x}$

 d. $-\frac{x^2+2}{4x}$

479. Simplify: $\frac{4ab^2-b^2}{8a^2+2a-1}$

 a. $\frac{b^2}{2a+1}$

 b. $-\frac{b^2}{2a+1}$

 c. $\frac{b^2}{2a-1}$

 d. $-\frac{b^2}{2a-1}$

480. $\frac{(2x-5)(x+4)(-(2x-5)(x+1)}{9(2x-5)} =$

 a. $\frac{1}{3(2x-5)}$

 b. $\frac{1}{9(2x-5)}$

 c. $\frac{1}{3}$

 d. $\frac{1}{9}$

Set 31 (Answers begin on page 198)

This problem set focuses on adding and subtracting rational expressions.

481. Compute and simplify: $\frac{4x-45}{x-9} + \frac{2x-9}{x-9} - \frac{3x+1}{x-9}$

 a. $\frac{3x-55}{x-9}$

 b. $\frac{3x-53}{x-9}$

 c. $\frac{3x-55}{3x-27}$

 d. $\frac{3x-53}{3x-27}$

482. Compute and simplify: $\frac{5a}{ab^3} + \frac{2a}{ab^3}$

 a. $\frac{7}{b^3}$

 b. $7b^3$

 c. $\frac{7}{ab^3}$

 d. $\frac{7a}{b^3}$

483. Compute and simplify: $\frac{3-2x}{(x+2)(x-1)} - \frac{2-x}{(x-1)(x+2)}$

 a. $\frac{-1}{(x-1)(x+2)}$

 b. $\frac{1}{(x-1)(x+2)}$

 c. $-\frac{1}{x-2}$

 d. $-\frac{1}{x+2}$

484. Compute and simplify: $\frac{4}{sr^3} + \frac{2}{rs^2}$

   a. $\frac{4(s+r^2)}{s^2r^3}$

   b. $\frac{2s+r^2}{s^2r^3}$

   c. $\frac{4s+r^2}{s^2r^3}$

   d. $\frac{2(2s+r^2)}{s^2r^3}$

485. Compute and simplify: $\frac{2}{x(x-2)} - \frac{5-2x}{(x-2)(x-1)}$

   a. $\frac{2x^2-3x+2}{x(x-1)(x-2)}$

   b. $\frac{2x^2+3x-2}{x(x-1)(x-2)}$

   c. $\frac{2x^2-3x-2}{x(x-1)(x-2)}$

   d. $\frac{2x^2+3x+2}{x(x-1)(x-2)}$

486. Compute and simplify: $\frac{4}{t(t+2)} - \frac{2}{t}$

   a. $\frac{-2t}{t+2}$

   b. $\frac{-2}{t+2}$

   c. $\frac{2}{t+2}$

   d. $\frac{2t}{t+2}$

487. Compute and simplify: $\frac{1}{x(x+1)} - \frac{2x}{(x+1)(x+2)} + \frac{3}{x}$

   a. $\frac{x^2+10x-8}{x(x+1)(x+2)}$

   b. $\frac{x^2+10x+8}{x(x+1)(x+2)}$

   c. $\frac{x^2-10x+8}{x(x+1)(x+2)}$

   d. $\frac{x^2-10x-8}{x(x+1)(x+2)}$

488. Compute and simplify: $\frac{x}{2x+1} - \frac{1}{2x-1} + \frac{2x^2}{4x^2-1}$

   a. $\frac{(4x-1)(x-1)}{(2x-1)(2x+1)}$

   b. $\frac{(4x+1)(x-1)}{(2x-1)(2x+1)}$

   c. $\frac{(4x+1)(x+1)}{(2x-1)(2x+1)}$

   d. $\frac{(4x-1)(x+1)}{(2x-1)(2x+1)}$

489. Compute and simplify:

$$\frac{3y+2}{(y-1)^2} - \frac{7y-3}{(y-1)(y+1)} + \frac{5}{y+1}$$

   a. $-\frac{y+4}{(y-1)(y+1)}$

   b. $-\frac{y+4}{(y-1)(y+1)^2}$

   c. $\frac{y+4}{(y-1)^2}$

   d. $\frac{y+4}{(y-1)(y+1)}$

490. Compute and simplify: $\left(\frac{6z+12}{4z+3} + \frac{2z-6}{4z+3}\right)^{-1}$

   a. $\frac{1}{2}$

   b. 2

   c. -2

   d. $-\frac{1}{2}$

491. Compute and simplify: $\frac{4}{x-3} + \frac{x+5}{3-x}$

   a. $-\frac{x+1}{x-3}$

   b. $-\frac{x+1}{x+3}$

   c. $\frac{x+1}{x-3}$

   d. $\frac{x-1}{x-3}$

492. Compute and simplify: $\frac{x}{x^2-10x+24} - \frac{3}{x-6} + 1$

   a. $\frac{x-6}{x+4}$

   b. $\frac{x+6}{x+4}$

   c. $\frac{x-6}{x-4}$

   d. $\frac{x+6}{x-4}$

493. Compute and simplify: $\frac{-x^2+5x}{(x-5)^2} + \frac{x+1}{x+5}$

   a. $-\frac{x(x-9)}{(x-5)(x+5)}$

   b. $\frac{x(x+9)}{(x-5)(x+5)}$

   c. $-\frac{x(x+9)}{(x-5)(x+5)}$

   d. $\frac{5-9x}{(x-5)(x+5)}$

494. Compute and simplify: $\frac{2x^2}{x^4-1} - \frac{1}{x^2-1} + \frac{1}{x^2+1}$

    a. $-\frac{2}{x^2+1}$

    b. $\frac{2}{x^2+1}$

    c. $-\frac{2}{(x-1)(x+1)}$

    d. $\frac{2}{(x-1)(x+1)}$

495. Compute and simplify: $\frac{x-1}{x-2} - \frac{3x-4}{x^2-2x}$

    a. $-\frac{x-2}{x}$

    b. $\frac{x+2}{x}$

    c. $\frac{x-2}{x}$

    d. $-\frac{x+2}{x}$

496. Compute and simplify: $1 + \frac{x-1}{x} - \frac{3x-3}{x^2+3x}$

    a. $\frac{2x+1}{x+3}$

    b. $-\frac{2(x+1)}{x+3}$

    c. $\frac{2(x+1)}{x+3}$

    d. $-\frac{2x+1}{x+3}$

Set 32 (Answers begin on page 199)

This problem set focuses on multiplying and dividing rational expressions.

497. Compute and simplify: $\frac{4x^3y^2}{z^3} \cdot \frac{y^3z^4}{2x^5}$

    a. $\frac{2y^6z^2}{x^2}$

    b. $\frac{2y^6z}{x^3}$

    c. $\frac{2y^5z}{x^2}$

    d. $\frac{2y^6z}{x^2}$

498. Compute and simplify: $\frac{8a^4}{9-a^2} \cdot \frac{5a^2+13a-6}{24a-60a^2}$

    a. $\frac{2a^3}{3(3-a)}$

    b. $\frac{2a^3}{3(3+a)}$

    c. $\frac{2a^3}{-3(3+a)}$

    d. $\frac{2a^3}{-3(3-a)}$

499. Compute and simplify: $\frac{9x-2}{8-4x} \cdot \frac{10-5x}{2-9x}$

    a. $-\frac{5(2-x)^2}{(9x-2)}$

    b. $\frac{5}{-4}$

    c. $-\frac{9(2-x)^2}{20(9x-2)^2}$

    d. $\frac{9(2-x)^2}{20(9x-2)^2}$

500. Compute and simplify: $\frac{12x^2y}{-18xy} \cdot \frac{-24xy^2}{56y^3}$

    a. $\frac{2x^2}{7y}$

    b. $\frac{2x}{7y}$

    c. $\frac{2x^2}{7y^2}$

    d. $\frac{2x}{7y^2}$

501. Compute and simplify: $\frac{x^2-x-12}{3x^2-x-2} \div (3x^2-10x-8)$

    a. $\frac{x+3}{(3x+2)^2(x-1)}$

    b. $\frac{(x-4)^2(x+3)}{(x-1)}$

    c. $\frac{x+3}{(3x+2)(x-1)^2}$

    d. $\frac{(x+3)^2}{(3x+2)^2(x-1)}$

502. Compute and simplify: $\frac{x-3}{2x^3} \div \frac{x^2-3x}{4r}$

    a. $-\frac{(x-3)^2}{8x^3}$

    b. $\frac{(x-3)^2}{8x^3}$

    c. $\frac{2}{x^3}$

    d. $\frac{(x-3)}{x^3}$

503. Compute and simplify: $\frac{x^2-64}{x^2-9} \div \frac{6x^2+48x}{2x-6}$

    a. $\frac{x-8}{3(x+3)}$

    b. $\frac{x+8}{3(x+3)}$

    c. $-\frac{x-8}{3x(x+3)}$

    d. $\frac{x-8}{3x(x+3)}$

504. Compute and simplify: $\frac{2(x-6)^2}{x+5} \cdot \frac{-(5+x)}{4(x-6)}$

    a. $\frac{-(x-6)}{2}$

    b. $\frac{(x-6)}{2}$

    c. $\frac{-x-6}{2}$

    d. $\frac{x+6}{2}$

505. Compute and simplify: $\frac{9x^2y^3}{14x} \cdot \frac{21y}{15xy^2} \cdot \frac{10x}{12y^3}$

   a. $\frac{9x}{4y}$

   b. $\frac{3x}{4y}$

   c. $\frac{3x}{16y}$

   d. $\frac{9x}{16y^3}$

506. Compute and simplify: $\frac{4x^2+4x+1}{4x^2-4x} \div \frac{2x^2+3x+1}{2x^2-2x}$

   a. $\frac{(2x+1)(x+2)}{2(x-1)(x+1)}$

   b. $\frac{(2x+1)}{2(x-1)(x+1)}$

   c. $\frac{(2x+1)(x-2)}{2(x^2+1)}$

   d. $\frac{(2x-1)(x-2)}{2(x^2+1)}$

507. Compute and simplify: $\frac{x^2-1}{x^2+x} \cdot \frac{2x+2}{1-x^2} \cdot \frac{x^2+x-2}{x^2-x}$

   a. $\frac{2(x+2)}{x^2}$

   b. $\frac{-2(x+2)}{x^2}$

   c. $\frac{2x+2}{x^2}$

   d. $\frac{-2x+2}{x^2}$

508. Compute and simplify:

$(4x^2-8x-5) \div \left[\frac{-(x-3)}{x+1} \cdot \frac{2x^2-3x-5}{x-3}\right]$

   a. $\frac{2x+1}{-(2x-5)}$

   b. $2x+1$

   c. $-(2x+1)$

   d. $\frac{2x+1}{2x-5}$

509. Compute and simplify:

$\frac{a^2-b^2}{2a^2-3ab+b^2} \cdot \left[\frac{2a^2-7ab+3b^2}{a^2+ab} \div \frac{ab-3b^2}{a^2+2ab+b^2}\right]$

   a. $\frac{(a+b)^2}{ab}$

   b. $\frac{a^2+b^2}{ab}$

   c. $\frac{2(a+b)}{ab}$

   d. $\frac{a-3b}{ab}$

510. Compute and simplify: $(x-3) \div \frac{x^2+3x-18}{x}$

   a. $\frac{x}{x+6}$

   b. $\frac{x}{x-3}$

   c. $\frac{(x-3)^2(x+6)}{x}$

   d. $\frac{x-3}{x+6}$

511. Compute and simplify: $\left[\frac{x^2-x}{4y} \cdot \frac{10xy^2}{2x-2}\right] \div \frac{3x^2+3x}{15x^2y^2}$

   a. $\frac{5x^3y^3}{2(x+1)}$

   b. $\frac{25x^3y^3}{4(x+1)}$

   c. $\frac{5x^2y^2}{4(x-1)}$

   d. $\frac{25x^2y^3}{4(x-1)}$

512. Compute and simplify:

$\frac{x+2}{x^2+5x+6} \cdot \frac{2x^2+7x+3}{4x^2+4x+1} \cdot \frac{6x^2+5x+1}{3x^2+x} \cdot \frac{x^2-4}{x^2+2x}$

   a. $\frac{x+2}{x}$

   b. $\frac{-x+2}{x}$

   c. $\frac{x+2}{x^2}$

   d. $\frac{x-2}{x^2}$

Set 33 (Answers begin on page 200)

This problem set focuses on simplifying complex fractions and performing multiple operations involving rational expressions.

513. Compute and simplify: $1 - \frac{\frac{3}{4}}{\frac{9}{16}} \cdot \frac{1}{4} + \left[\frac{5}{2} - \frac{1}{4}\right]^2$

   a. $\frac{275}{36}$

   b. $\frac{275}{48}$

   c. $\frac{245}{48}$

   d. $\frac{245}{36}$

514. Compute and simplify: $\dfrac{\frac{2}{3}+\frac{3}{4}}{\frac{3}{4}-\frac{1}{2}}$

    a. $\dfrac{16}{3}$

    b. $\dfrac{17}{3}$

    c. $\dfrac{17}{6}$

    d. $\dfrac{15}{4}$

515. Compute and simplify:

$$\left[\dfrac{3x^2+6x}{x-5}+\dfrac{2+x}{5-x}\right]\div\dfrac{3x-1}{25-x^2}$$

    a. $(x+2)(x+5)$

    b. $-(x+2)(x+5)$

    c. $-(x-2)(x+5)$

    d. $(x+2)(x-5)$

516. Compute and simplify: $\left[\dfrac{1}{(x+h)^2}-\dfrac{1}{x^2}\right]\div h$

    a. $\dfrac{2x+h}{x^2(x+h)}$

    b. $\dfrac{-(2x+h)^2}{x^2(x+h)}$

    c. $\dfrac{2x+h}{x^2(x+h)^2}$

    d. $\dfrac{-(2x+h)}{x^2(x+h)^2}$

517. Compute and simplify: $\dfrac{a+\frac{1}{b}}{b+\frac{1}{a}}$

    a. $\dfrac{a}{ab}+1$

    b. ab

    c. $\dfrac{ab+1}{ba-1}$

    d. $\dfrac{a}{b}$

518. Compute and simplify: $\dfrac{\frac{3}{x}-\frac{1}{2}}{\frac{5}{4x}-\frac{1}{2x}}$

    a. $\dfrac{2(6-x)}{3}$

    b. $\dfrac{6-x}{3}$

    c. $-\dfrac{6-x}{3}$

    d. $\dfrac{6+x}{6}$

519. Compute and simplify:

$$\dfrac{\frac{5}{(x-1)^3}-\frac{2}{(x-1)^2}}{\frac{2}{(x-1)^3}-\frac{5}{(x-1)^4}}$$

    a. $-(x+1)$

    b. $(x-1)$

    c. $-(x-1)$

    d. $\dfrac{1}{(x-1)}$

520. Compute and simplify: $1-\dfrac{\frac{x}{5}}{1+\frac{x}{5}}$

    a. $\dfrac{5}{x+5}$

    b. $\dfrac{1}{x+5}$

    c. 1

    d. 0

521. Compute and simplify:

$$\dfrac{\frac{a-2}{a+2}-\frac{a+2}{a-2}}{\frac{a-2}{a+2}+\frac{a+2}{a-2}}$$

    a. $\dfrac{4a^2}{(a+2)(a-2)}$

    b. $\dfrac{-4a}{(a+2)(a-2)}$

    c. $\dfrac{4a}{a^2+4}$

    d. $-\dfrac{4a}{a^2+4}$

522. Compute and simplify:

$$\dfrac{\frac{4}{4-x^2}-1}{\frac{1}{x+2}+\frac{1}{x-2}}$$

    a. $-\dfrac{x}{2}$

    b. $\dfrac{x}{2}$

    c. $-\dfrac{x}{x-4}$

    d. $\dfrac{x+4}{x-4}$

523. Compute and simplify: $(a^{-1} + b^{-1})^{-1}$

    a. $\frac{ab}{b+a}$

    b. $\frac{b+a}{ab}$

    c. ab

    d. $a+b$

524. Compute and simplify: $\frac{x^{-1} - y^{-1}}{x^{-1} + y^{-1}}$

    a. 0

    b. $\frac{y-x}{y+x}$

    c. $\frac{y+x}{y-x}$

    d. 1

525. Compute and simplify:

$$\left[\frac{x^2+4x-5}{2x^2+x-3} \cdot \frac{2x+3}{x+1}\right] - \frac{2}{x+2}$$

    a. $\frac{x^2-5x+8}{(x+1)(x+2)}$

    b. $\frac{x^2+5x+8}{(x+1)(x+2)}$

    c. $\frac{x^2-5x-8}{(x+1)(x+2)}$

    d. $-\frac{x^2-5x+8}{(x+1)(x+2)}$

526. Compute and simplify: $\left[\frac{x+5}{x-3} - x\right] \div \frac{1}{x-3}$

    a. $(x+5)(x-1)$

    b. $(x-5)(x+1)$

    c. $-(x-5)(x+1)$

    d. $-(x+5)(x+1)$

527. Compute and simplify: $\left[3 + \frac{1}{x+3}\right] \cdot \frac{x+3}{x-2}$

    a. $\frac{x+3}{3x+10}$

    b. $\frac{3x+10}{x-2}$

    c. $\frac{3x+10}{x+3}$

    d. $\frac{x+3}{x-2}$

528. Compute and simplify: $1 - \left(\frac{2}{x} - \left(\frac{3}{2x} - \frac{1}{6x}\right)\right)$

    a. $\frac{x+2}{x}$

    b. $\frac{x-2}{x}$

    c. $\frac{3x+2}{3x}$

    d. $\frac{3x-2}{3x}$

Set 34 (Answers begin on page 202)

This problem set focuses on solving rational equations.

529. Solve: $\frac{3}{x} = 2 + x$

    a. -3 and 1

    b. -3 only

    c. 1 only

    d. There are no solutions.

530. Solve: $\frac{2}{3} - \frac{3}{x} = \frac{1}{2}$

    a. $\frac{7}{18}$

    b. $\frac{18}{7}$

    c. -18

    d. 18

531. Solve: $\frac{2t}{t-7} + \frac{1}{t-1} = 2$

    a. -2

    b. 2

    c. $\frac{7}{5}$

    d. $\frac{5}{7}$

532. Solve: $\frac{x+8}{x+2} + \frac{12}{x^2+2x} = \frac{2}{x}$

    a. -4 only

    b. 4 only

    c. -4 and -2

    d. 4 and -2

533. Solve: $\frac{x}{x-3} + \frac{2}{x} = \frac{3}{x-3}$

    a. 3

    b. 2 and 3

    c. -2

    d. -2 and 3

534. Solve: $\frac{3}{x+2} + 1 = \frac{6}{(2-x)(2+x)}$
   a. −4 and 1
   b. 1 and 4
   c. −1 and 4
   d. −4 and −1

535. Solve: $\frac{10}{(2x-1)^2} = 4 + \frac{3}{2x-1}$
   a. $-\frac{9}{8}$ only
   b. $\frac{-1}{2}$ and $\frac{9}{8}$
   c. $\frac{1}{2}$ and $\frac{9}{8}$
   d. $\frac{1}{2}$ and $-\frac{9}{8}$

536. Solve for q: $\frac{1}{f} = (k-1)\left[\frac{1}{pq} + \frac{1}{q}\right]$
   a. $q = \frac{f(k-1)}{p}$
   b. $q = \frac{f(k-1)(1+p)}{p}$
   c. $q = \frac{f(k+1)(1+p)}{p}$
   d. $q = \frac{f(k+1)}{p}$

537. Solve: $\frac{x-1}{x-5} = \frac{4}{x-5}$
   a. 3 only
   b. −5 only
   c. 5 only
   d. There are no solutions.

538. Solve: $\frac{22}{2p^2-9p-5} - \frac{3}{2p+1} = \frac{2}{p-5}$
   a. −5 and 1
   b. −5 only
   c. 5 only
   d. There are no solutions.

539. Solve: $\frac{x+1}{x^3-9x} - \frac{1}{2x^2+x-21} = \frac{1}{2x^2+13x+21}$
   a. $\frac{7}{9}$ only
   b. $-\frac{7}{9}$ only
   c. $-\frac{9}{7}$ only
   d. There are no solutions.

540. Solve: $\frac{x}{x+1} - \frac{3}{x+4} = \frac{3}{x^2+5x+4}$
   a. −3 and −2
   b. 2 and 3
   c. −3 and 2
   d. −2 and 3

541. Solve: $1 + \frac{2}{x-3} = \frac{4}{x^2-4x-3}$
   a. −1 and 1
   b. 3 only
   c. −1 only
   d. −1 and 3

542. Solve: $\frac{3}{x+2} = \frac{x-3}{x-2}$
   a. 2 and 4
   b. 4 only
   c. 0 only
   d. 0 and 4

543. Solve: $\frac{t+1}{t-1} = \frac{4}{t^2-1}$
   a. −1 and −3
   b. 1 only
   c. −3 only
   d. −1 only

544. Solve for v_1: $v = \frac{v_1 + v_2}{1 + \frac{v_1 v_2}{c^2}}$
   a. $v_1 = \frac{c^2(v_2 - v)}{vv_2 - c^2}$
   b. $v_1 = \frac{c^2(v - v_2)}{vv_2 - c^2}$
   c. $v_1 = \frac{c^2(v - v_2)}{vv_2}$
   d. $v_1 = -\frac{c^2(v_2 + v_2)}{c^2 - vv_2}$

Set 35 (Answers begin on page 206)

This problem set focuses on solving rational inequalities.

545. Determine the solution set for the inequality
$\frac{(x-1)(x+2)}{(x+3)^2} \leq 0.$
a. $(-2,1)$
b. $[-2,1]$
c. $(-3,-2]$
d. $(-3,-2] \cup [1,\infty)$

546. Determine the solution set for the inequality
$\frac{x^2+9}{x^2-2x-3} > 0.$
a. $(-1,\infty)$
b. $(-1,3)$
c. $(-\infty,-1) \cup (3,\infty)$
d. $(-\infty,3)$

547. Determine the solution set for the inequality
$\frac{-x^2-1}{6x^4-x^3-2x^2} \geq 0.$
a. $(-\frac{1}{2},0) \cup (0,\frac{2}{3})$
b. $[-\frac{1}{2},\frac{2}{3}]$
c. $(-\infty,-\frac{1}{2}] \cup [\frac{2}{3},\infty)$
d. $(-\frac{1}{2},\frac{2}{3})$

548. Determine the solution set for the inequality
$\frac{\frac{1}{x}-\frac{1}{x+1}}{x+2} \geq 0.$
a. $(-\infty,-2) \cup (-1,0)$
b. $(-\infty,-2] \cup (-1,0)$
c. $(-2,-1) \cup (0,\infty)$
d. $[-2,-1) \cup [0,\infty)$

549. Determine the solution set for the inequality
$\frac{2z^2+5z+3}{z^2-3z-4} \leq 0.$
a. $(-\infty,-\frac{3}{2}) \cup (4,\infty)$
b. $[-\frac{3}{2},-1) \cup (-1,4)$
c. $(-\frac{3}{2},4)$
d. $[-\frac{3}{2},4)$

550. Determine the solution set for the inequality
$\frac{25(-x)^4}{x(5x^2)^2} \leq 0.$
a. the empty set
b. the set of all real numbers
c. $(0,\infty)$
d. $(-\infty,0)$

551. Determine the solution set for the inequality
$\frac{z^3-16z}{8z-32} < 0.$
a. $(0,4)$
b. $(-\infty,-4)$
c. $(4,\infty)$
d. $(-4,0)$

552. Determine the solution set for the inequality
$\frac{y^2-64}{8-y} \leq 0.$
a. $[-8,8]$
b. $[-8,8) \cup (8,\infty)$
c. $(-8,8)$
d. $(-8,8) \cup (8,\infty)$

553. Determine the solution set for the inequality
$\frac{x^2+8x}{x^3-64x} > 0.$
a. $(8,\infty)$
b. $[8,\infty)$
c. $(-\infty,8)$
d. $(-\infty,8]$

554. Determine the solution set for the inequality

$$\frac{5x^2(x-1) - 3x(x-1) - 2(x-1)}{10x^2(x-1) + 9x(x-1) + 2(x-1)} \le 0.$$

 a. $(-\frac{1}{2}, -\frac{2}{5}) \cup (-\frac{2}{5}, 1)$

 b. $[-\frac{1}{2}, -\frac{2}{5}) \cup (-\frac{2}{5}, 1]$

 c. $(-\infty, -\frac{1}{2}] \cup (-\frac{2}{5}, 1)$

 d. $(-\infty, -\frac{1}{2}] \cup [-\frac{2}{5}, \infty)$

555. Determine the solution set for the inequality

$$\frac{6x^3 - 24x}{24x^2} \ge 0.$$

 a. $[-2, 0] \cup [2, \infty)$

 b. $(-\infty, 0] \cup [2, \infty)$

 c. $[-2, 0) \cup [2, \infty)$

 d. $[0, 2]$

556. Determine the solution set for the inequality

$$\frac{(2x-5)(x+4) - (2x-5)(x+1)}{9(2x-5)} < 0.$$

 a. $(\frac{5}{2}, \infty)$

 b. $(-\infty, \frac{5}{2})$

 c. the empty set

 d. the set of all real numbers

557. Determine the solution set for the inequality

$$\frac{3-2x}{(x+2)(x-1)} - \frac{2-x}{(x-1)(x+2)} \ge 0.$$

 a. $(-\infty, -2]$

 b. $(-\infty, 1]$

 c. $(-\infty, -2)$

 d. $[-2, 1]$

558. Determine the solution set for the inequality

$$\left[\frac{x+5}{x-3} - x \right] \div \frac{1}{x-3} > 0.$$

 a. $(-1, 5)$

 b. $(-1, 3) \cup (3, 5)$

 c. $(-\infty, 3) \cup (3, \infty)$

 d. $(-\infty, 3) \cup (3, 5)$

559. Determine the solution set for the inequality

$$\frac{x}{2x+1} - \frac{1}{2x-1} + \frac{2x^2}{4x^2-1} \le 0.$$

 a. $[-\frac{1}{2}, -\frac{1}{4}] \cup (\frac{1}{2}, 1]$

 b. $[-\frac{1}{2}, -\frac{1}{4}]$

 c. $[-\frac{1}{2}, -\frac{1}{4}] \cup [\frac{1}{2}, 1]$

 d. $(-\frac{1}{2}, -\frac{1}{4}] \cup (\frac{1}{2}, 1]$

560. Determine the solution set for the inequality

$$\frac{3y+2}{(y-1)^2} - \frac{7y-3}{(y-1)(y+1)} + \frac{5}{y+1} < 0.$$

 a. $(-\infty, 4]$

 b. $(-\infty, -4]$

 c. $(-\infty, 4)$

 d. $(-\infty, -4)$

5 ▶ RADICAL EXPRESSIONS AND QUADRATIC EQUATIONS

An algebraic expression involving a term raised to a fractional exponent is a *radical expression*. The arithmetic of such expressions is a direct application of the familiar exponent rules. Sometimes, raising a negative real number to a fractional exponent results in a complex number of the form $a + bi$, where a and b are real numbers and $i = \sqrt{-1}$; the arithmetic of complex numbers resembles the algebra of binomials.

Various methods can be used to solve quadratic equations, and the solutions often involve radical terms. These topics are reviewed in the seven problem sets in this section.

Set 36 (Answers begin on page 211)

The definition of fractional powers and the simplification of expressions involving radicals are the focus of this problem set.

561. −5 is a third root of what real number?
   a. $\sqrt[3]{-5}$
   b. 25
   c. −125
   d. −625

562. Which of the following are second roots (i.e., square roots) of 49?
   a. 7 only
   b. −7 only
   c. 7 and −7
   d. none of the above

563. Which of the following is the principal fourth root of 625?
   a. 5
   b. −5
   c. 25
   d. −25

564. Simplify: $\sqrt[5]{-32}$
   a. $2\sqrt[5]{2}$
   b. $-2\sqrt[5]{2}$
   c. 2
   d. −2

565. Which of the following is a value of b that satisfies the equation $\sqrt[3]{b} = 4$?
   a. 64
   b. $\sqrt[3]{4}$
   c. 16
   d. none of the above

566. Simplify: $\sqrt[4]{3^{12}}$
   a. 27
   b. 9
   c. 81
   d. 243

567. Simplify: $\sqrt[5]{5^{15}}$
   a. 5
   b. 15
   c. 125
   d. 625

568. Find a number b that satisfies the following: $\sqrt[4]{(2^b)^4} = 8$
   a. 2
   b. 3
   c. 4
   d. There is no such value of b.

569. Simplify: $64^{\frac{1}{6}}$
   a. $2\sqrt[6]{2}$
   b. 2
   c. $\frac{64}{6}$
   d. none of the above

570. Simplify: $49^{\frac{5}{2}}$
   a. $\frac{245}{2}$
   b. 343
   c. 35
   d. 16,807

571. Simplify: $81^{-\frac{3}{4}}$
   a. $\frac{1}{27}$
   b. $-\frac{243}{4}$
   c. 27
   d. 9

572. Simplify: $32^{\frac{3}{5}}$
   a. $\frac{96}{5}$
   b. $\frac{1}{8}$
   c. 8
   d. 64

573. Simplify: $\left(\frac{8}{27}\right)^{-\frac{2}{3}}$
   a. $-\frac{16}{135}$
   b. $\frac{4}{9}$
   c. $\frac{9}{4}$
   d. $\frac{3}{2}$

574. Simplify: $(-64)^{-\frac{1}{3}}$
   a. $-\frac{1}{4}$
   b. -4
   c. -16
   d. $-\frac{1}{16}$

575. Simplify: $(4x^{-4})^{-\frac{1}{2}}$
   a. x^2
   b. $2x^2$
   c. $\frac{x^2}{2}$
   d. $2x^{-2}$

576. Simplify: $4\sqrt{x^{144}}$
   a. x^{36}
   b. $4x^{72}$
   c. $4x^{36}$
   d. $2x^{72}$

Set 37 (Answers begin on page 211)

The simplification of more complicated radical expressions is the focus of this problem set.

577. Simplify: $\sqrt[3]{9} \cdot \sqrt[3]{-3}$
   a. 3
   b. -3
   c. $\sqrt[3]{-12}$
   d. $\sqrt[3]{-81}$

578. Simplify, assuming $x > 0$: $\frac{\sqrt{x^5}}{\sqrt{x^7}}$
   a. x
   b. $\frac{1}{x}$
   c. $\frac{1}{x^2}$
   d. x^2

579. Simplify: $a^3\sqrt{a^3}$
   a. $a^4\sqrt{a}$
   b. a^5
   c. $a^5\sqrt{a}$
   d. a^6
   e. a^9

580. Simplify: $\frac{4\sqrt{g}}{\sqrt{4g}}$
   a. 2
   b. 4
   c. $\sqrt{g}$
   d. $\frac{2\sqrt{g}}{g}$
   e. $2\sqrt{g}$

581. Simplify: $\frac{\sqrt[3]{27y^3}}{\sqrt{27y^2}}$
   a. $\frac{\sqrt{3}}{3}$
   b. $\sqrt{3}$
   c. $\frac{y\sqrt{3}}{3}$
   d. y
   e. $y\sqrt{3}$

582. Simplify: $\dfrac{\sqrt{a^2b} \cdot \sqrt{ab^2}}{\sqrt{ab}}$
   a. $\dfrac{\sqrt{ab}}{ab}$
   b. $\sqrt{ab}$
   c. ab
   d. $ab\sqrt{ab}$
   e. a^2b^2

583. Simplify: $\sqrt{(4g^2)^3\,(g^4)}$
   a. $8g^3$
   b. $8g^4$
   c. $8g^5$
   d. $8g^{10}$
   e. $8g^{12}$

584. Simplify: $\dfrac{9pr}{(pr)^{-\frac{3}{2}}}$
   a. $\sqrt{3pr}$
   b. $\dfrac{3}{pr}$
   c. $3\sqrt{pr}$
   d. $3pr$
   e. $3p^2r^2$

585. If $n=20$, what is the value of $\dfrac{\sqrt{n+5}}{\sqrt{n}}\left(\dfrac{n}{2}\sqrt{5}\right)$?
   a. 5
   b. $\dfrac{5\sqrt{5}}{2}$
   c. 10
   d. $5\sqrt{5}$
   e. 25

586. Simplify: $\sqrt{\dfrac{125}{9}}$
   a. $\dfrac{5}{3}$
   b. $\dfrac{\sqrt{5}}{3}$
   c. $\dfrac{5\sqrt{5}}{3}$
   d. $\dfrac{\sqrt{5}}{9}$

587. Simplify: $\dfrac{\sqrt[4]{243}}{\sqrt[4]{3}}$
   a. $\sqrt[4]{9}$
   b. $3\sqrt[4]{3}$
   c. $\sqrt[4]{3}$
   d. 3

588. Simplify: $\sqrt{x^2+4x+4}$
   a. $\sqrt{x+2}$
   b. $x+2\sqrt{x+1}$
   c. $x+2\sqrt{x+2}$
   d. $x+2$

589. Simplify: $\sqrt[4]{32x^8}$
   a. $x^2\sqrt[4]{8}$
   b. $x^2\sqrt[4]{4}$
   c. $2x\sqrt[4]{4}$
   d. $2x^2\sqrt[4]{2}$

590. Simplify: $\sqrt[4]{x^{21}}$
   a. $x^4\sqrt[4]{x}$
   b. $x^5\sqrt[4]{x}$
   c. $x^4\sqrt[4]{x^3}$
   d. $x^3\sqrt[4]{x^3}$

591. Simplify: $\sqrt[3]{54x^5}$
   a. $2x\sqrt[3]{3x^2}$
   b. $3x\sqrt[3]{2x^2}$
   c. $3x^2\sqrt[3]{2x}$
   d. $2x^2\sqrt[3]{3x}$

592. Simplify: $\sqrt{x^3+40x^2+400x}$
   a. $(x+20)\sqrt{x}$
   b. $x\sqrt{x}+2x\sqrt{10}+20\sqrt{x}$
   c. $x\sqrt{x+20}$
   d. This radical expression cannot be simplified further.

Set 38 (Answers begin on page 212)

This problem set focuses on the arithmetic of radical expressions, including those involving complex numbers.

593. Simplify: $\sqrt{-25}$
- **a.** 5
- **b.** $5i$
- **c.** -5
- **d.** $-5i$

594. Simplify: $\sqrt{-32}$
- **a.** $4i\sqrt{2}$
- **b.** $-4i\sqrt{2}$
- **c.** $-3i\sqrt{2}$
- **d.** $3i\sqrt{2}$

595. Simplify: $-\sqrt{48} + 2\sqrt{27} - \sqrt{75}$
- **a.** $-3\sqrt{3}$
- **b.** $-3\sqrt{5}$
- **c.** $5\sqrt{3}$
- **d.** $-5\sqrt{5}$

596. Simplify: $3\sqrt{3} + 4\sqrt{5} - 8\sqrt{3}$
- **a.** $-\sqrt{8}$
- **b.** $-4\sqrt{3} + 4\sqrt{5}$
- **c.** $4\sqrt{3} - 3\sqrt{5}$
- **d.** $-5\sqrt{3} + 4\sqrt{5}$

597. Simplify: $xy\sqrt{8xy^2} + 3y^2\sqrt{18x^3}$
- **a.** $11y\sqrt{2x}$
- **b.** $11xy\sqrt{2xy}$
- **c.** $11x^2y^2\sqrt{2}$
- **d.** $11xy^2\sqrt{2x}$

598. Simplify: $\sqrt{\frac{18}{25}} + \sqrt{\frac{32}{9}}$
- **a.** $\frac{29\sqrt{2}}{5}$
- **b.** $\frac{23\sqrt{2}}{15}$
- **c.** $\frac{29\sqrt{2}}{15}$
- **d.** $\frac{32\sqrt{2}}{15}$

599. Simplify: $(5 - \sqrt{3})(7 + \sqrt{3})$
- **a.** $32 - 2\sqrt{3}$
- **b.** $32 + 2\sqrt{3}$
- **c.** $16 - 4\sqrt{3}$
- **d.** $16 + 4\sqrt{3}$

600. Simplify: $(4 + \sqrt{6})(6 - \sqrt{15})$
- **a.** $24 + 4\sqrt{15} + 6\sqrt{6} - 3\sqrt{10}$
- **b.** $24 - 4\sqrt{15} + 6\sqrt{6} - 3\sqrt{10}$
- **c.** $24 + 4\sqrt{15} - 6\sqrt{6} - 3\sqrt{10}$
- **d.** $24 + 4\sqrt{15} - 6\sqrt{6} + 3\sqrt{10}$

601. Simplify: $\frac{-10 + \sqrt{-25}}{5}$
- **a.** $-2 + i$
- **b.** $2 + i$
- **c.** $2 - i$
- **d.** $-2 - i$

602. Simplify: $(4 + 2i)(4 - 2i)$
- **a.** 12
- **b.** 16
- **c.** 20
- **d.** $20i$

603. Simplify: $(4 + 2i)^2$
- **a.** $12 - 16i$
- **b.** $16 + 16i$
- **c.** $16 - 16i$
- **d.** $12 + 16i$

604. Simplify: $\sqrt{21}\left(\dfrac{\sqrt{3}}{\sqrt{7}} + \dfrac{\sqrt{7}}{\sqrt{3}}\right)$

    a. $10\sqrt{3}$

    b. 10

    c. $10\sqrt{7}$

    d. $10\sqrt{21}$

605. Simplify: $(2 + \sqrt{3x})^2$

    a. $4\sqrt{3x} + 7x$

    b. $4\sqrt{3x} + 7 + x$

    c. $4 + 4\sqrt{3x} + x\sqrt{3}$

    d. $4 + 4\sqrt{3x} + 3x$

606. Simplify: $(\sqrt{3} + \sqrt{7})(2\sqrt{3} - 5\sqrt{7})$

    a. $-29 - 3\sqrt{21}$

    b. $-29 + 3\sqrt{21}$

    c. $29 - 3\sqrt{21}$

    d. $29 + 3\sqrt{21}$

607. Simplify: $\dfrac{1}{3 - 5\sqrt{2}}$

    a. $\dfrac{3 - 5\sqrt{2}}{41}$

    b. $-\dfrac{3 - 5\sqrt{2}}{41}$

    c. $\dfrac{3 + 5\sqrt{2}}{41}$

    d. $-\dfrac{3 + 5\sqrt{2}}{41}$

608. Simplify by rationalizing the denominator: $\dfrac{\sqrt{2x}}{2 - 3\sqrt{x}}$

    a. $-\dfrac{2\sqrt{2x} + 6x}{4 + 9x}$

    b. $\dfrac{\sqrt{2x} + 2x}{2 - 3x}$

    c. $\dfrac{2\sqrt{2x} - 6x}{4 - 9x}$

    d. $\dfrac{2\sqrt{2x} + 3x\sqrt{2}}{4 - 9x}$

Set 39 (Answers begin on page 213)

This problem set focuses on solving equations involving radicals.

609. Solve: $\sqrt{7 + 3x} = 4$

    a. 3 only

    b. −3 only

    c. $3i$

    d. There is no solution.

610. Solve: $\sqrt{4x + 33} = 2x - 1$

    a. 4 only

    b. −2 only

    c. −2 and 4

    d. There is no solution.

611. If $a^{\frac{2}{3}} = 6$, then $a^{\frac{4}{3}} =$

    a. $\sqrt{3}$

    b. $\sqrt{6}$

    c. $3\sqrt{6}$

    d. $6\sqrt{6}$

    e. 36

612. If $q^{-3} = -\dfrac{1}{2}$, which of the following is the value of q?

    a. −8

    b. $-\dfrac{1}{8}$

    c. $-\sqrt[3]{\dfrac{1}{2}}$

    d. $\sqrt[3]{-2}$

613. Solve: $\sqrt[3]{5x - 8} = 3$

    a. 49

    b. −7

    c. $7i$

    d. 7

614. Solve: $\sqrt[3]{7 - 3x} = -2$

    a. −5

    b. 5

    c. $5i$

    d. $-5i$

615. Solve: $(x-3)^2 = -28$
  a. $3 \pm 2i\sqrt{7}$
  b. $-3 \pm 2i\sqrt{7}$
  c. $2 \pm 3i\sqrt{7}$
  d. $-2 \pm 3i\sqrt{7}$

616. Solve: $\sqrt{10-3x} = x-2$
  a. -2 only
  b. -2 and -3
  c. 3 only
  d. There is no solution.

617. Solve: $\sqrt{3x+4} + x = 8$
  a. 4 and 15
  b. 15 only
  c. -4 and 15
  d. 4 only

618. Solve: $(x-1)^2 + 16 = 0$
  a. $1 \pm 2i$
  b. $1 \pm 4i$
  c. $-1 \pm 4i$
  d. $-1 \pm 2i$

619. Solve: $x^3 = -27$
  a. $3i$
  b. -3
  c. $-3i$
  d. 3

620. Solve: $x^2 = 225$
  a. $15i$
  b. $-15i$
  c. ± 15
  d. $\pm 5\sqrt{5}$

621. Solve: $x^3 = -125$
  a. -5
  b. 5
  c. $5i$
  d. $-5i$

622. Solve: $(x+4)^2 = 81$
  a. -13 only
  b. 5 only
  c. $-13, 5$
  d. There is no solution.

623. Solve: $x^2 + 1 = 0$
  a. ± 1
  b. $-1, -i$
  c. $1, i$
  d. $\pm i$

624. Solve: $x^2 + 81 = 0$
  a. ± 9
  b. $\pm 9i$
  c. $-9, -9i$
  d. $9, 9i$

Set 40 (Answers begin on page 214)

Solving quadratic equations using the quadratic formula is the topic of this problem set.

625. Solve using the quadratic formula: $x^2 - 7 = 0$
  a. $\pm 7i$
  b. ± 7
  c. $\pm i\sqrt{7}$
  d. $\pm \sqrt{7}$

626. Solve using the quadratic formula: $2x^2 - 1 = 0$
  a. $\pm \frac{\sqrt{2}}{2}$
  b. $\pm \frac{i\sqrt{2}}{2}$
  c. $\pm \sqrt{2}$
  d. $\pm i\sqrt{2}$

627. Solve using the quadratic formula: $4x^2 + 3x = 0$
  a. $0, -\frac{3}{4}$
  b. $-\frac{3}{4}$ only
  c. $-\frac{4}{3}$ only
  d. $\frac{3}{4}$ only

628. Solve using the quadratic formula: $-5x^2 + 20x = 0$
   a. -4 only
   b. $0, 4$
   c. $4, -4$
   d. $0, -4$

629. Solve using the quadratic formula:
   $x^2 + 4x + 4 = 0$
   a. 2
   b. $2i$
   c. -2
   d. $-2i$

630. Solve using the quadratic formula:
   $x^2 - 5x - 6 = 0$
   a. $-2, -3$
   b. $1, -6$
   c. $-1, 6$
   d. $3, 2$

631. Solve using the quadratic formula:
   $3x^2 + 5x + 2 = 0$
   a. $-1, \frac{2}{3}$
   b. $-1, -\frac{2}{3}$
   c. $1, -\frac{2}{3}$
   d. $1, \frac{2}{3}$

632. Solve using the quadratic formula: $5x^2 - 24 = 0$
   a. $\pm\frac{2\sqrt{30}}{5}$
   b. $\pm\frac{2i\sqrt{30}}{5}$
   c. $\pm 2\sqrt{6}$
   d. $\pm 2i\sqrt{6}$

633. Solve using the quadratic formula:
   $2x^2 = -5x - 4$
   a. $\frac{-5 \pm i\sqrt{7}}{4}$
   b. $\frac{5 \pm i\sqrt{7}}{4}$
   c. $\frac{-4 \pm i\sqrt{7}}{-5}$
   d. $\frac{-7 \pm i\sqrt{5}}{4}$

634. Solve using the quadratic formula:
   $x^2 - 2\sqrt{2}x + 3 = 0$
   a. $\sqrt{2} \pm i$
   b. $\sqrt{2} \pm 1$
   c. $1 \pm i\sqrt{2}$
   d. $\pm i\sqrt{2}$

635. Solve using the quadratic formula: $x^2 = -2x$
   a. $2, 0$
   b. $-2, 0$
   c. $2i, -2i$
   d. 0 only

636. Solve using the quadratic formula: $(3x - 8)^2 = 45$
   a. $\frac{-3 \pm 8\sqrt{5}}{3}$
   b. $\frac{-3 \pm \sqrt{5}}{8}$
   c. $\frac{-8 \pm 3\sqrt{5}}{3}$
   d. $\frac{-8 \pm \sqrt{5}}{3}$

637. Solve using the quadratic formula:
   $0.20x^2 - 2.20x + 2 = 0$
   a. $0.01, 0.1$
   b. $10, 100$
   c. $0.1, 1$
   d. $1, 10$

638. Solve using the quadratic formula:
$x^2 - 3x - 3 = 0$

  a. $\frac{3 \pm 3\sqrt{7}}{2}$

  b. $\frac{3 \pm 7\sqrt{3}}{2}$

  c. $\frac{-3 \pm \sqrt{21}}{2}$

  d. $\frac{3 \pm \sqrt{21}}{2}$

639. Solve using the quadratic formula:
$\frac{1}{6}x^2 - \frac{5}{3}x + 1 = 0$

  a. $-5 \pm \sqrt{19}$

  b. $5 \pm \sqrt{19}$

  c. $5 \pm i\sqrt{19}$

  d. $-5 \pm i\sqrt{19}$

640. Solve using the quadratic formula:
$(x - 3)(2x + 1) = x(x - 4)$

  a. $\frac{-1 \pm \sqrt{13}}{2}$

  b. $\frac{1 \pm \sqrt{13}}{2}$

  c. $\frac{1 \pm i\sqrt{13}}{2}$

  d. $\frac{-1 \pm i\sqrt{13}}{2}$

Set 41 (Answers begin on page 216)

Solving quadratic equations using radical and graphical methods is the focus of this problem set.

641. Solve using radical methods: $4x^2 = 3$

  a. $\pm \frac{\sqrt{3}}{2}$

  b. $\pm \sqrt{\frac{2}{3}}$

  c. $\pm \frac{i\sqrt{3}}{2}$

  d. $\pm i\sqrt{\frac{2}{3}}$

642. Solve using radical methods: $-3x^2 = -9$

  a. $\pm 3i$

  b. ± 3

  c. $\pm \sqrt{3}$

  d. $\pm i\sqrt{3}$

643. Solve using radical methods: $(4x + 5)^2 = -49$

  a. $\frac{5 \pm 7i}{4}$

  b. $\frac{-5 \pm 7i}{4}$

  c. $\frac{-7 \pm 5i}{4}$

  d. $\frac{7 \pm 5i}{4}$

644. Solve using radical methods: $(3x - 8)^2 = 45$

  a. $\frac{-8 \pm 3i\sqrt{5}}{3}$

  b. $\frac{-8 \pm 3\sqrt{5}}{3}$

  c. $\frac{8 \pm 3\sqrt{5}}{3}$

  d. $\frac{8 \pm 3i\sqrt{5}}{3}$

645. Solve using radical methods: $(-2x + 1)^2 - 50 = 0$

  a. $\frac{-1 \pm 5\sqrt{2}}{2}$

  b. $\frac{1 \pm 5i\sqrt{2}}{2}$

  c. $\frac{1 \pm 5\sqrt{2}}{2}$

  d. $\frac{-1 \pm 5i\sqrt{2}}{2}$

646. Solve using radical methods: $-(1 - 4x)^2 - 121 = 0$

  a. $\frac{-1 \pm 11i}{4}$

  b. $\frac{1 \pm 11i}{4}$

  c. $\frac{1 \pm i\sqrt{11}}{4}$

  d. $\frac{-1 \pm i\sqrt{11}}{4}$

647. Find the real solutions of the following equation, if they exist, using graphical methods:
$5x^2 - 24 = 0$
a. $\approx \pm 2.191$
b. ± 4.8
c. ≈ 2.191 only
d. The solutions are imaginary.

648. Find the real solutions of the following equation, if they exist, using graphical methods: $2x^2 = -5x - 4$
a. $0.5, 1.5$
b. $-1.5, 0$
c. $-0.5, 0.5$
d. The solutions are imaginary.

649. Find the real solutions of the following equation, if they exist, using graphical methods: $4x^2 = 20x - 24$
a. $2, 3$
b. $16, 36$
c. $-2, -3$
d. $16, -36$

650. Find the real solutions of the following equation, if they exist, using graphical methods: $12x - 15x^2 = 0$
a. $0, -0.8$
b. $-0.8, 0.8$
c. $0, 0.8$
d. The solutions are imaginary.

651. Find the real solutions of the following equation, if they exist, using graphical methods: $(3x - 8)^2 = 45$
a. $\approx -3.875, 3.875$
b. $-3, 5$
c. $\approx 3.875, 4.903$
d. The solution are imaginary.

652. Find the real solutions of the following equation, if they exist, using graphical methods: $0.20x^2 - 2.20x + 2 = 0$
a. $-10, -1$
b. $1, 10$
c. $-1, 10$
d. The solutions are imaginary.

653. Find the real solutions of the following equation, if they exist, using graphical methods: $x^2 - 3x - 3 = 0$
a. $\approx -0.791, 3.791$
b. $1, 3$
c. $-1, -3$
d. The solutions are imaginary.

654. Find the real solutions of the following equation, if they exist, using graphical methods: $x^2 = -2x$
a. $0, 2$
b. $-2, 0$
c. $-2, 2$
d. The solutions are imaginary.

655. Find the real solutions of the following equation, if they exist, using graphical methods: $\frac{1}{6}x^2 - \frac{5}{3}x + 1 = 0$
a. $\approx 0.51, 10.51$
b. $\approx 0.641, 9.359$
c. $1, 4.2$
d. The solutions are imaginary.

656. Find the real solutions of the following equation, if they exist, using graphical methods: $(2x + 1)^2 - 2(2x + 1) - 3 = 0$
a. $-1, -1$
b. $1, 1$
c. $-1, 1$
d. The solutions are imaginary.

Set 42 (Answers begin on page 220)

Solving equations that can be put in quadratic form via substitution is the focus of this problem set.

657. Solve: $b^4 - 7b^2 + 12 = 0$
   a. $\pm\sqrt{2}, \pm\sqrt{3}$
   b. $\pm 2, \pm\sqrt{3}$
   c. $\pm\sqrt{2}, \pm 3$
   d. $\pm 2, \pm 3$

658. Solve: $(3b^2 - 1)(1 - 2b^2) = 0$
   a. $\pm\frac{\sqrt{2}}{2}, \pm\frac{\sqrt{3}}{3}$
   b. $\pm\frac{\sqrt{2}}{3}, \pm\frac{\sqrt{3}}{2}$
   c. $\pm\frac{\sqrt{3}}{2}, \pm\frac{2}{3}$
   d. $\pm\frac{\sqrt{2}}{3}, \pm\frac{\sqrt{3}}{3}$

659. Solve: $4b^4 + 20b^2 + 25 = 0$
   a. $\pm i\sqrt{\frac{10}{3}}$
   b. $\pm\sqrt{\frac{10}{3}}$
   c. $\pm\frac{\sqrt{10}}{2}$
   d. $\pm i\left(\frac{\sqrt{10}}{2}\right)$

660. Solve: $16b^4 - 1 = 0$
   a. $\pm i\sqrt{\frac{1}{2}}, \pm\sqrt{\frac{1}{2}}$
   b. $\pm i\left(\frac{1}{2}\right), \pm\frac{1}{2}$
   c. $\pm i\left(\frac{1}{2}\right), \pm\sqrt{\frac{1}{2}}$
   d. $\pm i\sqrt{\frac{1}{2}}, \pm\frac{1}{2}$

661. Solve: $x + 21 = 10x^{\frac{1}{2}}$
   a. $-49, -9$
   b. $-49, 9$
   c. $9, 49$
   d. $-9, 49$

662. Solve: $16 - 56\sqrt{x} + 49x = 0$
   a. $\frac{16}{49}$
   b. $\frac{49}{16}$
   c. $-\frac{16}{49}$
   d. $-\frac{49}{16}$

663. Solve: $x - \sqrt{x} = 6$
   a. 9
   b. -9
   c. 16
   d. -16

664. Solve: $2x^{\frac{1}{6}} - x^{\frac{1}{3}} = 1$
   a. -1
   b. i
   c. 1
   d. $-i$

665. Solve: $3 + x^{-\frac{1}{4}} - x^{-\frac{1}{2}} = 0$
   a. $\frac{-16}{(1+\sqrt{13})^4}, \frac{-16}{(1-\sqrt{13})^4}$
   b. $\frac{16}{(1+\sqrt{13})^4}, \frac{16}{(1-\sqrt{13})^4}$
   c. $\frac{-16i}{(1+\sqrt{13})^4}, \frac{-16i}{(1-\sqrt{13})^4}$
   d. $\frac{-16i}{(1+\sqrt{13})^4}, \frac{16i}{(1-\sqrt{13})^4}$

666. Solve: $(x^3 + 5)^2 - 5(x^3 + 5) + 6 = 0$
   a. $\sqrt[3]{2}, \sqrt[3]{-3}$
   b. $\sqrt[3]{-2}, \sqrt[3]{3}$
   c. $\sqrt[3]{2}, \sqrt[3]{3}$
   d. $\sqrt[3]{-2}, \sqrt[3]{-3}$

667. Solve: $4x^6 + 1 = 5x^3$
   a. $-\sqrt[3]{\frac{1}{4}}, -1$
   b. $-\sqrt[3]{\frac{1}{4}}, 1$
   c. $\sqrt[3]{\frac{1}{4}}, 1$
   d. $\sqrt[3]{\frac{1}{4}}, -1$

668. Solve: $\left(x^2 + x\right)^2 + 12 = 8\left(x^2 + x\right)$

   a. 1, 2, 3

   b. −3, −2, 1, 2

   c. −3, −2, −1, 2

   d. −3, 1, 2, 3

669. Solve: $2\left(1 + \sqrt{w}\right)^2 = 13\left(1 + \sqrt{w}\right) - 6$

   a. 25 only

   b. 25 and $-\frac{1}{4}$

   c. $\frac{1}{2}$ and 6

   d. $\frac{1}{2}$ only

670. Solve: $(r - \frac{3}{r})^2 - (r - \frac{3}{r}) - 6 = 0$

   a. $-3, 1, \frac{3 \pm \sqrt{21}}{2}$

   b. $-1, 3, \frac{3 \pm \sqrt{21}}{2}$

   c. $-3, 1, \frac{3 \pm i\sqrt{21}}{2}$

   d. $-3, 1, \frac{-3 \pm \sqrt{21}}{2}$

671. Solve: $6\sqrt{x} - 13\sqrt[4]{x} + 6 = 0$

   a. $-\frac{16}{81}, -\frac{81}{16}$

   b. $\frac{16}{81}, \frac{81}{16}$

   c. $-\frac{16}{81}, \frac{81}{16}$

   d. $-\frac{81}{16}, \frac{16}{81}$

672. Solve: $2a^{\frac{2}{3}} - 11a^{\frac{1}{3}} + 12 = 0$

   a. $-64, -\frac{27}{8}$

   b. $-64, \frac{27}{8}$

   c. $64, \frac{27}{8}$

   d. $64, -\frac{27}{8}$

6 ▶ ELEMENTARY FUNCTIONS

The functions you typically first encounter are those described by sets of ordered pairs that can be visualized in the Cartesian plane. Such functions are generally described using either algebraic expressions or graphs, and are denoted using letters, such as f or g. When we want to emphasize the input-output defining relationship of a function, an expression of the form $y = f(x)$ is often used. The arithmetic of real-valued functions is performed using the arithmetic of real numbers and algebraic expressions.

The *domain* of a function can be thought of as the set of all possible x-values for which there corresponds an output y. From the graphical viewpoint, an x-value belongs to the domain of f if an ordered pair with that x-value belongs to the graph of f. When an algebraic expression is used to describe a function $y = f(x)$, it is convenient to view the domain as the set of all values of x that can be substituted into the expression and yield a meaningful output. The *range* of a function is the set of all possible y-values attained at some member of the domain. Basic functions and their properties are reviewed in the ten problem sets that make up this section.

Set 43 (Answers begin on page 225)

The problems in this set focus on the notions of domain and range and the basic arithmetic of elementary functions.

673. In the following graph of $f(x)$, for how many values of x does $f(x) = 3$?

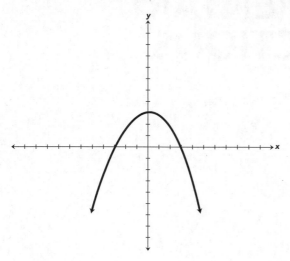

a. 0
b. 1
c. 2
d. 3
e. 4

674. In the following graph of $f(x)$, for how many values of x does $f(x) = 0$?

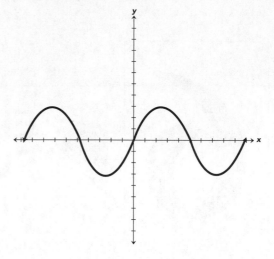

a. 2
b. 3
c. 4
d. 5
e. 8

675. In the following graph of $f(x)$, for how many values of x does $f(x) = 10$?

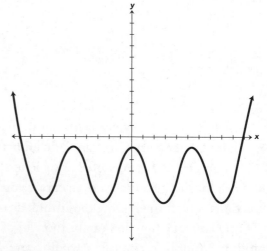

a. 0
b. 2
c. 4
d. 5
e. 8

676. What is the range of the function $f(x) = x^2 - 4$?
  a. the set of all real numbers
  b. the set of all real numbers excluding 2 and –2
  c. the set of all real numbers greater than or equal to 0
  d. the set of all real numbers greater than or equal to 4
  e. the set of all real numbers greater than or equal to –4

677. Which of the following is true of $f(x) = -\frac{1}{2}x^2$?
  a. The range of the function is the set of all real numbers less than or equal to 0.
  b. The range of the function is the set of all real numbers less than 0.
  c. The range of the function is the set of all real numbers greater than or equal to 0.
  d. The domain of the function is the set of all real numbers greater than or equal to 0.
  e. The domain of the function is the set of all real numbers less than or equal to $\sqrt{2}$.

For questions 678 through 680, refer to the functions f and g, both defined on $[-5,5]$, whose graphs are shown here.

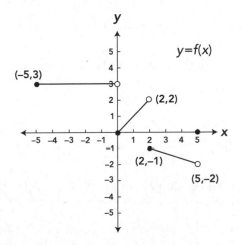

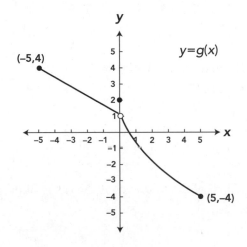

678. The range of f is which of the following?
  a. $[-2,2] \cup \{3\}$
  b. $(-2,-1] \cup [0,3]$
  c. $(-2,-1] \cup [0,2) \cup \{3\}$
  d. $[-2,2) \cup \{3\}$

679. The range of g is which of the following?
  a. $[-4,4]$
  b. $[-4,2] \cup (2,4]$
  c. $[-4,1) \cup (1,4]$
  d. none of the above

680. $2 \cdot f(0) + [f(2) \cdot g(5)]^2 =$
  a. 18
  b. 10
  c. 8
  d. 16

For questions 681 through 684, refer to the graph of the following fourth-degree polynomial function $y = p(x)$.

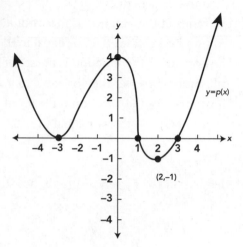

681. The zeros of $p(x)$ are $x =$
   a. −3, 0, 2
   b. −3, 1, 3
   c. −3, 0, 1, 2, 3
   d. none of the above

682. Which of the following is the range of $p(x)$?
   a. $\mathbb{R}$
   b. $[-1,4]$
   c. $[-1,°)$
   d. $[-4,4]$

683. Which of the following is the domain of $p(x)$?
   a. $\mathbb{R}$
   b. $[-1,4]$
   c. $[-1,\infty)$
   d. $[-4,4]$

684. Which of the following is the solution set for the inequality $-1 < p(x) \le 0$?
   a. $(1,3)$
   b. $[1,3]\cup\{-3\}$
   c. $(1,2)\cup(2,3)$
   d. $[1,2)\cup(2,3]\cup\{-3\}$

685. The range of the function $f(x) = \frac{2x+1}{1-x}$ is which of the following?
   a. $(-°,1)\cup(1,°)$
   b. $(-°,-2)\cup(-2,°)$
   c. $(-°,-\frac{1}{2})\cup(-\frac{1}{2},°)$
   d. $\mathbb{R}$

For questions 686 through 688, use the following functions:

$$f(x) = -(2x-(-1-x^2)) \quad g(x) = 3(1+x) \quad h(x) = \frac{1}{1+x^2}$$

686. Which of the following is equivalent to $\frac{9f(x)}{g(x)}$?
   a. $-g(x)$
   b. $g(x)$
   c. $-3g(x)$
   d. $3g(x)$

687. What is the domain of the function $2g(x)h(x)$?
   a. $(-\infty,-1)\cup(-1,1)\cup(1,\infty)$
   b. $(-\infty,-1)\cup(1,\infty)$
   c. $(-\infty,-1)\cup(-1,\infty)$
   d. the set of all real numbers

688. Which of the following is equivalent to $3f(x) - 2xg(x) - \frac{1}{h(x)}$?
   a. $-10x^2 + 6x + 2$
   b. $-2(5x^2 + 6x + 2)$
   c. $10x^2 + 12x + 4$
   d. $2(5x +2)(x + 1)$

Set 44 (Answers begin on page 226)

This problem set focuses on compositions of functions, the simplification involved therein, and the general principles of the graph of a function.

For questions 689 through 692, use the following diagrams:

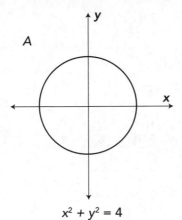

$$x^2 + y^2 = 4$$

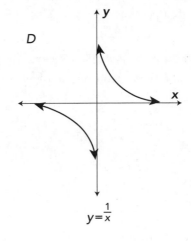

$$y = \frac{1}{x}$$

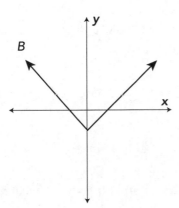

$$y = |x| - 3$$

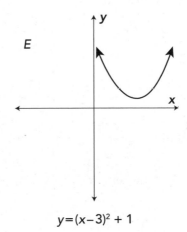

$$y = (x-3)^2 + 1$$

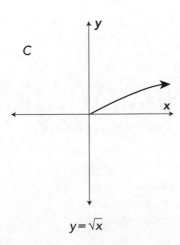

$$y = \sqrt{x}$$

689. Which of the coordinate planes shows the graph of an equation that is not a function?
   a. A
   b. B
   c. C
   d. D
   e. A and D

690. Which of the coordinate planes shows the graph of a function that has a range that contains negative values?
  a. A
  b. B
  c. D
  d. B and D
  e. A, B, and D

691. Which of the coordinate planes shows the graph of a function that has a domain of all real numbers?
  a. B
  b. D
  c. E
  d. B and D
  e. B and E

692. Which of the coordinate planes shows the graph of a function that has the same range as its domain?
  a. B and C
  b. C and D
  c. B and D
  d. B and E
  e. D and E

693. Simplify $f(-\frac{2}{x})$ when $f(x) = -\frac{1}{x^3}$.
  a. $\frac{x^3}{8}$
  b. $8x^3$
  c. $-\frac{8}{x^3}$
  d. $-\frac{x^3}{8}$

694. Simplify $f(2y - 1)$ when $f(x) = x^2 + 3x - 2$.
  a. $4y^2 + 2y - 4$
  b. $4y^2 + 6y - 2$
  c. $4y^2 + 6y - 3$
  d. $2y^2 + 6y - 4$

695. Simplify $f(x + h) - f(x)$ when $f(x) = -(x - 1)^2 + 3$.
  a. h
  b. $f(h)$
  c. $-h(2x + h - 2)$
  d. $-2hx + h^2 - 2h$

696. Compute $(g \circ h)(4)$ when $g(x) = 2x^2 - x - 1$ and $h(x) = x - 2\sqrt{x}$.
  a. 0
  b. 1
  c. -1
  d. 4

697. Simplify $(f \circ f \circ f)(2x)$ when $f(x) = -x^2$.
  a. $16x$
  b. $-16x$
  c. $64x^8$
  d. $-256x^8$

698. If $f(x) = 3x + 2$ and $g(x) = 2x - 3$, what is the value of $g(f(-2))$?
  a. -19
  b. -11
  c. -7
  d. -4
  e. -3

699. If $f(x) = 2x + 1$ and $g(x) = x - 2$, what is the value of $f(g(f(3)))$?
  a. 1
  b. 3
  c. 5
  d. 7
  e. 11

700. If $f(x) = 6x + 4$ and $g(x) = x^2 - 1$, which of the following is equivalent to $g(f(x))$?
  a. $6x^2 - 2$
  b. $36x^2 + 16$
  c. $36x^2 + 48x + 15$
  d. $36x^2 + 48x + 16$
  e. $6x^3 + 4x^2 - 6x - 4$

For questions 701 and 702, refer to the functions f and g, both defined on $[-5,5]$, whose graphs are shown here.

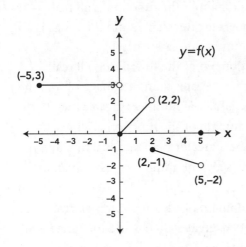

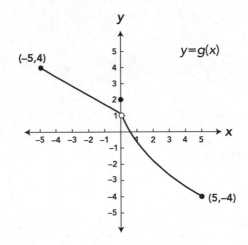

701. $(f \circ g)(0) =$
 a. $\frac{1}{2}$
 b. -1
 c. 2
 d. undefined

702. $\big(f(f(f(f(5))))\big) =$
 a. 0
 b. -1
 c. 3
 d. undefined

703. If $f(x) = \sqrt{x^2 - 4x}$, then $f(x + 2) =$
 a. $\sqrt{x^2 - 4x + 2}$
 b. $\sqrt{x^2 - 4x - 4}$
 c. $\sqrt{x^2 - 4}$
 d. $|x - 2|$

704. If $f(x) = \sqrt{-3x}$ and $g(x) = \sqrt{2x^2 + 18}$, then the domain of $g \circ f$ is
 a. $[0, \infty)$
 b. $(-\infty, 0]$
 c. $\mathbb{R}$
 d. none of the above

Set 45 (Answers begin on page 229)

This problem set explores some basic features of common elementary functions.

705. Determine the domain of the function
 $f(x) = \sqrt{-x}$.
 a. $(-\infty, 0]$
 b. $[0, \infty)$
 c. $(-\infty, 0)$
 d. $(0, \infty)$

706. Determine the domain of the function

$g(x) = \dfrac{1}{\sqrt[3]{-1-x}}$.

  a. the set of all real numbers

  b. $(-1,\infty)$

  c. $(-\infty,-1)$

  d. $(-\infty,-1)\cup(-1,\infty)$

707. Which of the following is true of the function $f(x) = 2$?

  a. It is not a function.

  b. Its range is $\{2\}$.

  c. It has no domain.

  d. It has a slope of 2.

  e. It has no y-intercept.

708. Which of the following is true about the function $f(x) = |x|$?

  a. It has one y-intercept and two x-intercepts.

  b. It has one y-intercept and one x-intercept.

  c. There exists precisely one x-value for which $f(x) = 1$.

  d. $f(x) > 0$, for all real numbers x.

709. Which of the following is true about the function $f(x) = x^3$?

  a. $f(x) > 0$, for all real numbers x.

  b. The graph of $y = f(x)$ crosses the line $y = a$ precisely once, for any real number a.

  c. The graph of $y = f(x)$ is decreasing on its domain.

  d. The range is $[0,°)$.

710. Which of the following is true about the function $f(x) = \frac{1}{x}$?

  a. $f(x) \geq 0$, for all real numbers x.

  b. The graph has one x-intercept.

  c. The graph of $y = f(x)$ is decreasing on the interval $(0,\infty)$.

  d. The range is $(0,\infty)$.

711. Which of the following is true about the function $f(x) = \sqrt{4x-1}$?

  a. The domain of the function is all real numbers greater than $\frac{1}{4}$ and the range is all real numbers greater than 0.

  b. The domain of the function is all real numbers greater than or equal to $\frac{1}{4}$ and the range is all real numbers greater than 0.

  c. The domain of the function is all real numbers greater than or equal to $\frac{1}{4}$ and the range is all real numbers greater than or equal to 0.

  d. The domain of the function is all real numbers greater than 0 and the range is all real numbers greater than or equal to $\frac{1}{4}$.

  e. The domain of the function is all real numbers greater than or equal to 0 and the range is all real numbers greater than or equal to $\frac{1}{4}$.

712. Consider the graphs of $f(x) = x^2$ and $g(x) = x^4$. Which of the following statements is true?

  a. $f(x) \leq g(x)$, for all real numbers x.

  b. The graphs of $y = f(x)$ and $y = g(x)$ do not intersect.

  c. The range of both f and g is $[0,\infty)$.

  d. The graphs of both f and g are increasing on their entire domains.

713. Which of the following is the domain of the function $f(x) = \dfrac{1}{(2-x)^{\frac{2}{5}}}$?

  a. $(-\infty,2)$

  b. $(2,\infty)$

  c. $(-\infty,2)\cup(2,\infty)$

  d. none of the above

714. How many x-intercepts does the function $f(x) = 1 - |2x-1|$ have?

  a. 0

  b. 1

  c. 2

  d. more than 2

715. How many points of intersection are there of the graphs of $f(x) = x^2$ and $g(x) = x^4$?
   a. 0
   b. 1
   c. 2
   d. more than 2

716. How many points of intersection are there of the graphs of $f(x) = 2x$ and $g(x) = 4x^3$?
   a. 0
   b. 1
   c. 2
   d. more than 2

717. How many points of intersection are there of the graphs of $f(x) = \frac{3}{4}x^2$ and $g(x) = \frac{5}{16}x^2$?
   a. 0
   b. 1
   c. 2
   d. more than 2

718. What is the y-intercept of the function $f(x) = \frac{-2 - |2 - 3x|}{4 - 2x^2 \, |-x|}$?
   a. $(0,0)$
   b. $(0, -1)$
   c. $(-1, 0)$
   d. There is no y-intercept.

719. What can you conclude about the graph of $y = f(x)$ if you know that the equation $f(x) = 0$ does not have a solution?
   a. The graph has no x-intercept.
   b. The graph has no y-intercept.
   c. The graph has neither an x-intercept nor a y-intercept.
   d. There is not enough information to conclude anything about the graph of f.

720. What can you conclude about the graph of $y = f(x)$ if you know that the equation $f(x) = 3$ does not have a solution?
   a. 3 is not in the domain of f.
   b. 3 is not in the range of f.
   c. The graph of the function cannot have y-values larger than 3.
   d. The graph of the function cannot be defined for x-values larger than 3.

Set 46 (Answers begin on page 231)

This problem set focuses on properties of more sophisticated functions, including monotonicity, asymptotes, and the existence of inverse functions.

721. The domain of $f(x) = \frac{2x}{x^3 - 4x}$ is
   a. $(-\infty, -2) \cup (2, \infty)$
   b. $(-\infty, 2) \cup (2, \infty)$
   c. $(-\infty, -2) \cup (-2, 0) \cup (0, 2) \cup (2, \infty)$
   d. $(-\infty, -2) \cup (-2, 2) \cup (2, \infty)$

722. Which of the following are the vertical and horizontal asymptotes for the function $f(x) = \frac{(x - 3)(x^2 - 16)}{(x^2 + 9)(x - 4)}$?
   a. $x = -3, x = 4$
   b. $x = -3, x = 4, y = 1$
   c. $x = 4, y = 1$
   d. $y = 1$

723. On what intervals is the graph of the following fourth degree polynomial function $y = p(x)$ increasing?

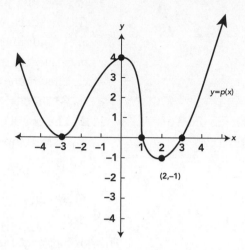

a. $(-3,0) \cup (2,\infty)$
b. $(-3,0) \cup (3,\infty)$
c. $(-\infty,-3) \cup (0,1)$
d. $(-\infty,-3) \cup (0,2)$

724. Consider the functions f and g, both defined on $[-5,5]$, whose graphs are shown here.

c. g does not have an inverse on the interval $(-5,-1)$.
d. All of the above statements are false.

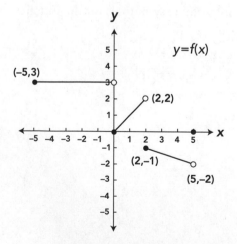

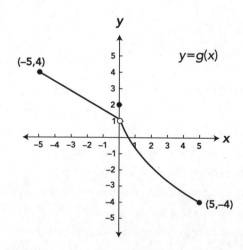

Which of the following statements is true?
a. f has an inverse on the interval $(0,5)$.
b. f has an inverse on the interval $(-5,2)$.

725. Which of the following is the inverse function for $f(x) = \frac{x-1}{5x+2}, x \neq -\frac{2}{5}$?
   a. $f^{-1}(y) = \frac{2y-1}{5y+1}, y \neq -\frac{1}{5}$
   b. $f^{-1}(y) = \frac{2y+1}{5y-1}, y \neq \frac{1}{5}$
   c. $f^{-1}(y) = \frac{-2y-1}{5y-1}, y \neq \frac{1}{5}$
   d. $f^{-1}(y) = \frac{2y+1}{5y+1}, y \neq -\frac{1}{5}$

726. Assume that the function f has an inverse f^{-1}, and the point $(1, 4)$ is on the graph of $y = f(x)$. Which of the following statements is true?
   a. If the range of f^{-1} is $[1,\infty)$, then $f(0)$ is not defined.
   b. $f^{-1}(4) = 1$
   c. The point $(4, 1)$ must lie on the graph of $y = f^{-1}(x)$.
   d. All of the above statements are true.

727. Which of the following is the inverse of $f(x) = x^3 + 2$?
   a. $f^{-1}(y) = \sqrt[3]{y-2}$
   b. $f^{-1}(y) = \sqrt[3]{y} - 2$
   c. $f^{-1}(y) = \sqrt[3]{2-y}$
   d. $f^{-1}(y) = \sqrt[3]{2} - \sqrt[3]{y}$

728. Which of the following are characteristics of the graph of $f(x) = 2 - \frac{x^2+1}{x-1}$?
   I. The function is equivalent to the linear function $g(x) = 2 - (x + 1)$ with a hole at $x = 1$.
   II. There is one vertical asymptote, no horizontal asymptote, and an oblique asymptote.
   III. There is one x-intercept and one y-intercept.
   a. I only
   b. II only
   c. II and III only
   d. I and III only

729. Which of the following are characteristics of the graph of $f(x) = \frac{(2-x)^2(x+3)}{x(x-2)^2}$?
   I. The graph has a hole at $x = 2$.
   II. $y = 1$ is a horizontal asymptote and $x = 0$ is a vertical asymptote.
   III. There is one x-intercept and one y-intercept.
   a. I and III only
   b. I and II only
   c. I only
   d. none of these choices

730. Which of the following functions is decreasing on $(-\infty,0)$?
   a. $f(x) = x^3$
   b. $f(x) = 2x + 5$
   c. $f(x) = \frac{1}{x}$
   d. $f(x) = 3$

731. Which of the following functions is increasing on $(0,\infty)$?
   a. $f(x) = x^3$
   b. $f(x) = 2x + 5$
   c. $f(x) = |x|$
   d. all of the above

732. Which of the following statements is false?
   a. The domain of any polynomial function is the set of all real numbers.
   b. There exists a rational function whose domain is the set of all real numbers.
   c. A rational function must have both a vertical and a horizontal asymptote.
   d. All of the statements are true.

733. Which of the following statements is false?
  a. There exists a polynomial whose graph is increasing everywhere.
  b. A polynomial with degree greater than 1 must have at least one turning point.
  c. There exists a polynomial whose graph remains below the *x*-axis on its entire domain.
  d. All of the statements are true.

734. Which of the following statements is true?
  a. Linear functions with positive slopes are increasing.
  b. There exists a rational function whose graph intersects both Quadrants I and II.
  c. All quadratic functions are decreasing on one side of the vertex and increasing on the other side of the vertex.
  d. All of the statements are true.

735. Determine the *x*-values of the points of intersection of the graphs of $f(x) = -4x$ and $g(x) = 2\sqrt{x}$.
  a. 0, 4
  b. $0, \frac{1}{4}$
  c. ± 2
  d. $\pm \frac{1}{2}$

736. Determine the *x*-values of the points of intersection of the graphs of $f(x) = \sqrt{x}$ and $g(x) = 3\sqrt{x}$.
  a. 0
  b. 0,9
  c. 0,3
  d. The graphs do not intersect.

Set 47 (Answers begin on page 235)

This problem set focuses on translations and reflections of known graphs.

737. Which of the following sequence of shifts would you perform in order to obtain the graph of $f(x) = (x + 2)^3 - 3$ from the graph of $g(x) = x^3$?
  a. Shift the graph of *g* up 3 units and then left two units.
  b. Shift the graph of *g* down 3 units and then right two units.
  c. Shift the graph of *g* up 3 units and then right two units.
  d. Shift the graph of *g* down 3 units and then left two units.

738. Which of the following parabolas has its turning point in the second quadrant of the coordinate plane?
  a. $y = (x + 1)^2 - 2$
  b. $y = (x - 1)^2 - 2$
  c. $y = -(x + 1)^2 - 2$
  d. $y = -(x + 2)^2 + 1$
  e. $y = (x - 2)^2 + 1$

739. Compared to the graph of $y = x^2$, the graph of $y = (x - 2)^2 - 2$ is
  a. shifted 2 units right and 2 units down.
  b. shifted 2 units left and 2 units down.
  c. shifted 2 units right and 2 units up.
  d. shifted 2 units left and 2 units up.

740. Which of the following sequence of shifts would you perform in order to obtain the graph of $f(x) = (x - 4)^3 + 1$ from the graph of $g(x) = x^3$?

a. Shift the graph of g up 1 unit and then left 4 units.

b. Shift the graph of g down 1 unit and then right 4 units.

c. Shift the graph of g up 1 unit and then right 4 units.

d. Shift the graph of g down 4 units and then right 1 unit.

741. Which of the following sequence of shifts would you perform in order to obtain the graph of $f(x) = (x - 2)^2 - 4$ from the graph of $g(x) = x^2$?

a. Shift the graph of g up 4 units and then left 2 units.

b. Shift the graph of g down 4 units and then left 2 units.

c. Shift the graph of g up 4 units and then right 2 units.

d. Shift the graph of g down 4 units and then right 2 units.

742. Which of the following sequence of shifts would you perform in order to obtain the graph of $f(x) = (x - 2)^3 - 1$ from the graph of $g(x) = x^3$?

a. Shift the graph of g up 1 unit and then left 2 units.

b. Shift the graph of g down 1 unit and then right 2 units.

c. Shift the graph of g up 1 unit and then right 2 units.

d. Shift the graph of g down 2 units and then left 1 unit.

743. Which of the following sequence of shifts would you perform in order to obtain the graph of $f(x) = \sqrt{x - 5} - 3$ from the graph of $g(x) = \sqrt{x}$?

a. Shift the graph of g up 3 units and then left 5 units.

b. Shift the graph of g down 3 units and then left 5 units.

c. Shift the graph of g down 5 units and then left 3 units.

d. Shift the graph of g down 3 units and then right 5 units.

744. Which of the following sequence of shifts would you perform in order to obtain the graph of $f(x) = 2\sqrt{x + 3}$ from the graph of $g(x) = 2\sqrt{x}$?

a. Shift the graph of g up 3 units.

b. Shift the graph of g down 3 units.

c. Shift the graph of g right 3 units.

d. Shift the graph of g left 3 units.

745. Which of the following sequence of shifts would you perform in order to obtain the graph of $f(x) = |x + 6| + 4$ from the graph of $g(x) = |x|$?

a. Shift the graph of g up 4 units and then left 6 units.

b. Shift the graph of g down 4 units and then right 6 units.

c. Shift the graph of g up 6 units and then right 4 units.

d. Shift the graph of g down 6 units and then left 4 units.

746. Which of the following sequence of shifts would you perform in order to obtain the graph of $f(x) = -|x - 1| + 5$ from the graph of $g(x) = |x|$?

   a. Shift the graph of g left 1 unit, then reflect over the x-axis, and then up 5 units.
   b. Shift the graph of g right 1 unit, then reflect over the x-axis, and then up 5 units.
   c. Shift the graph of g left 1 unit, then reflect over the x-axis, and then down 5 units.
   d. Shift the graph of g right 1 unit, then reflect over the x-axis, and then down 5 units.

747. Which of the following sequence of shifts would you perform in order to obtain the graph of $f(x) = -(x + 3)^3 + 5$ from the graph of $g(x) = x^3$?

   a. Shift the graph of g left 3 units then reflect over the x-axis, and then up 5 units.
   b. Shift the graph of g right 3 units, then reflect over the x-axis, and then up 5 units.
   c. Shift the graph of g left 3 units, then reflect over the x-axis, and then down 5 units.
   d. Shift the graph of g right 3 units, then reflect over the x-axis, and then down 5 units.

748. The graph of which of the following functions can be obtained by shifting the graph of $g(x) = x^4$ right 5 units and then reflecting it over the x-axis?

   a. $f(x) = -x^4 + 5$
   b. $f(x) = -x^4 - 5$
   c. $f(x) = -(x + 5)^4$
   d. $f(x) = -(x - 5)^4$

749. The graph of which of the following functions can be obtained by shifting the graph of $g(x) = \sqrt{x}$ right 5 units and then up 2 units?

   a. $f(x) = \sqrt{x - 5} + 2$
   b. $f(x) = \sqrt{x + 5} + 2$
   c. $f(x) = \sqrt{x - 2} + 5$
   d. $f(x) = \sqrt{x + 2} - 5$

750. The graph of which of the following functions can be obtained by shifting the graph of $g(x) = |x|$ left 3 units, then reflecting it over the x-axis, and then shifting it down 2 units?

   a. $f(x) = -|x - 3| + 2$
   b. $f(x) = -|x + 2| - 3$
   c. $f(x) = -|x + 3| - 2$
   d. $f(x) = -|x - 2| + 3$

751. The graph of which of the following functions can be obtained by reflecting the graph of $g(x) = \frac{1}{x}$ over the x-axis, and then shifting it up 2 units?

   a. $f(x) = 2 + \frac{1}{x}$
   b. $f(x) = 2 - \frac{1}{x}$
   c. $f(x) = -\frac{1}{x + 2}$
   d. $f(x) = -\frac{1}{x - 2}$

752. The graph of which of the following functions can be obtained by shifting the graph of $g(x) = \frac{1}{x^2}$ right 2 units and then reflecting it over the x-axis?

   a. $f(x) = -\frac{1}{(x + 2)^2}$
   b. $f(x) = -\frac{1}{x^2} + 2$
   c. $f(x) = -\frac{1}{x^2} - 2$
   d. $f(x) = -\frac{1}{(x - 2)^2}$

Set 48 (Answers begin on page 236)

This problem set focuses on the basic computations and graphs involving exponentials, as well as application of the exponent rules.

753. If $e^x = 2$ and $e^y = 3$, then $e^{3x-2y} =$
   a. $\frac{9}{8}$
   b. $\frac{8}{9}$
   c. 1
   d. -1

754. Simplify: $2^x \cdot 2^{x+1}$
   a. 2^{x^2+x}
   b. 2^{2x+1}
   c. $\frac{2^{x^2}}{2^{-x}}$
   d. $\frac{2^{-x^2}}{2^x}$

755. Simplify: $(4^{x-1})^2 \cdot 16$
   a. 4^{2x+2}
   b. 4^x
   c. 4^{2x}
   d. 4^{-2x}

756. Simplify: $\left(\frac{5^{4x}}{5^{2x-6}}\right)^{\frac{1}{2}}$
   a. 5^{x-3}
   b. 5^{3x}
   c. 5^{x+3}
   d. 5^{-3x}

757. Simplify: $(e^x + e^{-x})^2$
   a. $e^{2x} + e^{-2x}$
   b. $e^{2x} + e^{-2x} + 2$
   c. $e^{x^2} + e^{-x^2}$
   d. $e^{x^2} + e^{-x^2} + 2$

758. Simplify: $\frac{(5^{3x-1})^3 \cdot 5^{x-1}}{5^{2x}}$
   a. 5^{8x-1}
   b. 125^{3x-2}
   c. 25^{4x-1}
   d. 625^{2x-1}

759. Simplify: $e^x(e^x - 1) - e^{-x}(e^x - 1)$
   a. $e^{2x} - e^x - 1 + e^{-x}$
   b. $e^{2x} + e^x - 1 + e^{-x}$
   c. $e^{2x} + 2e^x - 1$
   d. $e^{2x} + 1 - 2e^{-x}$

760. Simplify: $\frac{e^x(e^x - e^{-x}) + e^{-x}(e^x + e^{-x})}{e^{-2x}}$
   a. $e^{2x} + 1$
   b. $e^{4x} + 1$
   c. $e^{-4x} + 1$
   d. $e^{-2x} + 1$

761. Which of the following statements is true?
   a. If $b > 1$, the graph of $y = b^x$ gets very close to the x-axis as the x-values move to the right.
   b. If $b > 1$, the graph of $y = b^x$ gets very close to the x-axis as the x-values move to the left.
   c. If $b > 1$, the graph of $y = b^x$ grows without bound as the x-values move to the left.
   d. If $b > 1$, the y-values associated with the graph of $y = b^x$ grow very rapidly as the x-values move to the left.

762. Which of the following statements is true?
 a. If $0 < b < 1$, the graph of $y = b^x$ gets very close to the x-axis as the x-values move to the left, and the y-values grow very rapidly as the x-values move to the left.
 b. If $0 < b < 1$, the graph of $y = b^x$ gets very close to the x-axis as the x-values move to the right, and the y-values grow very rapidly as the x-values move to the right.
 c. If $0 < b < 1$, the graph of $y = b^x$ gets very close to the x-axis as the x-values move to the right, and the y-values grow very rapidly as the x-values move to the right.
 d. If $0 < b < 1$, the graph of $y = b^x$ gets very close to the x-axis as the x-values move to the right, and the y-values grow very rapidly as the x-values move to the left.

763. Which of the following statements is true?
 a. If $0 < b < 1$, then the equation $b^x = -1$ has a solution.
 b. If $b > 1$, then the equation $b^x = 1$ has two solutions.
 c. If $b > 0$, then the equation $b^x = 0$ has no solution.
 d. If $b > 0$, then only negative x-values can be solutions to the equation $b^x = 0$.

764. Which of the following statements is true?
 a. $2^x > 3^x$, for all $x > 0$.
 b. $\left(\frac{1}{2}\right)^x < \left(\frac{1}{3}\right)^x$, for all $x > 0$.
 c. $\left(\frac{1}{2}\right)^{-x} > 0$, for any real number x.
 d. All of the above statements are true.

765. What is the solution set for the inequality $1 - 3^x \le 0$?
 a. $[1, \infty)$
 b. $[0, \infty)$
 c. the empty set
 d. the set of all real numbers

766. What is the solution set for $\left(-\frac{2}{3}\right)^{2x} \le 0$?
 a. $(-\infty, 0)$
 b. $(-\infty, 0]$
 c. the empty set
 d. the set of all real numbers

767. Which of the following is a true characterization of the graph of $f(x) = -\left(\frac{3}{4}\right)^x$?
 a. The graph has one x-intercept and one y-intercept.
 b. There exists an x-value for which $f(x) = 1$.
 c. The graph is increasing as the x-values move from left to right.
 d. The graph is decreasing as the x-values move from left to right.

768. Which of the following is a true characterization of the graph of $f(x) = -\left(\frac{15}{7}\right)^{3x}$?
 a. The graph has one x-intercept and one y-intercept.
 b. There exists an x-value for which $f(x) = 1$.
 c. The graph is increasing as the x-values move from left to right.
 d. The graph is decreasing as the x-values move from left to right.

Set 49 (Answers begin on page 238)

This problem set focuses on additional features of exponential functions, including solving equations involving exponential expressions.

769. The range of the function $f(x) = 1 - 2e^x$ is which of the following?
 a. $(-\infty, 1]$
 b. $(-\infty, 1)$
 c. $(1, \infty)$
 d. $[1, \infty)$

770. Determine the values of x that satisfy the equation $2^{7x^2-1} = 4^{3x}$.

  a. $x = \dfrac{-3 \pm \sqrt{37}}{14}$

  b. $x = \dfrac{3 \pm \sqrt{37}}{14}$

  c. $x = -\dfrac{1}{7}$ and $x = 1$

  d. $x = \dfrac{1}{7}$ and $x = -1$

771. Determine the values of x, if any, that satisfy the equation $5^{\sqrt{x+1}} = \dfrac{1}{25}$.

  a. 3

  b. -3

  c. $\dfrac{1}{3}$

  d. no solution

772. Which of the following are characteristics of the graph of $f(x) = -e^{2-x} - 3$?

  a. The graph of f lies below the x-axis.

  b. $y = -3$ is the horizontal asymptote for the graph of f.

  c. The domain is $\mathbb{R}$.

  d. all of the above

773. Solve: $2^{x-5} = 8$

  a. -3

  b. 3

  c. 2

  d. 8

774. Solve: $3^{2x} = 9 \cdot 3^{x-1}$

  a. 1

  b. 0

  c. -1

  d. none of the above

775. Solve: $4^{2x-3} = \dfrac{1}{4^x}$

  a. -1

  b. 0

  c. 1

  d. none of the above

776. Solve: $125^x = 25$

  a. $\dfrac{3}{2}$

  b. $-\dfrac{3}{2}$

  c. $-\dfrac{2}{3}$

  d. $\dfrac{2}{3}$

777. Solve: $(e^x)^{x-3} = e^{10}$

  a. -5 and -2

  b. -2 and 5

  c. -5 and 2

  d. 2 and 5

778. Solve: $16^{3x-1} = 4^{2x+3}$

  a. $-\dfrac{5}{4}$

  b. $-\dfrac{4}{5}$

  c. $\dfrac{4}{5}$

  d. $\dfrac{5}{4}$

779. Solve: $4^{x+1} = \left(\dfrac{1}{2}\right)^{2x}$

  a. -2^{-2}

  b. $\dfrac{1}{4}$

  c. $-\dfrac{1}{2}$

  d. -2

780. Solve: $x \cdot 3^x + 5 \cdot 3^x = 0$

  a. -5 only

  b. 5 only

  c. 0 only

  d. 0 and -5

781. Solve: $\left(10^{x+1}\right)^{2x} = 100$

  a. $\dfrac{-1 \pm \sqrt{5}}{2}$

  b. $\dfrac{1}{3}$ only

  c. $-\dfrac{1}{3}$ only

  d. $\pm\dfrac{1}{3}$

782. Solve: $2^{\sqrt{x}} \cdot 2 = 8$
    a. 16
    b. 8
    c. 4
    d. 2

783. Solve: $2x^2 \cdot e^x - 7x \cdot e^x + 6 \cdot e^x = 0$
    a. $\frac{2}{3}$ and 2
    b. $\frac{3}{2}$ and -2
    c. $\frac{3}{2}$ and 2
    d. $-\frac{2}{3}$ and 2

784. Solve: $e^{2x} + 5e^x - 6 = 0$
    a. 2 and 3
    b. 0 only
    c. -3 and -2
    d. $-3, -2$, and 0

Set 50 (Answers begin on page 240)

This problem set focuses on basic computations involving logarithms.

785. $\log_3 27 =$
    a. -2
    b. 2
    c. -3
    d. 3

786. $\log_3 \left(\frac{1}{9} \right) =$
    a. 2
    b. -2
    c. 3
    d. -3

787. $\log_{\frac{1}{2}} 8 =$
    a. 16
    b. 3
    c. -3
    d. 4

788. $\log_7 \sqrt{7} =$
    a. 0
    b. 1
    c. $\frac{1}{2}$
    d. -1

789. $\log_5 1 =$
    a. 0
    b. $\frac{1}{5}$
    c. 1
    d. $-\frac{1}{5}$

790. $\log_{16} 64 =$
    a. $\frac{2}{3}$
    b. $-\frac{2}{3}$
    c. 4
    d. $\frac{3}{2}$

791. If $\log_6 x = 2$, then $x =$
    a. 6
    b. 12
    c. 36
    d. -36

792. If $5\sqrt{a} = x$, then $\log_a x =$
    a. $\log_a 5 - \frac{1}{2}$
    b. $\log_a 5 + \frac{1}{2}$
    c. $\log_a 5 + 2$
    d. $\log_a 5 - 2$

793. $\log_3 (3^4 \cdot 9^3) =$
    a. 8
    b. 10
    c. 6
    d. 12

794. If $5^{3x-1} = 7$, then $x =$
    a. $\frac{1}{3}(1 - \log_5 7)$
    b. $-3(1 + \log_5 7)$
    c. $-\frac{1}{3}(1 - \log_5 7)$
    d. $\frac{1}{3}(1 + \log_5 7)$

795. If $\log_a x = 2$ and $\log_a y = -3$, then $\log_a \left(\frac{x}{y^3}\right) =$

  a. $-\frac{2}{3}$

  b. 11

  c. $-\frac{3}{2}$

  d. 8

796. $3^{\log_3 2} =$

  a. 2

  b. 1

  c. 0

  d. −1

797. $\log_a(a^x) =$

  a. ax

  b. 0

  c. x

  d. x^a

798. If $3 \ln\left(\frac{1}{x}\right) = \ln 8$, then $x =$

  a. $\frac{1}{2}$

  b. 2

  c. $-\frac{1}{2}$

  d. −2

799. $e^{-\frac{1}{2}\ln 3} =$

  a. $-\frac{\sqrt{3}}{3}$

  b. $\frac{\sqrt{3}}{3}$

  c. $\frac{3}{\sqrt{3}}$

  d. $-\frac{3}{\sqrt{3}}$

800. If $\ln x = 3$ and $\ln y = 2$, then $\ln\left(\frac{e^2 y}{\sqrt{x}}\right) =$

  a. $\frac{2}{5}$

  b. $-\frac{2}{5}$

  c. $\frac{5}{2}$

  d. $-\frac{5}{2}$

Set 51 (Answers begin on page 241)

This problem set focuses on basic features of logarithmic functions, and simplifying logarithmic expressions using the logarithm rules.

801. Which of the following is equivalent to $3 \ln (xy^2) - 4 \ln(x^2 y) + \ln(xy)$?

  a. $\ln \left[\frac{y^3}{x^4}\right]$

  b. $\ln \left[y^3 x^4\right]$

  c. $\ln \left[\frac{x^4}{y^3}\right]$

  d. $\ln \left[y^3\right] + \ln \left[x^4\right]$

802. Simplify: $\log_8 2 + \log_8 4$

  a. 1

  b. −1

  c. $2 \log_8 2$

  d. 3

803. Simplify: $4 \log_9 3$

  a. 8

  b. −8

  c. 2

  d. −2

804. Which of the following is equivalent to $\ln (18x^3) - \ln (6x)$?

  a. $\ln (3x^2)$

  b. $2 \ln (3x)$

  c. $\ln (3x)^2$

  d. $\ln (108x^4)$

805. Simplify: $\log_7 \frac{2}{49} - \log_7 \frac{2}{7}$

  a. 2

  b. −1

  c. $\frac{2}{49}$

  d. $-\frac{2}{49}$

806. Simplify: $3 \log_4 \frac{2}{3} + \log_4 27$

    a. $-\frac{2}{3}$

    b. $\frac{2}{3}$

    c. $-\frac{3}{2}$

    d. $\frac{3}{2}$

807. Which of the following is equivalent to $\log (2x^3)$?

    a. $\log 2 - 3 \log x$

    b. $-\log 2 + 3 \log x$

    c. $\log 2 + 3 \log x$

    d. $-\log 2 - 3 \log x$

808. Which of the following is equivalent to $\log_3 \left(\frac{8yz^4}{x^2} \right)$?

    a. $3 + \log_2 y - 4 \log_2 z - 2 \log_2 x$

    b. $3 - \log_2 y - 4 \log_2 z + 2 \log_2 x$

    c. $3 + \log_2 y + 4 \log_2 z + 2 \log_2 x$

    d. $3 + \log_2 y + 4 \log_2 z - 2 \log_2 x$

809. Which of the following is equivalent to $\frac{3}{2} \log_2 4 - \frac{2}{3} \log_2 8 + \log_2 2$?

    a. $\log_2 2$

    b. -1

    c. -2

    d. 2

810. Which of the following is equivalent to $3 \log_b (x + 3)^{-1} - 2 \log_b x + \log_b (x + 3)^3$?

    a. $2 \log_b x$

    b. $-\log_b x$

    c. $\log_b [x(x + 3)]$

    d. $\log_b \left(\frac{1}{x^2} \right)$

811. Which of the following is equivalent to $\ln \left[(2\sqrt{x + 1})(x^2 + 3)^4 \right]$?

    a. $\frac{1}{2} \ln 2(x + 1) - 4 \ln (x^2 + 3)$

    b. $\ln 2 - \frac{1}{2} \ln (x + 1) - 4 \ln (x^2 + 3)$

    c. $\ln 2 + \frac{1}{2} \ln (x + 1) + 4 \ln (x^2 + 3)$

    d. $\frac{1}{2} \ln 2(x + 1) + 4 \ln (x^2 + 3)$

812. Which of the following is equivalent to $\log_3 \left(\frac{x^2 \sqrt{2x - 1}}{(2x + 1)^{\frac{3}{2}}} \right)$?

    a. $2 \log_3 (x) + \frac{1}{2} \log_3 (2x - 1)(2x + 1)$

    b. $2 \log_3 (x) - \frac{3}{2} \log_3 (2x - 1)(2x + 1)$

    c. $2 \log_3 (x) + \frac{1}{2} \log_3 (2x - 1) - \frac{3}{2} \log_3 (2x + 1)$

    d. $2 \log_3 (x) - \frac{3}{2} \log_3 \frac{2x - 1}{2x + 1}$

813. Which of the following is a true characterization of the graph of $f(x) = \ln x$?

    a. As the x-values decrease toward zero, the y-values plunge downward very sharply.

    b. As the x-values decrease toward zero, the y-values shoot upward very sharply.

    c. As the x-values move to the right, the y-values decrease very slowly.

    d. As the x-values move to the left, the y-values increase very slowly.

814. What is the domain of $k(x) = \log_3(-x)$?

    a. $(-\infty, 0)$

    b. $(0, \infty)$

    c. $[0, \infty)$

    d. the set of all real numbers

815. What is the domain of $b(x) = \log_5(x^2 + 1)$?

    a. $(-\infty, 0)$

    b. $(0, \infty)$

    c. $[0, \infty)$

    d. the set of all real numbers

816. What is the x-intercept of $f(x) = \log_2 x$?

    a. $(2,0)$

    b. $(0,1)$

    c. $(1,0)$

    d. This function does not have an x-intercept.

Set 52 (Answers begin on page 243)

This problem set focuses on additional features of logarithmic functions and solving equations and inequalities involving logarithms.

817. The range of the function $f(x) = \ln(2x - 1)$ is which of the following?

    a. $\left(-\infty, -\frac{1}{2}\right)$

    b. $\left[\frac{1}{2}, \infty\right)$

    c. $\left(\frac{1}{2}, \infty\right)$

    d. $\mathbb{R}$

818. Which of the following, if any, are x-intercepts of the function $f(x) = \ln (x^2 - 4x + 4)$?

    a. $(1,0)$

    b. $(3,0)$

    c. both **a** and **b**

    d. neither **a** nor **b**

819. The domain of the function $f(x) = \ln (x^2 - 4x + 4)$ is which of the following?

    a. $(2, \infty)$

    b. $(-\infty, 2)$

    c. $(-\infty, 2) \cup (2, \infty)$

    d. $\mathbb{R}$

820. Which of the following is a characteristic of the graph of $f(x) = \ln(x + 1) + 1$?

    a. The y-intercept is $(e, 1)$.

    b. $x = -1$ is a vertical asymptote of f.

    c. There is no x-intercept.

    d. $y = 1$ is a horizontal asymptote.

821. Which of the following choices for f and g are inverses?

    a. $f(x) = e^{-x}$ and $g(x) = \ln \sqrt{x}, x > 0$

    b. $f(x) = e^{2x}$ and $g(x) = \ln \sqrt{x}, x > 0$

    c. $f(x) = e^{-x}$ and $g(x) = \ln \sqrt{-x}, x > 0$

    d. $f(x) = e^{2x}$ and $g(x) = \ln \sqrt{2x}, x > 0$

822. Solve: $\ln(x - 2) - \ln(3 - x) = 1$

    a. $\frac{3e + 2}{e + 1}$

    b. $\frac{3(e + 2)}{e + 1}$

    c. 2

    d. 3

823. Determine the solution set for the inequality $5 \le 4e^{2 - 3x} + 1 < 9$.

    a. $\left(\frac{2 - \ln 2}{3}, \frac{e - 2}{3}\right]$

    b. $\left(\frac{2 - \ln 2}{3}, \frac{2}{3}\right]$

    c. $\left[\frac{2}{3}, \frac{-2 + \ln 2}{3}\right]$

    d. $\left[\frac{e - 2}{3}, \frac{-2 + \ln 2}{3}\right)$

824. Determine the solution set for the inequality $\ln(1 - x^2) \le 0$.

    a. $(-1, 0) \cup (0, 1)$

    b. $(-\infty, -1) \cup ((1, \infty)$

    c. $(-1, 1)$

    d. $[-1, 1]$

825. Solve: $\log x + \log(x + 3) = 1$

    a. -2 and 5

    b. 2 and 5

    c. -5 and 2

    d. 2 only

826. Solve: $\log_2(2x) + \log_2(x+1) = 2$
 a. 0 only
 b. 1 only
 c. 0 and 1
 d. no solution

827. Solve: $\log(x-2) = 2 + \log(x+3)$
 a. $\frac{302}{99}$
 b. 3
 c. 2
 d. There is no solution to this equation.

828. Assuming that $b > 1$, solve: $b^{3\log_b x} = 1$
 a. b
 b. b^2
 c. 0
 d. 1

829. Solve for x: $y = e^{-a(b+x)}$
 a. $x = -\frac{1}{a}(ab + \ln y)$
 b. $x = -b + \ln y$
 c. $x = \frac{1}{a}(ab - \ln y)$
 d. $x = \frac{-ab + \ln y}{a}$

830. Solve for y: $3\ln 4y + \ln A = \ln B$
 a. $y = \frac{1}{4}e^{\frac{1}{3}(\ln B - \ln A)}$
 b. $y = \frac{1}{4}e^{\frac{1}{3}(\ln B + \ln A)}$
 c. $y = -\frac{1}{4}e^{\frac{1}{3}(\ln \frac{B}{A})}$
 d. $y = -\frac{1}{4}e^{\frac{1}{3}(\ln AB)}$

831. Solve for x: $1 + \ln(x \cdot y) = \ln z$
 a. $x = e^{\ln z + \ln y - 1}$
 b. $x = e^{\ln z - \ln y + 1}$
 c. $x = e^{\ln z - \ln y - 1}$
 d. $x = e^{-(\ln z + \ln y - 1)}$

832. Solve for t: $P = P_0 e^{-kt}$
 a. $t = -\frac{1}{k}\ln\left(\frac{P}{P_0}\right)$
 b. $t = -k\ln\left(\frac{P}{P_0}\right)$
 c. $t = \frac{1}{k}\ln\left(-\frac{P}{P_0}\right)$
 d. $t = -\frac{1}{k}\ln\left(\frac{P_0}{P}\right)$

7 ▶ MATRIX ALGEBRA

Systems of linear equations can also be solved using Cramer's rule, which involves the use of matrices. Matrix operations, including matrix arithmetic, computing determinants and inverses, and applying back substitution and Cramer's rule to solve systems of linear equations, are reviewed in the seven problem sets in this section.

Set 53 (Answers begin on page 245)

Basic features of matrices and the arithmetic of matrices are explored in this problem set.

833. What are the dimensions of the matrix
[1 2 −1 0]?
a. 4×4
b. 1×4
c. 4×1
d. 1×1

834. What are the dimensions of the matrix
$$\begin{bmatrix} 0 & -2 \\ 0 & 1 \\ 0 & -2 \\ 0 & 0 \end{bmatrix}?$$
a. 2×4
b. 4×2
c. 2×2
d. 4×4

835. Which of the following matrices has dimensions 3×2?
a. $\begin{bmatrix} 2 & 1 \\ 3 & 5 \\ 2 & 0 \end{bmatrix}$

b. $\begin{bmatrix} -1 & -1 & 0 \\ 0 & -3 & 1 \end{bmatrix}$

c. $\begin{bmatrix} 3 & 3 \\ 3 & 3 \end{bmatrix}$

d. none of the above

836. Compute, if possible: $-3\begin{bmatrix} -1 & -1 & 0 \\ 0 & -3 & 1 \end{bmatrix}$
a. $\begin{bmatrix} -3 & -3 & 0 \\ 0 & -9 & -3 \end{bmatrix}$

b. $\begin{bmatrix} -3 & 3 & 0 \\ 0 & 9 & 3 \end{bmatrix}$

c. $\begin{bmatrix} 3 & 3 & 0 \\ 0 & 9 & -3 \end{bmatrix}$

d. This computation is not well-defined.

837. Compute, if possible: $[1 \ -1 \ -2 \ 2] + \begin{bmatrix} 1 \\ 0 \\ 1 \\ -5 \end{bmatrix}$
a. $[2 \ -1 \ -1 \ -3]$

b. $\begin{bmatrix} 2 \\ -1 \\ -1 \\ -3 \end{bmatrix}$

c. $\begin{bmatrix} 1 & -1 & -2 & 1 \\ 0 & 0 & 0 & 0 \\ 1 & 0 & 0 & 0 \\ -5 & 0 & 0 & 0 \end{bmatrix}$

d. This computation is not well-defined.

838. Compute, if possible: $2\begin{bmatrix} -3 & -1 \\ 0 & 1 \end{bmatrix} - 3\begin{bmatrix} 2 & -1 \\ 2 & 2 \end{bmatrix}$
a. $\begin{bmatrix} -4 & -6 \\ 1 & -12 \end{bmatrix}$

b. $\begin{bmatrix} 12 & -1 \\ 6 & 4 \end{bmatrix}$

c. $\begin{bmatrix} -12 & 1 \\ -6 & -4 \end{bmatrix}$

d. This computation is not well-defined.

839. Compute, if possible: $\frac{2}{5}\begin{bmatrix} -1 & 0 & 1 \\ 1 & 0 & 1 \\ 0 & 1 & -1 \end{bmatrix}$

a.
$$\begin{bmatrix} \frac{2}{5} & 0 & -\frac{2}{5} \\ -\frac{2}{5} & 0 & -\frac{2}{5} \\ 0 & -\frac{2}{5} & \frac{2}{5} \end{bmatrix}$$

b.
$$\begin{bmatrix} \frac{2}{5} & 0 & \frac{2}{5} \\ \frac{2}{5} & 0 & \frac{2}{5} \\ 0 & \frac{2}{5} & \frac{2}{5} \end{bmatrix}$$

c.
$$\begin{bmatrix} -\frac{2}{5} & 0 & \frac{2}{5} \\ \frac{2}{5} & 0 & \frac{2}{5} \\ 0 & \frac{2}{5} & -\frac{2}{5} \end{bmatrix}$$

d. This computation is not well-defined.

840. Determine the values of x, if any exist, that make the following equality true:

$$\begin{bmatrix} x & -2 \\ 0 & 2 \end{bmatrix} = \frac{1}{2}\begin{bmatrix} 6 & -4 \\ 0 & 4 \end{bmatrix}$$

a. 3
b. −3
c. 6
d. There is no such x-value.

841. Determine the values of x, if any exist, that make the following equality true:

$$\begin{bmatrix} -1 & x^2 \\ 3x & -1 \end{bmatrix} = \begin{bmatrix} -1 & 4 \\ 6 & -1 \end{bmatrix}$$

a. −1 and 1
b. −2 and 2
c. −2 only
d. 2 only

842. How many ordered pairs (x,y) make the following equality true:

$$3\begin{bmatrix} 0 & 2 \\ 1 & 1 \\ 0 & x \end{bmatrix} = -1\begin{bmatrix} 0 & -6 \\ -3 & -3 \\ 1 & 6y \end{bmatrix}?$$

a. 0
b. 1
c. 2
d. infinitely many

843. Determine an ordered pair (x,y) that makes the following equality true:

$$-4\begin{bmatrix} 3x-2 & 0 & -2 \\ -2 & 2y & -1 \end{bmatrix} - 2\begin{bmatrix} 4x+2 & -5 & 1 \\ 2 & 4-3y & -1 \end{bmatrix} = \begin{bmatrix} 2x & 10 & 6 \\ 4 & 4 & 6 \end{bmatrix}$$

a. $(\frac{-2}{11}, \frac{-8}{3})$
b. $(\frac{2}{11}, \frac{-8}{3})$
c. $(\frac{2}{11}, \frac{8}{3})$
d. $(\frac{-2}{11}, \frac{8}{3})$

844. Determine an ordered triple (x,y,z) if one exists, that makes the following equality true:

$$\begin{bmatrix} x & 2y \\ 3z & 4 \end{bmatrix} - \begin{bmatrix} 2x & 3y \\ 4z & 4 \end{bmatrix} = \begin{bmatrix} 3x & 4y \\ -2z & 0 \end{bmatrix}$$

a. (1,1,1)
b. (0,1,1)
c. (0,0,0)
d. (0,1,0)

845. Which of the following statements is true?
a. The sum of two 4×2 matrices must be a 4×2 matrix.
b. The sum of a 4×2 matrix and a 2×4 matrix is well-defined.
c. A constant multiple of a 3×1 matrix need not be a 3×1 matrix.
d. All of the above statements are true.

846. Which of the following statements is true?

  a. $0\begin{bmatrix} -1 & 2 \\ 2 & -1 \end{bmatrix} = 0$

  b. $0\begin{bmatrix} 2 \\ -1 \end{bmatrix} + \begin{bmatrix} -1 \\ 3 \end{bmatrix} = \begin{bmatrix} 2 & -1 \\ -1 & 3 \end{bmatrix}$

  c. $3[-1\ 0\ 0] - 2[0\ 1\ 0] + [0\ 0\ -1] = -[3\ 2\ 1]$

  d. All of the above statements are false.

847. Which of the following statements is true?

  a. $\begin{bmatrix} -1 & -1 \\ -1 & -1 \end{bmatrix} + 1 = \begin{bmatrix} 0 & 0 \\ 0 & 0 \end{bmatrix}$

  b. There is an X-value that makes the following equation true: $-3\begin{bmatrix} X & 1 & 0 & 0 \\ 0 & X & 1 & 0 \\ 0 & 0 & X & 1 \\ 0 & 0 & 0 & X \end{bmatrix} = \begin{bmatrix} 15 & 1 & 0 & 0 \\ 0 & 15 & 1 & 0 \\ 0 & 0 & 15 & 1 \\ 0 & 0 & 0 & 15 \end{bmatrix}$

  c. $\begin{bmatrix} 1 & 1 \\ 1 & 1 \end{bmatrix} - \begin{bmatrix} 1 & 1 \end{bmatrix} - \begin{bmatrix} 1 & 1 \end{bmatrix} = \begin{bmatrix} 0 & 0 \\ 0 & 0 \end{bmatrix}$

  d. All of the above statements are false.

848. How many ordered triples (x,y,z) (where x, y, and z are real numbers) make the following equation true?

$$\begin{bmatrix} 1 & x-2 & -1 & -1 \\ -3 & -1 & 2\sqrt{y} & 1 \\ -2 & 1 & 1 & 4z^2 \\ 0 & -3 & -4 & 0 \end{bmatrix} = \begin{bmatrix} 1 & -x^2 & -1 & -1 \\ -3 & -1 & y^2 & 1 \\ -2 & 1 & 1 & 8z \\ 0 & -3 & -4 & 0 \end{bmatrix}?$$

  a. 8

  b. 4

  c. 2

  d. 0

Set 54 (Answers begin on page 247)

The multiplication of matrices and matrix computations involving multiple operations are the focus of this problem set.

For questions 849 through 864, use the following matrices:

$$A = \begin{bmatrix} -1 & 2 \\ 0 & 2 \\ -1 & -1 \end{bmatrix} \qquad F = \begin{bmatrix} 0 \\ 0 \end{bmatrix}$$

$$B = \begin{bmatrix} 1 & -2 & -1 \\ 3 & 5 & 0 \end{bmatrix} \qquad G = \begin{bmatrix} -2 & -1 & 0 & 1 \\ -1 & -2 & -1 & 0 \\ 1 & -1 & -2 & -1 \end{bmatrix}$$

$$C = \begin{bmatrix} 0 & 1 \\ 1 & -4 \end{bmatrix} \qquad H = \begin{bmatrix} 3 & 1 & -1 \\ 1 & -2 & 1 \\ 0 & 0 & -2 \\ -2 & 1 & 0 \end{bmatrix}$$

$$D = \begin{bmatrix} 3 & 2 & 1 \\ 0 & 1 & 2 \\ -1 & -1 & 0 \end{bmatrix} \qquad I = \begin{bmatrix} 2 \\ 2 \\ 1 \end{bmatrix}$$

$$E = \begin{bmatrix} -4 & -2 & 0 \end{bmatrix}$$

849. Express as a single matrix, if possible: *CF*

    a. $\begin{bmatrix} 0 & 0 \end{bmatrix}$

    b. $\begin{bmatrix} 0 \\ 0 \end{bmatrix}$

    c. $\begin{bmatrix} 0 & 0 \\ 0 & 0 \end{bmatrix}$

    d. not possible

850. Express as a single matrix, if possible: $(2G)(-3E)$

    a. $\begin{bmatrix} -1 & 1 & 0 \\ 0 & 1 & 0 \\ 1 & 0 & 1 \end{bmatrix}$

    b. $\begin{bmatrix} 1 & 0 \\ 0 & 1 \end{bmatrix}$

    c. $\begin{bmatrix} -1 & -1 & 1 \end{bmatrix}$

    d. not possible

851. Express as a single matrix, if possible: AB

a. $\begin{bmatrix} 5 & 12 \\ 6 & 10 \end{bmatrix}$

b. $\begin{bmatrix} 12 & 1 \\ -3 & 1 \end{bmatrix}$

c. $\begin{bmatrix} 5 & 12 & 1 \\ 6 & 10 & 0 \\ -4 & -3 & 1 \end{bmatrix}$

d. not possible

852. Express as a single matrix, if possible: $4BA$

a. $\begin{bmatrix} 64 & -4 \\ -12 & 0 \end{bmatrix}$

b. $\begin{bmatrix} 0 & 4 \\ 12 & -64 \end{bmatrix}$

c. $\begin{bmatrix} 0 & -4 \\ -12 & 64 \end{bmatrix}$

d. not possible

853. Express as a single matrix, if possible: $(-2D)(3D)$

a. $\begin{bmatrix} -48 & -42 & -42 \\ 12 & 6 & -12 \\ 18 & 18 & 18 \end{bmatrix}$

b. $\begin{bmatrix} -6 & -7 & -7 \\ 2 & 1 & -2 \\ 3 & 3 & 3 \end{bmatrix}$

c. $\begin{bmatrix} 48 & 42 & 42 \\ -12 & -6 & 12 \\ -18 & -18 & -18 \end{bmatrix}$

d. not possible

854. Express as a single matrix, if possible: FF

a. $\begin{bmatrix} 0 & 0 \\ 0 & 0 \end{bmatrix}$

b. $\begin{bmatrix} 0 \\ 0 \end{bmatrix}$

c. $\begin{bmatrix} 0 & 0 \end{bmatrix}$

d. not possible

855. Express as a single matrix, if possible: $IE + D$

a. $\begin{bmatrix} 5 & 2 & -1 \\ 8 & 3 & -2 \\ 5 & 3 & 0 \end{bmatrix}$

b. $\begin{bmatrix} -5 & -2 & 1 \\ -8 & -3 & 2 \\ -5 & -3 & 0 \end{bmatrix}$

c. $\begin{bmatrix} 1 & -2 & 5 \\ -8 & -3 & 2 \\ 1 & -3 & -5 \end{bmatrix}$

d. not possible

856. Express as a single matrix, if possible: $(BG)H$

a. $\begin{bmatrix} -3 & -7 & -3 \\ -52 & 18 & 8 \end{bmatrix}$

b. $\begin{bmatrix} -3 & -52 \\ -7 & 18 \\ -3 & 8 \end{bmatrix}$

c. $\begin{bmatrix} 3 & 52 \\ 7 & -18 \\ 3 & -8 \end{bmatrix}$

d. not possible

857. Express as a single matrix, if possible: (*EG*)(*HI*)
   a. 22
   b. 33
   c. 66
   d. not possible

858. Express as a single matrix, if possible:
   (*ED*)(*AC*)
   a. [36 −164]
   b. [−36 164]
   c. $\begin{bmatrix} 36 \\ -164 \end{bmatrix}$
   d. not possible

859. Express as a single matrix, if possible: *E*(*G* + *A*)
   a. $\begin{bmatrix} 1 & 0 \\ 1 & 0 \end{bmatrix}$
   b. $\begin{bmatrix} -2 \\ 1 \end{bmatrix}$
   c. −1
   d. not possible

860. Express as a single matrix, if possible: 4*B* − 3*EF*
   a. $\begin{bmatrix} 2 & -1 & 2 \\ 1 & 1 & -2 \end{bmatrix}$
   b. $\begin{bmatrix} 2 & 1 \\ -1 & 1 \\ 2 & -2 \end{bmatrix}$
   c. $\begin{bmatrix} 0 & 0 \\ 0 & 0 \end{bmatrix}$
   d. not possible

861. Express as a single matrix, if possible:
   (2*C*)(2*C*)(2*C*)*F*
   a. $\begin{bmatrix} 0 \\ 0 \end{bmatrix}$
   b. [0 0]
   c. $\begin{bmatrix} 0 & 0 \\ 0 & 0 \end{bmatrix}$
   d. not possible

862. Express as a single matrix, if possible:
   (*EAF*)(*CF*)
   a. $\begin{bmatrix} 0 & 0 \\ 0 & 0 \end{bmatrix}$
   b. [0 0]
   c. $\begin{bmatrix} 0 \\ 0 \end{bmatrix}$
   d. not possible

863. Express as a single matrix, if possible:
   3*D* − 2*AB* + *GH*
   a. $\begin{bmatrix} 10 & 17 & -2 \\ 17 & 14 & -7 \\ -9 & -5 & 0 \end{bmatrix}$
   b. $\begin{bmatrix} -10 & -17 & 2 \\ -17 & -14 & 7 \\ 9 & 5 & 0 \end{bmatrix}$
   c. $\begin{bmatrix} -10 & -17 & -2 \\ -17 & -14 & -7 \\ -9 & -5 & 0 \end{bmatrix}$
   d. not possible

864. Express as a single matrix, if possible:
$(2E)(-2F) + 2B$

a. $\begin{bmatrix} 2 & 1 \\ -1 & 0 \end{bmatrix}$

b. $\begin{bmatrix} -2 & -5 \\ 4 & 2 \end{bmatrix}$

c. $\begin{bmatrix} 3 & 2 \\ 1 & 5 \end{bmatrix}$

d. not possible

Set 55 (Answers begin on page 250)

This problem set is focused on computing determinants of square matrices.

865. Compute the determinant: $\begin{bmatrix} -3 & 7 \\ 1 & 5 \end{bmatrix}$
a. −38
b. −26
c. 22
d. −22

866. Compute the determinant: $\begin{bmatrix} a & 0 \\ 0 & b \end{bmatrix}$
a. 0
b. a
c. ab
d. b

867. Compute the determinant: $\begin{bmatrix} 1 & 2 \\ 2 & 3 \end{bmatrix}$
a. −1
b. 1
c. −3
d. 3

868. Compute the determinant: $\begin{bmatrix} 2 & 3 \\ 1 & 1 \end{bmatrix}$
a. −5
b. 5
c. 1
d. −1

869. Compute the determinant: $\begin{bmatrix} -1 & 2 \\ 2 & -4 \end{bmatrix}$
a. −10
b. −6
c. 6
d. 0

870. Compute the determinant: $\begin{bmatrix} 6 & 3 \\ 2 & 1 \end{bmatrix}$
a. 9
b. 0
c. 16
d. −16

871. Compute the determinant: $\begin{bmatrix} -3 & 4 \\ 4 & 2 \end{bmatrix}$
a. −20
b. 20
c. −22
d. 22

872. Compute the determinant: $\begin{bmatrix} 1 & -4 \\ 0 & 25 \end{bmatrix}$
a. 25
b. −25
c. −4
d. 4

873. Compute the determinant: $\begin{bmatrix} 3 & -1 \\ 1 & -2 \end{bmatrix}$
a. 1
b. −1
c. −5
d. 5

874. Compute the determinant: $\begin{bmatrix} -2 & 0 \\ -12 & 3 \end{bmatrix}$
a. 24
b. 36
c. −6
d. −24

875. Compute the determinant: $\begin{bmatrix} 0 & 1 \\ -2 & -1 \end{bmatrix}$
a. −2
b. 2
c. 1
d. −1

876. Compute the determinant: $\begin{bmatrix} -1 & 0 \\ 2 & -1 \end{bmatrix}$
 a. 2
 b. 1
 c. −2
 d. 0

877. Compute the determinant: $\begin{bmatrix} 3 & 2 \\ 3 & 2 \end{bmatrix}$
 a. 0
 b. 5
 c. 12
 d. −12

878. Compute the determinant: $\begin{bmatrix} 3 & -2 \\ 9 & -6 \end{bmatrix}$
 a. 0
 b. 15
 c. 48
 d. −15

879. Compute the determinant: $\begin{bmatrix} -1 & -1 \\ -1 & 0 \end{bmatrix}$
 a. 1
 b. 0
 c. −1
 d. 2

880. Compute the determinant: $\begin{bmatrix} 0 & 2 \\ 4 & 0 \end{bmatrix}$
 a. 2
 b. −2
 c. 8
 d. −8

Set 56 (Answers begin on page 251)

This problem set is focused on writing systems in matrix form.

881. Write this system in matrix form:
$$\begin{cases} -3x + 7y = 2 \\ x + 5y = 8 \end{cases}$$

 a. $\begin{bmatrix} 3 & -7 \\ 1 & 5 \end{bmatrix} \begin{bmatrix} x \\ y \end{bmatrix} = \begin{bmatrix} 2 \\ 8 \end{bmatrix}$

 b. $\begin{bmatrix} -3 & 7 \\ -1 & -5 \end{bmatrix} \begin{bmatrix} x \\ y \end{bmatrix} = \begin{bmatrix} 2 \\ 8 \end{bmatrix}$

 c. $\begin{bmatrix} -3 & 1 \\ 7 & 5 \end{bmatrix} \begin{bmatrix} x \\ y \end{bmatrix} = \begin{bmatrix} 2 \\ 8 \end{bmatrix}$

 d. $\begin{bmatrix} -3 & 7 \\ 1 & 5 \end{bmatrix} \begin{bmatrix} x \\ y \end{bmatrix} = \begin{bmatrix} 2 \\ 8 \end{bmatrix}$

882. Write this system in matrix form: $\begin{cases} x = a \\ y = b \end{cases}$

 a. $\begin{bmatrix} 1 & 0 \\ 0 & 1 \end{bmatrix} \begin{bmatrix} x \\ y \end{bmatrix} = \begin{bmatrix} a \\ b \end{bmatrix}$

 b. $\begin{bmatrix} 1 & 1 \\ 0 & 0 \end{bmatrix} \begin{bmatrix} x \\ y \end{bmatrix} = \begin{bmatrix} a \\ b \end{bmatrix}$

 c. $\begin{bmatrix} 0 & 0 \\ 1 & 1 \end{bmatrix} \begin{bmatrix} x \\ y \end{bmatrix} = \begin{bmatrix} a \\ b \end{bmatrix}$

 d. $\begin{bmatrix} 1 & 0 \\ 0 & 1 \end{bmatrix} \begin{bmatrix} x \\ y \end{bmatrix} = \begin{bmatrix} b \\ a \end{bmatrix}$

883. Write this system in matrix form:
$$\begin{cases} x + 2y = 4 \\ 2x + 3y = 2 \end{cases}$$

 a. $\begin{bmatrix} 3 & 2 \\ 2 & 1 \end{bmatrix} \begin{bmatrix} x \\ y \end{bmatrix} = \begin{bmatrix} 4 \\ 2 \end{bmatrix}$

 b. $\begin{bmatrix} 1 & 2 \\ 2 & 3 \end{bmatrix} \begin{bmatrix} x \\ y \end{bmatrix} = \begin{bmatrix} 4 \\ 2 \end{bmatrix}$

 c. $\begin{bmatrix} 1 & 2 \\ 2 & 3 \end{bmatrix} \begin{bmatrix} x \\ y \end{bmatrix} = \begin{bmatrix} 2 \\ 4 \end{bmatrix}$

 d. $\begin{bmatrix} 3 & 2 \\ 2 & 1 \end{bmatrix} \begin{bmatrix} x \\ y \end{bmatrix} = \begin{bmatrix} 2 \\ 4 \end{bmatrix}$

884. Write this system in matrix form:

$$\begin{bmatrix} 2 & 1 \\ 3 & 1 \end{bmatrix} \begin{bmatrix} x \\ y \end{bmatrix} = \begin{bmatrix} 1 \\ -2 \end{bmatrix}?$$

a. $\begin{bmatrix} 2 & 1 \\ 3 & 1 \end{bmatrix} \begin{bmatrix} x \\ y \end{bmatrix} = \begin{bmatrix} 1 \\ -2 \end{bmatrix}$

b. $\begin{bmatrix} 2 & 3 \\ 1 & 1 \end{bmatrix} \begin{bmatrix} x \\ y \end{bmatrix} = \begin{bmatrix} -2 \\ 1 \end{bmatrix}$

c. $\begin{bmatrix} 2 & 3 \\ 1 & 1 \end{bmatrix} \begin{bmatrix} x \\ y \end{bmatrix} = \begin{bmatrix} 1 \\ -2 \end{bmatrix}$

d. $\begin{bmatrix} 2 & 1 \\ 3 & 1 \end{bmatrix} \begin{bmatrix} x \\ y \end{bmatrix} = \begin{bmatrix} -2 \\ 1 \end{bmatrix}$

885. Write this system in matrix form:

$$\begin{cases} -x + 2y = 3 \\ 2x + 4y = -6 \end{cases} ?$$

a. $\begin{bmatrix} -4 & 2 \\ 2 & -1 \end{bmatrix} \begin{bmatrix} x \\ y \end{bmatrix} = \begin{bmatrix} 3 \\ -6 \end{bmatrix}$

b. $\begin{bmatrix} 2 & 3 \\ 1 & 1 \end{bmatrix} \begin{bmatrix} x \\ y \end{bmatrix} = \begin{bmatrix} -2 \\ 1 \end{bmatrix}$

c. $\begin{bmatrix} -1 & 2 \\ 2 & -4 \end{bmatrix} \begin{bmatrix} x \\ y \end{bmatrix} = \begin{bmatrix} 3 \\ -6 \end{bmatrix}$

d. $\begin{bmatrix} 2 & 1 \\ 3 & 1 \end{bmatrix} \begin{bmatrix} x \\ y \end{bmatrix} = \begin{bmatrix} -2 \\ 1 \end{bmatrix}$

886. Write this system in matrix form:

$$\begin{cases} 6x + 3y = 8 \\ 2x + y = 3 \end{cases} ?$$

a. $\begin{bmatrix} 6 & 3 \\ 2 & 1 \end{bmatrix} \begin{bmatrix} x \\ y \end{bmatrix} = \begin{bmatrix} 3 \\ 8 \end{bmatrix}$

b. $\begin{bmatrix} 6 & 2 \\ 3 & 1 \end{bmatrix} \begin{bmatrix} x \\ y \end{bmatrix} = \begin{bmatrix} 3 \\ 8 \end{bmatrix}$

c. $\begin{bmatrix} 6 & 2 \\ 3 & 1 \end{bmatrix} \begin{bmatrix} x \\ y \end{bmatrix} = \begin{bmatrix} 8 \\ 3 \end{bmatrix}$

d. $\begin{bmatrix} 6 & 3 \\ 2 & 1 \end{bmatrix} \begin{bmatrix} x \\ y \end{bmatrix} = \begin{bmatrix} 8 \\ 3 \end{bmatrix}$

887. Write this system in matrix form:

$$\begin{cases} -3x + 1 = 4y \\ 2y + 3 = -4x \end{cases} ?$$

a. $\begin{bmatrix} 2 & 4 \\ 4 & -3 \end{bmatrix} \begin{bmatrix} x \\ y \end{bmatrix} = \begin{bmatrix} 1 \\ -3 \end{bmatrix}$

b. $\begin{bmatrix} -3 & 4 \\ 4 & 2 \end{bmatrix} \begin{bmatrix} x \\ y \end{bmatrix} = \begin{bmatrix} -3 \\ 1 \end{bmatrix}$

c. $\begin{bmatrix} -3 & 4 \\ 4 & 2 \end{bmatrix} \begin{bmatrix} x \\ y \end{bmatrix} = \begin{bmatrix} 1 \\ -3 \end{bmatrix}$

d. $\begin{bmatrix} 2 & 4 \\ 4 & -3 \end{bmatrix} \begin{bmatrix} x \\ y \end{bmatrix} = \begin{bmatrix} -3 \\ 1 \end{bmatrix}$

888. Write this system in matrix form:
$$\begin{cases} -2 = x - 4y \\ 5y = \frac{1}{5} \end{cases}$$

a. $\begin{bmatrix} 1 & -4 \\ 0 & 5 \end{bmatrix} \begin{bmatrix} x \\ y \end{bmatrix} = \begin{bmatrix} 2 \\ \frac{1}{5} \end{bmatrix}$

b. $\begin{bmatrix} 1 & 0 \\ -4 & 5 \end{bmatrix} \begin{bmatrix} x \\ y \end{bmatrix} = \begin{bmatrix} -2 \\ \frac{1}{5} \end{bmatrix}$

c. $\begin{bmatrix} 1 & 0 \\ -4 & 5 \end{bmatrix} \begin{bmatrix} x \\ y \end{bmatrix} = \begin{bmatrix} \frac{1}{5} \\ -2 \end{bmatrix}$

d. $\begin{bmatrix} 1 & -4 \\ 0 & 5 \end{bmatrix} \begin{bmatrix} x \\ y \end{bmatrix} = \begin{bmatrix} \frac{1}{5} \\ -2 \end{bmatrix}$

889. Write this system in matrix form:
$$\begin{cases} x - 2 = 4x - y + 3 \\ 6 - 2y = -3 - x \end{cases}$$

a. $\begin{bmatrix} -2 & 1 \\ 1 & -3 \end{bmatrix} \begin{bmatrix} x \\ y \end{bmatrix} = \begin{bmatrix} 5 \\ -9 \end{bmatrix}$

b. $\begin{bmatrix} -3 & 1 \\ 1 & -2 \end{bmatrix} \begin{bmatrix} x \\ y \end{bmatrix} = \begin{bmatrix} 5 \\ -9 \end{bmatrix}$

c. $\begin{bmatrix} -3 & 1 \\ 1 & -2 \end{bmatrix} \begin{bmatrix} x \\ y \end{bmatrix} = \begin{bmatrix} -9 \\ 5 \end{bmatrix}$

d. $\begin{bmatrix} -2 & 1 \\ 1 & -3 \end{bmatrix} \begin{bmatrix} x \\ y \end{bmatrix} = \begin{bmatrix} -9 \\ 5 \end{bmatrix}$

890. Write this system in matrix form:
$$\begin{cases} -1 = -3 - 2x \\ 2 - 3y = 6(1 - 2x) \end{cases}$$

a. $\begin{bmatrix} -3 & 0 \\ 12 & 2 \end{bmatrix} \begin{bmatrix} x \\ y \end{bmatrix} = \begin{bmatrix} -2 \\ 4 \end{bmatrix}$

b. $\begin{bmatrix} 2 & 0 \\ 12 & -3 \end{bmatrix} \begin{bmatrix} x \\ y \end{bmatrix} = \begin{bmatrix} -2 \\ 4 \end{bmatrix}$

c. $\begin{bmatrix} 2 & 12 \\ 0 & -3 \end{bmatrix} \begin{bmatrix} x \\ y \end{bmatrix} = \begin{bmatrix} -2 \\ 4 \end{bmatrix}$

d. $\begin{bmatrix} 2 & 12 \\ 0 & -3 \end{bmatrix} \begin{bmatrix} x \\ y \end{bmatrix} = \begin{bmatrix} 4 \\ -2 \end{bmatrix}$

891. Write this system in matrix form:
$$\begin{cases} 2x - 3y + 5 = 1 + 2x - 4y \\ 3 - x - 2y = -y + x + 3 \end{cases}$$

a. $\begin{bmatrix} 0 & 1 \\ -2 & -1 \end{bmatrix} \begin{bmatrix} x \\ y \end{bmatrix} = \begin{bmatrix} -4 \\ 0 \end{bmatrix}$

b. $\begin{bmatrix} -1 & 1 \\ -2 & 0 \end{bmatrix} \begin{bmatrix} x \\ y \end{bmatrix} = \begin{bmatrix} -4 \\ 0 \end{bmatrix}$

c. $\begin{bmatrix} 0 & 1 \\ -2 & -1 \end{bmatrix} \begin{bmatrix} x \\ y \end{bmatrix} = \begin{bmatrix} 0 \\ -4 \end{bmatrix}$

d. $\begin{bmatrix} 0 & -2 \\ 1 & -1 \end{bmatrix} \begin{bmatrix} x \\ y \end{bmatrix} = \begin{bmatrix} -4 \\ 0 \end{bmatrix}$

892. Which of the following systems can be written in the matrix form

$$\begin{bmatrix} -1 & 0 \\ 2 & -1 \end{bmatrix} \begin{bmatrix} x \\ y \end{bmatrix} = \begin{bmatrix} -2 \\ 1 \end{bmatrix}?$$

a. $\begin{cases} -x = -2 \\ -2x + y = 1 \end{cases}$

b. $\begin{cases} x = -2 \\ 2x - y = 1 \end{cases}$

c. $\begin{cases} -x = -2 \\ 2x - y = 1 \end{cases}$

d. $\begin{cases} -x = -2 \\ 2x + y = 1 \end{cases}$

893. Which of the following systems can be written in the matrix form

$$\begin{bmatrix} 3 & 2 \\ 3 & 2 \end{bmatrix} \begin{bmatrix} x \\ y \end{bmatrix} = \begin{bmatrix} -2 \\ 1 \end{bmatrix}?$$

a. $\begin{cases} 2x - 3y = -2 \\ 2x + 3y = 1 \end{cases}$

b. $\begin{cases} 2x + 3y = -2 \\ 2x + 3y = 1 \end{cases}$

c. $\begin{cases} 3x - 2y = -2 \\ 3x - 2y = 1 \end{cases}$

d. $\begin{cases} 3x + 2y = -2 \\ 3x + 2y = 1 \end{cases}$

894. Which of the following systems can be written in the matrix form

$$\begin{bmatrix} 3 & -2 \\ 9 & -6 \end{bmatrix} \begin{bmatrix} x \\ y \end{bmatrix} = \begin{bmatrix} 4 \\ 12 \end{bmatrix}?$$

a. $\begin{cases} -36 + 2x = 4 \\ -9y + 6x = 12 \end{cases}$

b. $\begin{cases} -3x + 2x = 4 \\ -9y + 6x = 12 \end{cases}$

c. $\begin{cases} 3x - 2y = 4 \\ 9y - 6x = 12 \end{cases}$

d. $\begin{cases} 3x - 2x = 4 \\ 9y - 6x = 12 \end{cases}$

895. Which of the following systems can be written in the matrix form ?

$$\begin{bmatrix} -1 & -1 \\ -1 & 0 \end{bmatrix} \begin{bmatrix} x \\ y \end{bmatrix} = \begin{bmatrix} -1 \\ 1 \end{bmatrix}?$$

a. $\begin{cases} x - y = -1 \\ x = 1 \end{cases}$

b. $\begin{cases} -x - y = -1 \\ -x = 1 \end{cases}$

c. $\begin{cases} -x + y = -1 \\ x = 1 \end{cases}$

d. $\begin{cases} -y + x = -1 \\ -y = 1 \end{cases}$

896. Which of the following systems can be written in the matrix form

$$\begin{bmatrix} 0 & 2 \\ 4 & 0 \end{bmatrix} \begin{bmatrix} x \\ y \end{bmatrix} = \begin{bmatrix} 14 \\ -20 \end{bmatrix}?$$

a. $\begin{cases} 2x = 14 \\ 4y = -20 \end{cases}$

b. $\begin{cases} 2y = 14 \\ 4x = -20 \end{cases}$

c. $\begin{cases} 2y = 14 \\ 4y = -20 \end{cases}$

d. $\begin{cases} 2x = 14 \\ 4x = -20 \end{cases}$

Set 57 (Answers begin on page 254)

This problem set is focused on computing inverse matrices.

897. Compute the inverse, if it exists: $\begin{bmatrix} -3 & 7 \\ 1 & 5 \end{bmatrix}$

a. $\begin{bmatrix} \frac{5}{22} & -\frac{7}{22} \\ -\frac{1}{22} & -\frac{3}{22} \end{bmatrix}$

b. $\begin{bmatrix} -\frac{5}{22} & \frac{7}{22} \\ \frac{1}{22} & \frac{3}{22} \end{bmatrix}$

c. $\begin{bmatrix} 5 & -7 \\ -1 & -3 \end{bmatrix}$

d. The inverse does not exist.

898. Assume that a and b are not zero. Compute the inverse, if it exists:

$$\begin{bmatrix} a & 0 \\ 0 & b \end{bmatrix}$$

a. $\begin{bmatrix} \frac{1}{a} & 0 \\ 0 & \frac{1}{b} \end{bmatrix}$

b. $\begin{bmatrix} -\frac{1}{a} & 0 \\ 0 & -\frac{1}{b} \end{bmatrix}$

c. $\begin{bmatrix} -a & 0 \\ 0 & -b \end{bmatrix}$

d. The inverse does not exist.

899. Compute the inverse, if it exists: $\begin{bmatrix} 1 & 2 \\ 2 & 3 \end{bmatrix}$

a. $\begin{bmatrix} -3 & 2 \\ 2 & -1 \end{bmatrix}$

b. $\begin{bmatrix} 3 & -2 \\ -2 & 1 \end{bmatrix}$

c. $\begin{bmatrix} -3 & -2 \\ -2 & 1 \end{bmatrix}$

d. The inverse does not exist.

900. Compute the inverse, if it exists: $\begin{bmatrix} 2 & 3 \\ 1 & 1 \end{bmatrix}$

a. $\begin{bmatrix} -1 & 3 \\ 1 & -2 \end{bmatrix}$

b. $\begin{bmatrix} 1 & -3 \\ -1 & 2 \end{bmatrix}$

c. $\begin{bmatrix} -1 & 3 \\ 1 & 2 \end{bmatrix}$

d. The inverse does not exist.

901. Compute the inverse, if it exists: $\begin{bmatrix} -1 & 2 \\ 2 & -4 \end{bmatrix}$

  a. $\begin{bmatrix} 1 & -2 \\ -2 & 4 \end{bmatrix}$

  b. $\begin{bmatrix} 1 & 2 \\ 2 & 4 \end{bmatrix}$

  c. $\begin{bmatrix} -4 & 2 \\ 2 & -1 \end{bmatrix}$

  d. The inverse does not exist.

902. Compute the inverse, if it exists: $\begin{bmatrix} 6 & 3 \\ 2 & 1 \end{bmatrix}$

  a. $\begin{bmatrix} 1 & 3 \\ 2 & 6 \end{bmatrix}$

  b. $\begin{bmatrix} 6 & -3 \\ -2 & 1 \end{bmatrix}$

  c. $\begin{bmatrix} 1 & -3 \\ -2 & 6 \end{bmatrix}$

  d. The inverse does not exist.

903. Compute the inverse, if it exists: $\begin{bmatrix} -3 & 4 \\ 4 & 2 \end{bmatrix}$

  a. $\begin{bmatrix} -\frac{1}{11} & \frac{2}{11} \\ \frac{2}{11} & \frac{3}{22} \end{bmatrix}$

  b. $\begin{bmatrix} 2 & 4 \\ 4 & -3 \end{bmatrix}$

  c. $\begin{bmatrix} 2 & -4 \\ -4 & -3 \end{bmatrix}$

  d. The inverse does not exist.

904. Compute the inverse, if it exists: $\begin{bmatrix} 1 & -4 \\ 0 & 25 \end{bmatrix}$

  a. $\begin{bmatrix} 1 & \frac{4}{25} \\ 0 & \frac{1}{25} \end{bmatrix}$

  b. $\begin{bmatrix} 1 & 0 \\ -4 & 25 \end{bmatrix}$

  c. $\begin{bmatrix} 25 & 4 \\ 0 & 1 \end{bmatrix}$

  d. The inverse does not exist.

905. Compute the inverse, if it exists: $\begin{bmatrix} 3 & -1 \\ 1 & -2 \end{bmatrix}$

  a. $\begin{bmatrix} -2 & -1 \\ 1 & 3 \end{bmatrix}$

  b. $\begin{bmatrix} -2 & 1 \\ -1 & 3 \end{bmatrix}$

  c. $\begin{bmatrix} -\frac{2}{5} & -\frac{1}{5} \\ -\frac{1}{5} & -\frac{3}{5} \end{bmatrix}$

  d. The inverse does not exist.

906. Compute the inverse, if it exists: $\begin{bmatrix} 2 & 0 \\ 12 & -3 \end{bmatrix}$

  a. $\begin{bmatrix} 3 & 0 \\ -12 & -2 \end{bmatrix}$

  b. $\begin{bmatrix} \frac{1}{2} & 0 \\ 2 & -\frac{1}{3} \end{bmatrix}$

  c. $\begin{bmatrix} 3 & 0 \\ 12 & -2 \end{bmatrix}$

  d. The inverse does not exist.

907. Compute the inverse, if it exists: $\begin{bmatrix} 0 & 1 \\ -2 & -1 \end{bmatrix}$

a. $\begin{bmatrix} -\frac{1}{2} & -\frac{1}{2} \\ 1 & 0 \end{bmatrix}$

b. $\begin{bmatrix} -1 & 1 \\ -2 & 0 \end{bmatrix}$

c. $\begin{bmatrix} -1 & -1 \\ 2 & 0 \end{bmatrix}$

d. The inverse does not exist.

908. Compute the inverse, if it exists: $\begin{bmatrix} -1 & 0 \\ 2 & -1 \end{bmatrix}$

a. $\begin{bmatrix} 1 & 0 \\ 2 & 1 \end{bmatrix}$

b. $\begin{bmatrix} 1 & 2 \\ 0 & 1 \end{bmatrix}$

c. $\begin{bmatrix} -1 & 0 \\ -2 & -1 \end{bmatrix}$

d. The inverse does not exist.

909. Compute the inverse, if it exists: $\begin{bmatrix} 3 & 2 \\ 3 & 2 \end{bmatrix}$

a. $\begin{bmatrix} 2 & 2 \\ 3 & 3 \end{bmatrix}$

b. $\begin{bmatrix} 2 & -2 \\ -3 & 3 \end{bmatrix}$

c. $\begin{bmatrix} 2 & -3 \\ -2 & 3 \end{bmatrix}$

d. The inverse does not exist.

910. Compute the inverse, if it exists: $\begin{bmatrix} 3 & -2 \\ 9 & -6 \end{bmatrix}$

a. $\begin{bmatrix} -6 & -2 \\ 9 & 3 \end{bmatrix}$

b. $\begin{bmatrix} -6 & 9 \\ -2 & 3 \end{bmatrix}$

c. $\begin{bmatrix} -6 & 2 \\ -9 & 3 \end{bmatrix}$

d. The inverse does not exist.

911. Compute the inverse, if it exists: $\begin{bmatrix} -1 & -1 \\ -1 & 0 \end{bmatrix}$

a. $\begin{bmatrix} 0 & -1 \\ -1 & 1 \end{bmatrix}$

b. $\begin{bmatrix} 1 & -1 \\ -1 & 0 \end{bmatrix}$

c. $\begin{bmatrix} 1 & 1 \\ 1 & 0 \end{bmatrix}$

d. The inverse does not exist.

912. Compute the inverse, if it exists: $\begin{bmatrix} 0 & 2 \\ 4 & 0 \end{bmatrix}$

a. $\begin{bmatrix} 0 & -2 \\ -4 & 0 \end{bmatrix}$

b. $\begin{bmatrix} 0 & \frac{1}{4} \\ \frac{1}{2} & 0 \end{bmatrix}$

c. $\begin{bmatrix} 0 & -\frac{1}{4} \\ -\frac{1}{2} & 0 \end{bmatrix}$

d. The inverse does not exist.

Set 58 (Answers begin on page 257)

This problem set is focused on solving matrix equations of the form $Ax = b$.

913. Solve this system by first converting to an equivalent matrix equation:

$$\begin{cases} -3x + 7y = 2 \\ x + 5y = 8 \end{cases}$$

a. $x = \frac{23}{11}, y = \frac{13}{11}$
b. $x = -\frac{23}{11}, y = -\frac{13}{11}$
c. There is no solution.
d. There are infinitely many solutions.

914. Solve this system by first converting to an equivalent matrix equation:

$$\begin{cases} x = a \\ y = b \end{cases}$$

a. There is no solution.
b. There are infinitely many solutions.
c. $x = a, y = b$
d. $x = b, y = a$

915. Solve this system by first converting to an equivalent matrix equation:

$$\begin{cases} x + 2y = 4 \\ 2x + 3y = 2 \end{cases}$$

a. $x = 8, y = -6$
b. $x = -8, y = 6$
c. There is no solution.
d. There are infinitely many solutions.

916. Solve this system by first converting to an equivalent matrix equation:

$$\begin{cases} 2x + 3y = 1 \\ x + y = -2 \end{cases}$$

a. $x = 1, y = -5$
b. $x = -7, y = 5$
c. There is no solution.
d. There are infinitely many solutions.

917. Solve this system by first converting to an equivalent matrix equation:

$$\begin{cases} -x + 2y = 3 \\ 2x - 4y = -6 \end{cases}$$

a. $x = -3, y = 2$
b. $x = 2, y = -3$
c. There is no solution.
d. There are infinitely many solutions.

918. Solve this system by first converting to an equivalent matrix equation:

$$\begin{cases} 6x + 3y = 8 \\ 2x + y = 3 \end{cases}$$

a. $x = -4, y = -6$
b. $x = 4, y = -6$
c. There are infinitely many solutions.
d. There is no solution.

919. Solve this system by first converting to an equivalent matrix equation:

$$\begin{cases} -3x = 1 - 4y \\ 2y + 3 = -4x \end{cases}$$

a. $x = -\frac{7}{11}, y = -\frac{5}{22}$
b. $x = \frac{7}{11}, y = \frac{5}{22}$
c. There is no solution.
d. There are infinitely many solutions.

920. Solve this system by first converting to an equivalent matrix equation:

$$\begin{cases} -2 = x - 4y \\ 5y = \frac{1}{5} \end{cases}$$

a. $x = -\frac{46}{25}, y = \frac{1}{25}$
b. $x = \frac{46}{25}, y = -\frac{1}{25}$
c. There is no solution.
d. There are infinitely many solutions.

921. Solve this system by first converting to an equivalent matrix equation:

$$\begin{cases} x - 2 = 4x - y + 3 \\ 6 - 2y = -3 - x \end{cases}$$

a. $x = \frac{19}{5}, y = -\frac{22}{5}$
b. $x = -\frac{19}{5}, y = \frac{22}{5}$
c. There is no solution.
d. There are infinitely many solutions.

922. Solve this system by first converting to an equivalent matrix equation:

$$\begin{cases} -1 = -3 - 2x \\ 2 - 3y = 6(1 - 2x) \end{cases}$$

a. There is no solution.
b. There are infinitely many solutions.
c. $x = -1, y = -\frac{16}{3}$
d. $x = 1, y = \frac{16}{3}$

923. Solve this system by first converting to an equivalent matrix equation:

$$\begin{cases} 2x - 3y + 5 = 1 + 2x - 4y \\ 3 - x - 2y = -y + x + 3 \end{cases}$$

a. There is no solution.
b. There are infinitely many solutions.
c. $x = 2, y = -4$
d. $x = -2, y = 4$

924. Solve this system:

$$\begin{bmatrix} -1 & 0 \\ 2 & -1 \end{bmatrix} \begin{bmatrix} x \\ y \end{bmatrix} = \begin{bmatrix} -2 \\ 1 \end{bmatrix}$$

a. $x = -2, y = -3$
b. $x = 2, y = 3$
c. There is no solution.
d. There are infinitely many solutions.

925. Solve this system: $\begin{bmatrix} 3 & 2 \\ 3 & 2 \end{bmatrix} \begin{bmatrix} x \\ y \end{bmatrix} = \begin{bmatrix} -2 \\ 1 \end{bmatrix}$

a. $x = -2, y = -1$
b. $x = 2, y = -1$
c. There are infinitely many solutions.
d. There is no solution.

926. Solve this system: $\begin{bmatrix} 3 & -2 \\ 9 & -6 \end{bmatrix} \begin{bmatrix} x \\ y \end{bmatrix} = \begin{bmatrix} 4 \\ 12 \end{bmatrix}$

a. $x = 9, y = -2$
b. $x = -9, y = 2$
c. There is no solution.
d. There are infinitely many solutions.

927. Solve this system:

$$\begin{bmatrix} -1 & -1 \\ -1 & 0 \end{bmatrix} \begin{bmatrix} x \\ y \end{bmatrix} = \begin{bmatrix} -1 \\ 1 \end{bmatrix}$$

a. There is no solution.
b. There are infinitely many solutions.
c. $x = -1, y = 2$
d. $x = 1, y = -2$

928. Solve this system: $\begin{bmatrix} 0 & 2 \\ 4 & 0 \end{bmatrix} \begin{bmatrix} x \\ y \end{bmatrix} = \begin{bmatrix} 14 \\ -20 \end{bmatrix}$

a. There is no solution.
b. There are infinitely many solutions.
c. $x = -5, y = 7$
d. $x = 5, y = -7$

Set 59 (Answers begin on page 264)

This problem set is focused on solving matrix equations of the form $Ax = b$ using Cramer's rule.

929. Solve this system using Cramer's rule:

$$\begin{cases} -3x + 7y = 2 \\ x + 5y = 8 \end{cases}$$

a. There is no solution.
b. There are infinitely many solutions.
c. $x = \frac{23}{11}, y = \frac{13}{11}$
d. $x = -\frac{23}{11}, y = -\frac{13}{11}$

930. Assume that a and b are nonzero. Solve this system using Cramer's rule:

$$\begin{cases} x = a \\ y = b \end{cases}$$

a. $x = b, y = a$
b. $x = a, y = b$
c. There is no solution.
d. There are infinitely many solutions.

931. Solve this system using Cramer's rule:

$$\begin{cases} x + 2y = 4 \\ 2x + 3y = 2 \end{cases}$$

a. $x = 8, y = -6$
b. $x = -8, y = 6$
c. There is no solution.
d. There are infinitely many solutions.

932. Solve this system using Cramer's rule:

$$\begin{cases} 2x + 3y = 1 \\ x + y = -2 \end{cases}$$

a. $x = -7, y = 5$
b. $x = 7, y = -5$
c. There is no solution.
d. There are infinitely many solutions.

933. Solve this system using Cramer's rule:

$$\begin{cases} x - 2 = 4x - y + 3 \\ 6 - 2y = -3 - x \end{cases}$$

a. $x = 2, y = 5$
b. $x = -2, y = -5$
c. There is no solution.
d. There are infinitely many solutions.

934. Solve this system using Cramer's rule:

$$\begin{cases} 6x + 3y = 8 \\ 2x + y = 3 \end{cases}$$

a. $x = -12, y = 4$
b. $x = 12, y = -4$
c. There are infinitely many solutions.
d. There is no solution.

935. Solve this system using Cramer's rule:

$$\begin{cases} -3y = 1 - 4y \\ 2y + 3 = -4x \end{cases}$$

a. $x = -\frac{7}{11}, y = -\frac{5}{22}$
b. $x = \frac{7}{11}, y = \frac{5}{22}$
c. There are infinitely many solutions.
d. There is no solution.

936. Solve this system using Cramer's rule:

$$\begin{cases} -2 = x - 4y \\ 5y = \frac{1}{5} \end{cases}$$

a. There is no solution.
b. There are infinitely many solutions.
c. $x = -\frac{46}{25}, y = \frac{1}{25}$
d. $x = \frac{46}{25}, y = -\frac{1}{25}$

937. Solve this system using Cramer's rule:

$$\begin{cases} x - 2 = 4x - y + 3 \\ 6 - 2y = -3 - x \end{cases}$$

    a. There is no solution.
    b. There are infinitely many solutions.
    c. $x = -\frac{1}{5}, y = \frac{22}{5}$
    d. $x = \frac{1}{5}, y = -\frac{22}{5}$

938. Solve this system using Cramer's rule:

$$\begin{cases} -1 = -3 - 2x \\ 2 - 3y = 6(1 - 2x) \end{cases}$$

    a. $x = 1, y = \frac{16}{3}$
    b. $x = -1, y = \frac{-16}{3}$
    c. There is no solution.
    d. There are infinitely many solutions.

939. Solve this system using Cramer's rule:

$$\begin{cases} 2x - 3y + 5 = 1 + 2x - 4y \\ 3 - x - 2y = -y + x + 3 \end{cases}$$

    a. $x = 2, y = -4$
    b. $x = -2, y = 4$
    c. There is no solution.
    d. There are infinitely many solutions.

940. Solve this system using Cramer's rule:

$$\begin{bmatrix} -1 & 0 \\ 2 & -1 \end{bmatrix} \begin{bmatrix} x \\ y \end{bmatrix} = \begin{bmatrix} -2 \\ 1 \end{bmatrix}$$

    a. $x = -2, y = -3$
    b. $x = 2, y = 3$
    c. There is no solution.
    d. There are infinitely many solutions.

941. Solve this system using Cramer's rule:

$$\begin{bmatrix} 3 & 2 \\ 3 & 2 \end{bmatrix} \begin{bmatrix} x \\ y \end{bmatrix} = \begin{bmatrix} -2 \\ 1 \end{bmatrix}$$

    a. $x = -5, y = -3$
    b. $x = 5, y = 3$
    c. There are infinitely many solutions.
    d. There is no solution.

942. Solve this system using Cramer's rule:

$$\begin{bmatrix} 3 & -2 \\ 9 & -6 \end{bmatrix} \begin{bmatrix} x \\ y \end{bmatrix} = \begin{bmatrix} 4 \\ 12 \end{bmatrix}$$

    a. $x = -1, y = 4$
    b. $x = 1, y = -4$
    c. There is no solution.
    d. There are infinitely many solutions.

943. Solve this system using Cramer's rule:

$$\begin{bmatrix} -1 & -1 \\ -1 & 0 \end{bmatrix} \begin{bmatrix} x \\ y \end{bmatrix} = \begin{bmatrix} -1 \\ 1 \end{bmatrix}$$

    a. $x = 1, y = -2$
    b. $x = -1, y = 2$
    c. There is no solution.
    d. There are infinitely many solutions.

944. Solve this system using Cramer's rule:

$$\begin{bmatrix} 0 & 2 \\ 4 & 0 \end{bmatrix} \begin{bmatrix} x \\ y \end{bmatrix} = \begin{bmatrix} 14 \\ -20 \end{bmatrix}$$

    a. $x = 5, y = -7$
    b. $x = -5, y = 7$
    c. There is no solution.
    d. There are infinitely many solutions.

SECTION 8 ▶ COMMON ALGEBRA ERRORS

It is common and expected for those who are learning algebra for the first time or reviewing the subject after having been away from it for a while to make errors. Several of the most typical errors made are explored in the four sets in this section.

For all of the questions in this section, identify the choice that best describes the error, if any, made in each scenario.

143

Set 61 (Answers begin on page 268)

Some common arithmetic and pre-algebra errors are explored in this problem set.

945. $(-3)^{-2} = 6$

  a. The answer should be -9 because $(-3)^{-2} = -3 \cdot 3$.

  b. The answer should be $\frac{1}{9}$ because $(-3)^{-2} = \frac{1}{(-3)\cdot(-3)} = \frac{1}{9}$.

  c. There is no error.

946. $(-\frac{2}{3})^0 = 1$

  a. The answer should be -1 because $(-\frac{2}{3})^0 = -(\frac{2}{3})^0 = -1$.

  b. The answer should be zero because you should multiply the base and exponent.

  c. There is no error.

947. $0.00013 = 1.3 \times 10^4$

  a. The statement should be $0.00013 = 1.3 \times 10^{-4}$ because the decimal point must move four places to the left in order to yield 0.00013.

  b. The statement should be $0.00013 = 1.3 \times 10^{-3}$ because there are three zeros before the decimal point.

  c. There is no error.

948. $-4^2 = -16$

  a. The answer should be -8 because $4^2 = 8$, and this is then multiplied by -1.

  b. The answer should be 16 because $-4^2 = (-4)(-4) = 16$.

  c. There is no error.

949. $\frac{a+2}{a} = 2$, for any nonzero value of a.

  a. This is incorrect because you cannot cancel terms in a sum; you can cancel only factors that are common to the numerator and denominator.

  b. The correct result should be 3 because $\frac{a}{a} = 1$ and $\frac{a+2}{a} = \frac{a}{a} + 2 = 1 + 2 = 3$.

  c. There is no error.

950. $\frac{3}{4} + \frac{a}{2} = \frac{3+a}{6}$, for any real number a.

  a. You must first get a common denominator before you add two fractions. The correct computation is: $\frac{3}{4} + \frac{a}{2} = \frac{3}{4} + \frac{2a}{4} = \frac{3+2a}{8}$.

  b. You must first get a common denominator before you add two fractions. The correct computation is: $\frac{3}{4} + \frac{a}{2} = \frac{3}{4} + \frac{2a}{4} = \frac{3+2a}{4}$.

  c. There is no error.

951. 4 is 200% of 8.

  a. There is no such quantity as "200%." You cannot exceed 100%.

  b. The placement of the quantities is incorrect. A correct statement would be "200% of 4 is 8."

  c. There is no error.

952. 0.50% of 10 is 0.05.

  a. In order to compute this percentage, you should multiply 0.50 times 10 to get 5, not 0.05.

  b. In order to compute this percentage, you should multiply 50.0 times 10 to get 500, not 0.05.

  c. There is no error.

953. $\sqrt{3} + \sqrt{6} = \sqrt{3+6} = \sqrt{9} = 3$

    a. You must first simplify $\sqrt{6}$ as $\sqrt{6} = \sqrt{2 \cdot 3}$ $= 2\sqrt{3}$, and then combine with $\sqrt{3}$ to conclude that $\sqrt{3} + \sqrt{6} = 3\sqrt{3}$.

    b. The sum $\sqrt{3} + \sqrt{6}$ cannot be simplified further because the radicands are different.

    c. There is no error.

954. $\sqrt{6} + \sqrt[3]{-38} = \sqrt[2+3]{6-38} = \sqrt[5]{-32} = -2$

    a. The calculation is correct until the last line; the fifth root of a negative number is not defined.

    b. The first equality is incorrect: the radicals cannot be combined since their indices are different.

    c. There is no error.

955. $\frac{1}{2+\sqrt{3}} = \frac{1}{2+\sqrt{3}} \cdot \frac{2+\sqrt{3}}{2+\sqrt{3}} = \frac{2+\sqrt{3}}{(2+\sqrt{3})^2} =$

    $\frac{2+\sqrt{3}}{2^2+(\sqrt{3})^2} = \frac{2+\sqrt{3}}{4+3} = \frac{2+\sqrt{3}}{7}$

    a. The third equality is incorrect because the binomial was not squared correctly. The correct denominator should be $2^2 + 2\sqrt{3} + (\sqrt{3})^2 = 7 + 2\sqrt{3}$.

    b. The first equality is wrong because multiplying by $\frac{2+\sqrt{3}}{2+\sqrt{3}}$ changes the value of the expression. The remaining equalities are correct.

    c. There is no error.

956. $(x^5)^2 = x^7$, for any real number x.

    a. The exponents should be multiplied, not added, so that the correct answer should be x^{10}.

    b. The correct answer should be x^{25} because $(x^5)^2 = x^{5^2}$.

    c. There is no error.

957. $\sqrt[3]{27x^3} = \sqrt[3]{27} \cdot \sqrt[3]{x^3} = 3x$, for any real number x.

    a. The first equality is wrong because the radical of a product is not the product of the radicals.

    b. The second equality holds only if x is not negative because you can only take the cube root of a non-negative real number.

    c. There is no error.

958. $\frac{\frac{3}{4}}{\frac{8}{5}} = \frac{3}{4} \cdot \frac{8}{5} = \frac{6}{5}$

    a. The first equality is wrong because you must multiply the numerator by the *reciprocal* of the denominator.

    b. The second equality is wrong because the fraction on the far right should be $\frac{24}{20}$, which cannot be simplified further.

    c. There is no error.

959. $\frac{x^{12}}{x^{-3}} = x^{\frac{12}{-3}} = x^{-4}$, for any non-negative real number x.

    a. The correct answer should be x^{36} because $\frac{x^{17}}{x^{-3}} = x^{12}x^3 = x^{12 \cdot 3} = x^{36}$.

    b. The correct answer should be x^{15} because $\frac{x^{12}}{x^{-3}} = x^{12}x^3 = x^{12+3} = x^{15}$.

    c. There is no error.

960. $(e^{4x})^2 = e^{4x+2}$

    a. The correct answer should be e^8x because $(e^{4x})^2 = e^{4x \cdot 2} = e^{8x}$.

    b. The correct answer should be e^{16x^2} because $(e^{4x})^2 = e^{(4x)^2} = e^{4^2x^2} = e^{16x^2}$.

    c. There is no error.

Set 62 (Answers begin on page 269)

Some common errors in solving equations and inequalities, as well as simplifying algebraic expressions, are explored in this problem set.

961. The solution set for the inequality $-6x > 24$ is $(-4, \infty)$.

   a. The inequality sign must be switched when multiplying both sides by a negative real number. The correct solution set should be $(\infty, -4)$.

   b. You should multiply both sides by -6, not divide by -6. The correct solution set should be $(-144, \infty)$.

   c. There is no error.

962. The solution set for the equation $|x - 1| = 2$ is $\{-1\}$.

   a. There are two solutions of this equation, namely $x = -1$ and $x = 3$.

   b. The solution of an absolute value equation cannot be negative. The only solution is $x = 3$.

   c. There is no error.

963. $\log_3 1 = 0$

   a. 1 is an invalid input for a logarithm. As such, the quantity $\log_3 1$ is undefined.

   b. The input and output are backward. The real statement should read $\log_3 0 = 1$.

   c. There is no error.

964. The solution of $-\frac{4}{x+7} = \frac{x+3}{x+7}$ is $x = -7$.

   a. The equation obtained after multiplying both sides by $x + 7$ was not solved correctly. The correct solution should be $x = 1$.

   b. $x = -7$ cannot be the solution because it makes the terms in the original equation undefined—you cannot divide by zero. As such, this equation has no solution.

   c. There is no error.

965. The solutions of the equation $\log_5 x + \log_5(5x^3) = 1$ are $x = -1$ and $x = 1$.

   a. Both solutions should be divided by $\sqrt{5}$; that is, the solutions should be $x = \pm\frac{1}{\sqrt{5}}$.

   b. While $x = 1$ satisfies the original equation, $x = -1$ cannot because negative inputs into a logarithm are not allowed.

   c. There is no error.

966. The complex solutions of the equation $x^2 + 5 = 0$ obtained using the quadratic formula are given by $x = \frac{0 \pm \sqrt{0^2 - 4(1)(5)}}{1} = \pm 2i\sqrt{5}$.

   a. The denominator in the quadratic formula is $2a$, which in this case is 2, not 1. As such, the complex solutions should be $x = \pm i\sqrt{5}$.

   b. There are no complex solutions to this equation because the graph of $y = x^2 + 5$ does not cross the x–axis.

   c. There is no error.

967. $x^2 - 4x - 21 = (x + 7)(x - 3)$

   a. This is incorrect because multiplying the binomials on the right side of the equality yields $x^2 - 21$, which is not the left side listed above.

   b. The signs used to define the binomials on the right side should be switched. The correct factorization is $(x - 7)(x + 3)$.

   c. There is no error.

968. Since taking the square root of both sides of the inequality $x^2 \leq 4$ yields the statement $x < \pm 2$. Since both statements must be satisfied simultaneously, the solution set is $(-\infty, -2]$.

  a. You must move all terms to one side of the inequality, factor (if possible), determine the values that make the factored expression equal to zero, and construct a sign chart to solve such an inequality. The correct solution set should be $[-2,2]$.

  b. When taking the square root of both sides of an equation, you use only the principal root. As such, the correct statement should be $x \leq 2$, so that the solution set is $(-\infty, 2]$.

  c. There is no error.

969. $(x - y)^2 = x^2 - y^2$

  a. The left side must be expanded by FOILing. The correct statement should be $(x - y)^2 = x^2 - 2xy + y^2$.

  b. The -1 must be squared. The correct statement should be $(x - y)^2 = x^2 + y^2$.

  c. There is no error.

970. The solution of the equation $\sqrt{x} = -2$ is $x = 4$, as seen by squaring both sides of the equation.

  a. The correct solution is $x = -4$ because when you square both sides of the equation, you do not square the -1.

  b. This equation has no real solutions because the output of a square root must be nonnegative.

  c. There is no error.

971. The solution set of the inequality $|x + 2| > 5$ is $(-\infty, -7) \cup (3, \infty)$.

  a. The interval $(-\infty, -7)$ should be deleted because an absolute value inequality cannot have negative solutions.

  b. You must include the values that make the left side equal to 5. As such, the solution set should be $(-\infty, -7] \cup [3, \infty)$.

  c. There is no error.

972. $x^2 + 25 = (x - 5)(x + 5)$

  a. The correct factorization of the left side is $x^2 + 25 = (x + 5)^2$.

  b. The left side is not a difference of squares. It cannot be factored further.

  c. There is no error.

973. $\dfrac{2x^1 - y^1}{y^1 + 4y^1} = \dfrac{2-1}{1+4} = \dfrac{1}{5}$

  a. Cancelling the terms $x-1$ and $y-1$ leaves 0 each time, not 1. So, the correct statement should be $\dfrac{2x^1 - y^1}{x^1 + 4y^1} = \dfrac{2}{4} = \dfrac{1}{2}$.

  b. You cannot cancel terms of a sum in the numerator and denominator. You can only cancel factors common to both. The complex fraction must first be simplified before any cancelation can occur. The correct statement is:

  $$\frac{2x^{-1} - y^{-1}}{x^{-1} + 4y^{-1}} = \frac{\frac{2}{x} - \frac{1}{y}}{\frac{1}{x} + \frac{4}{y}} = \frac{\frac{2y}{xy} - \frac{x}{xy}}{\frac{y}{xy} + \frac{4x}{xy}} = \frac{\frac{2y - x}{xy}}{\frac{y + 4x}{xy}}$$

  $$= \frac{2y - x}{xy} \cdot \frac{xy}{y + 4x} = \frac{2y - x}{y + 4x}$$

  c. There is no error.

974. $\ln(e^x + e^{2y}) = \ln(e^x) + \ln(e^{2y}) = x + 2y$

  a. The first equality is incorrect because the natural logarithm of a sum is the product of the natural logarithms. So, the statement should be $\ln(e^x + e^{2y}) = \ln(e^x) \cdot \ln(e^{2y}) = 2xy$.

  b. The first equality is incorrect because the natural logarithm of a sum is not the sum of the natural logarithms. In fact, the expression on the extreme left side of the string of equalities cannot be simplified further.

  c. There is no error.

975. $\log_5(5x^2) = 2\log_5(5x) = 2[\log_5(5) + \log_5(x)] = 2[1 + \log_5(x)]$

    a. The first equality is incorrect because $2\log_5(5x) = \log_5(5x)^2 = \log_5(25x^2)$. The other equalities are correct.

    b. The very last equality is incorrect because $\log_5 5 = 0$. The other equalities are correct.

    c. There is no error.

976. $\ln(4x^2 - 1) = \ln[(2x - 1)(2x + 1)] + \ln(2x - 1) + \ln(2x + 1)$

    a. The "natural logarithm of a difference rule" was not applied correctly. The correct statement should be $\ln(4x^2 - 1) = \ln(4x^2) - \ln(1) = \ln(4x^2) - 0 = \ln(4x^2)$. The last expression in this string of equalities cannot be simplified because the exponent 2 does not apply to the entire input of the logarithm.

    b. Using the fact that the natural logarithm of a difference is the quotient of the natural logarithms, we see that the expression $\ln(4x^2 - 1) = \frac{\ln(4x^2)}{\ln 1} = \frac{\ln(4x^2)}{0}$, so the expression is not well-defined.

    c. There is no error.

Set 63 (Answers begin on page 269)

This problem set highlights common errors made when working with functions.

977. The vertical asymptote for the graph of $f(x) = \frac{x+2}{x^2+4}$ is $y = 0$.

    a. The expression should be factored and simplified to obtain $f(x) = \frac{1}{x-2}$. Then, we can conclude that the vertical asymptote for f is $x = 2$.

    b. The line $y = 0$ is the horizontal asymptote for f.

    c. There is no error.

978. The line $x = a$ has a slope of zero, for any real number a.

    a. The line is vertical, so its slope is undefined.

    b. The statement is true except when $a = 0$. The y–axis cannot be described by such an equation.

    c. There is no error.

979. The point $(-2, 1)$ lies in Quadrant IV.

    a. The point is actually in Quadrant II.

    b. The point is actually in Quadrant III.

    c. There is no error.

980. The inverse of the function $f(x) = x^2$, where x is any real number, is the function $f^{-1}(x) = \sqrt{x}$.

    a. f cannot have an inverse because it doesn't pass the vertical line test.

    b. The domain of f must be restricted to $[0,\infty)$ in order for f to have an inverse. In such case, the given function $f^{-1}(x) = \sqrt{x}$ is indeed its inverse.

    c. There is no error.

981. The lines $y = 3x + 2$ and $y = -\frac{1}{3}x + 2$ are perpendicular.

    a. The lines are parallel since their slopes are negative reciprocals of each other.

    b. The lines cannot be perpendicular since the product of their slopes is not 1.

    c. There is no error.

982. The slope of a line passing through the points (a, b) and (c, d) is $m = \frac{b-d}{a-c}$, provided that $a \neq c$.

    a. The slope is actually equal to the quantity $m = \frac{a-c}{b-d}$, provided that $b \neq d$.

    b. The slope is actually equal to the quantity $m = \frac{b-a}{d-c}$, provided that $c \neq d$.

    c. There is no error.

983. The graph of the function $f(x) = \frac{8-x}{x^2-64}$ has an open hole at $x = 8$.
  a. The graph actually has a vertical asymptote at $x = 8$ because this value makes the denominator equal to zero.
  b. The graph actually has a horizontal asymptote at $x = 8$ because this value makes the denominator equal to zero.
  c. There is no error.

984. If $f(2) = 5$, then the point $(5, 2)$ must be on the graph of $y = f(x)$.
  a. The coordinates of the point that is known to lie on the graph of $y = f(x)$ are reversed; it should be $(2,5)$.
  b. The given information is insufficient to make any conclusion about a point being on the graph of $y = f(x)$. All that can be said is that 2 is in the range of f.
  c. There is no error.

985. The range of the function $f(x) = (x-1)^2$ is $[0,\infty)$.
  a. The graph of f is the graph of $g(x) = x^2$ shifted vertically up one unit. Since the range of g is $[0,\infty)$, it follows that the range of f must be $[1,\infty)$.
  b. The graph of f is the graph of $g(x) = x^2$ shifted vertically down one unit. Since the range of g is $[0,\infty)$, it follows that the range of f must be $[-1,\infty)$.
  c. There is no error.

986. If $f(x) = 5$ and $g(x) = \sqrt{x}$, it follows that $(f \circ g)(-2) = 5$.
  a. The composition was computed in the wrong order. The correct output should be $\sqrt{5}$.
  b. -2 is not in the domain of g, so that the composition is not defined at -2.
  c. There is no error.

987. The x–intercept of $f(x) = x^3 + 1$ is $(0, 1)$.
  a. The point $(0,1)$ is the y–intercept of f, not the x–intercept.
  b. There are no x–intercepts for this function because $x^3 + 1$ is always positive.
  c. There is no error.

988. The graph of $g(x) = 2^{-x}$ is increasing as x moves from left to right through the domain.
  a. The graph of g is actually decreasing as x moves from left to right through the domain.
  b. There are intervals on which the graph of g is increasing and others on which it is decreasing.
  c. There is no error.

989. The graph of $y = f(x + 3)$ is obtained by shifting the graph of $y = f(x)$ to the right 3 units.
  a. The graph of $y = f(x + 3)$ is actually obtained by shifting the graph of $y = f(x)$ to the left 3 units.
  b. The graph of $y = f(x + 3)$ is actually obtained by shifting the graph of $y = f(x)$ vertically up 3 units.
  c. There is no error.

990. The graph of $y = f(x) - 2$ is obtained by shifting the graph of $y = f(x)$ down 2 units.

    a. The graph of $y = f(x) - 2$ is obtained by shifting the graph of $y = f(x)$ to the left 2 units.

    b. The graph of $y = f(x) - 2$ is obtained by shifting the graph of $y = f(x)$ to the right 2 units.

    c. There is no error.

991. If $f(x) = x^4$, then $f(x - h) = f(x) - f(h) = x^4 - h^4$.

    a. You cannot distribute a function across parts of a single input. As such, the correct statement should be $f(x - h) = (x - h)^4$.

    b. The second equality is incorrect because you must also raise the -1 to the fourth power. As such, the correct statement should be $f(x - h) = f(x) - f(h) = x^4 + h^4$.

    c. There is no error.

992. The graph of $y = 5$ does not represent a function because it does not pass the horizontal line test.

    a. The graph of $y = 5$ passes the vertical line test, so it represents a function. It is, however, not invertible.

    b. The fact that $y = 5$ does not pass the horizontal line test does not imply it is not a function. However, since the range of a function must consist of more than a single value, we conclude that it $y = 5$ cannot represent a function.

    c. There is no error.

Set 64 (Answers begin on page 270)

This problem set highlights common errors made when dealing with linear systems of equations and matrix algebra.

993. The system $\begin{cases} 2x + 3y = 6 \\ -2x - 3y = 2 \end{cases}$ has infinitely many solutions.

    a. Since adding the two equations results in the false statement $0 = 8$, there can be no solution of this system.

    b. The slopes of the two lines comprising the system are negatives of each other. As such, the lines are perpendicular, so the system has a unique solution.

    c. There is no error.

994. The system $\begin{cases} 2x - 5y = -1 \\ 4x - 10y = -2 \end{cases}$ has no solution.

    a. Since multiplying the first equation by -2 and then adding the two equations results in the true statement $0 = 0$, there are infinitely many solutions of this system.

    b. The two lines comprising the system intersect at a point, so the system has a unique solution.

    c. There is no error.

995. The matrix equation $\begin{bmatrix} 2 & 1 \\ -4 & -2 \end{bmatrix} \begin{bmatrix} x \\ y \end{bmatrix} =$

$\begin{bmatrix} -2 \\ 4 \end{bmatrix}$ has infinitely many solutions.

a. Since the determinant of the coefficient

matrix $\begin{bmatrix} 2 & 1 \\ -4 & -2 \end{bmatrix}$ is zero, the system has

no solution.

b. The system has a unique solution given by

$\begin{bmatrix} 0 \\ 1 \end{bmatrix} + \begin{bmatrix} 0 & 1 \\ 0 & 0 \end{bmatrix} = \begin{bmatrix} -2 \\ 4 \end{bmatrix}$.

c. There is no error.

996. $\begin{bmatrix} 0 \\ 1 \end{bmatrix} + \begin{bmatrix} 0 & 1 \\ 0 & 0 \end{bmatrix} = \begin{bmatrix} 0 & 1 \\ 1 & 0 \end{bmatrix}$

a. The two matrices on the left side of the equality do not have the same dimension. As such, their sum is undefined.

b. The matrices were added incorrectly. The

right side should be $\begin{bmatrix} 0 & 0 & 1 \\ 1 & 0 & 0 \end{bmatrix}$.

c. There is no error.

997. $\begin{bmatrix} -1 & 2 \end{bmatrix} \cdot \begin{bmatrix} 2 & -1 \end{bmatrix} = \begin{bmatrix} -2 & 1 \\ 4 & -2 \end{bmatrix}$

a. The product should be a real number, namely $\begin{bmatrix} -1 & 2 \end{bmatrix} \cdot \begin{bmatrix} 2 & -1 \end{bmatrix} =$ $(-1)(2) + (2)(-1) = -4$.

b. The inner dimensions of the two matrices on the left side are not the same. As such, they cannot be multiplied.

c. There is no error.

998. $\det \begin{bmatrix} 4 & 2 \\ 1 & -1 \end{bmatrix} = (2)(1) - (4)(-1) = 6$

a. The wrong pairs of entries are being multiplied to compute the determinant. The

correct statement should be $\det \begin{bmatrix} 4 & 2 \\ 1 & -1 \end{bmatrix}$

$= (4)(2) - (1)(-1) = 9$.

b. The difference is computed in the wrong order. The correct statement should be

$\det \begin{bmatrix} 4 & 2 \\ 1 & -1 \end{bmatrix} = (4)(-1) - (2)(1) = -6$.

c. There is no error.

999. $\begin{bmatrix} 1 & 1 \\ 1 & 0 \end{bmatrix}^{-1} = -\begin{bmatrix} 0 & -1 \\ -1 & 1 \end{bmatrix} = \begin{bmatrix} 0 & 1 \\ 1 & -1 \end{bmatrix}$

a. The constant multiple on the right side of the first equality should be 1, not −1. Therefore, the inverse should be

$\begin{bmatrix} 0 & -1 \\ -1 & 1 \end{bmatrix}$.

b. The inverse does not exist because several of the entries are the same real number.

c. There is no error.

1000. $\begin{bmatrix} 1 & -2 \\ 2 & 3 \end{bmatrix} + 1 = \begin{bmatrix} 2 & -1 \\ 3 & 4 \end{bmatrix}$

    a. You cannot add a 2×2 matrix and a real number because their dimensions are different. Therefore, the sum is not well–defined.

    b. The 1 should be added only to the diagonal entries, so that the correct statement should be $\begin{bmatrix} 1 & -2 \\ 2 & 3 \end{bmatrix} + 1 = \begin{bmatrix} 2 & -2 \\ 2 & 4 \end{bmatrix}$.

    c. There is no error.

1001. $\begin{bmatrix} -1 \\ -1 \end{bmatrix} \cdot \begin{bmatrix} -1 & -1 & -1 & -1 \end{bmatrix} =$

$$\begin{bmatrix} 1 & 1 & 1 & 1 \\ 1 & 1 & 1 & 1 \end{bmatrix}$$

    a. The product is not well–defined because the matrices must have the same dimensions in order to be multiplied.

    b. The correct product should be $\begin{bmatrix} 1 & 1 & 1 & 1 \end{bmatrix}$.

    c. There is no error.

ANSWERS AND EXPLANATIONS ▶

Section 1— Pre-Algebra Fundamentals

Set 1 (Page 2)

1. b. Multiply the contents of each set of parentheses first. Then, multiply the resulting products: $(15 + 32)(56 - 39) = (47)(17) = 799$

2. b. Dividing 65,715 by 4 results in 16,428 with a remainder of 3. Since the hundreds place is not 5 or greater, rounding the quotient to the nearest thousand yields 16,000.

3. c. Approximate 7,404 by 7,400. The quotient $7,400 \div 74 = 100$ is a good approximation of the quotient $7,404 \div 74$.

4. a. Using the order of operations, compute the quantities within each set of parentheses first. Then, multiply left to right as such products arise. Finally, compute sums and differences from left to right as they arise, as follows: $12(84 - 5) - (3 \times 54) = 12(79) - (162) = 948 - 162 = 786$

5. d. Computing the sum $60,000 + 800 + 2$ yields 60,802.

6. c. Since $112 \div 7 = 16$ and $112 \div 8 = 14$, we conclude that 112 is divisible by both 7 and 8.

7. a. Rounding 162 to the nearest hundred yields 200 (since the tens place is greater than 5), and rounding 849 to the nearest hundred yields 800 (since the tens place is less than 5). Multiplying 200 times 800 yields a product of 160,000, which is an estimation of the product of 162 and 849.

8. d. Multiplying 5 times 5 yields 25. Then, multiplying this product by 5 results in 125. Thus, $5 \times 5 \times 5 = 125$.

9. c. By the definition of an exponent, we have $3^5 = 3 \times 3 \times 3 \times 3 \times 3 = 243$.

10. b. First, the following are the multiples of 6 between 0 and 180:
6, 12, 18, 24, 30, 36, 42, 48, 54, 60, 66, 72, 78, 84, 90, 96, 102, 108, 114, 120, 126, 132, 138, 144, 150, 156, 162, 168, 174, 180
Of these, the following are also factors of 180: 6, 12, 18, 30, 36, 60, 90, 180
There are eight possibilities for the whole number p.

11. c. The only choice that is a product of prime numbers equaling 90 is $2 \times 3 \times 3 \times 5$.

12. c. The factors of 12 are 1, 2, 3, 4, 6, 12. Of these, only 1 and 3 are not multiples of 2. Thus, the set of positive factors of 12 that are NOT multiples of 2 is {1,3}.

13. e. The sum of 13 and 12 is 25, which is an odd number. Each of the other operations produces an even number: $20 \times 8 = 160$, $37 + 47 = 84$,
$7 \times 12 = 84$, $36 + 48 = 84$

14. d. By the definition of an exponent, $2^4 = 2 \times 2 \times 2 \times 2 = 16$.

15. b. Applying the order of operations, we first perform exponentiation, then subtract from left to right to obtain $9 - 2^2 = 9 - 4 = 5$.

16. c. The only choice that is divisible by only 1 and itself is 11. Each of the other choices has factors other than 1 and itself.

Set 2 (Page 3)

17. c. Begin by simplifying the absolute value quantity. Then, divide left to right:
$-25 \div |4 - 9| = -25 \div |-5| = -25 \div 5 = -5$.

18. a. Since there is an odd number of negative signs, the product will be negative. Computing this product yields $-4 \times -2 \times -6 \times 3 = -144$.

19. c. Applying the order of operations, first simplify the quantity enclosed in parentheses, then square it, then multiply left to right, and finally compute the resulting difference:
$5 - (-17 + 7)^2 \times 3 = 5 - (-10)^2 \times 3 = 5 - 100 \times 3 = 5 - 300 = -295$

20. a. Applying the order of operations, first compute the quantities enclosed in parentheses. Then, compute the resulting difference:
$(49 \div 7) - (48 \div (-4)) = (7) - (-12) = 7 + 12 = 19$

21. d. Note that substituting the values 1, 2, and 3 in for p in the equation $y = 6p - 23$ yields $-17, -11$, and -5, respectively. However, substituting 4 in for p results in the positive number 1. So, of the choices listed, the least value of p for which y is positive is 4.

22. b. Applying the order of operations, first compute the quantities enclosed in parentheses. Then, compute the difference from left to right:
$-(5 \cdot 3) + (12 \div (-4)) = -(15) + (-3) = -15 - 3 = -18$

23. c. Applying the order of operations, compute both exponentiated quantities first. Then, multiply from left to right as products arise. Finally, compute the resulting difference:
$-2(-2)^2 - 2^2 = -2(4) - 4 = -8 - 4 = -12$

24. a. Applying the order of operations, first perform the exponentiation. Then, compute the quantities enclosed with parentheses. Finally, compute the resulting quotient:
$(3^2 + 6) \div (-24 \div 8) = (9 + 6) \div (-3) = (15) \div (-3) = -5$

25. d. This one is somewhat more complicated since we have an expression consisting of terms within parentheses which are, in turn, enclosed within parentheses, and the whole thing is raised to a power. Proceed as follows:

$$(-2[1 - 2(4 - 7)])^2 = (-2[1 - 2(-3)])^2$$
$$= (-2[1 - (-6)])^2$$
$$= (-2[7])^2$$
$$= (-14)^2$$
$$= 196$$

26. b. Applying the order of operations, first compute quantities enclosed within parentheses and exponentiated terms on the same level from left to right. Repeat this until all such quantities are simplified. Then, multiply from left to right. Finally, compute the resulting difference:

$$3(5 - 3)^2 - 3(5^2 - 3^2) = 3(2)^2 - 3(25 - 9) =$$
$$3(4) - 3(16) = 12 - 48 = -36$$

27. a. Here we have an expression consisting of terms within parentheses which are, in turn, enclosed within parentheses. Proceed as follows:

$$-(-2 - (-11 - (-3^2 - 5) - 2)) =$$
$$-(-2 - (-11 - (-9 - 5) - 2))$$
$$= -(-2 - (-11 - (-14) - 2))$$
$$= -(-2 - (-11 + 14 - 2))$$
$$= -(-2 - (1))$$
$$= -(-3)$$
$$= 3$$

28. c. Since $h < 0$, it follows that $-h > 0$. Since we are also given that $g > 0$, we see that $g - h = g + (-h)$ is the sum of two positive numbers and hence, is itself positive.

29. c. Observe that $-g - h = -(g + h)$. Since $g < 0$ and $h < 0$, it follows that $g + h < 0$, so $-(g + h)$ is positive.

30. d. First, note that since $g < 0$ and $h < 0$, it follows that $g + h$ must be negative, so $-g - h = -(g + h)$ is positive. As such, we know that $-(g + h)$ is larger than $g + h$. Next, each of the sums $-g + h$ and $-g - h$ consists of one positive integer and one negative integer. Thus, while it is possible for one of them to be positive, its value cannot exceed that of $-g - h$ since this sum consists of two positive integers. As such, we conclude that $-g - h$ is the largest of the four expressions provided.

31. c. First, note that since $g < 0$ and $h < 0$, it follows that $g + h$ must be negative and so, $-g - h = -(g + h)$ is positive. As such, we know that $-(g + h)$ is larger than $g + h$. Next, each of the sums $-g + h$ and $g - h$ consists of one positive integer and one negative integer. Thus, while it is possible for one of them to be negative, its value cannot be smaller than $g + h$ since this sum consists of two negative integers. As such, we conclude that $g + h$ is the smallest of the four expressions provided.

32. d. First, note that since we are given that $g < -2$, it follows that both g and $-g^2$ are negative, while both $-g$ and $(-g)^2$ are positive. Moreover, $-g$ is an integer larger than 2 (which follows by multiplying both sides of the given inequality by -1). Squaring an integer larger than 2 produces an even larger integer. As such, we conclude that $(-g)^2$ is the largest of the four expressions provided.

Set 3 (Page 5)

33. a. Express both fractions using the least common denominator, then add: $\frac{5}{9} - \frac{1}{4} =$
$\frac{5 \cdot 4}{9 \cdot 4} - \frac{1 \cdot 9}{4 \cdot 9} = \frac{20}{36} - \frac{9}{36} = \frac{20 - 9}{36} = \frac{11}{36}$

34. b. First, rewrite all fractions using the least common denominator, which is 30. Then, add:

$$\frac{2}{15} + \frac{1}{5} + \frac{1}{6} + \frac{3}{10} = \frac{2 \cdot 2}{15 \cdot 2} + \frac{1 \cdot 6}{5 \cdot 6} + \frac{1 \cdot 5}{6 \cdot 5} + \frac{3 \cdot 3}{10 \cdot 3}$$

$$= \frac{4}{30} + \frac{6}{30} + \frac{5}{30} + \frac{9}{30}$$

$$= \frac{4 + 6 + 5 + 9}{30}$$

$$= \frac{24}{30}$$

$$= \frac{4}{5}$$

35. d. The square is divided into 8 congruent parts, 3 of which are shaded. Thus, $\frac{3}{8}$ of the figure is shaded.

36. c. First, rewrite both fractions using the least common denominator, which is 60. Then, subtract:

$$\frac{17}{20} - \frac{5}{6} = \frac{17 \cdot 3}{20 \cdot 3} - \frac{5 \cdot 10}{6 \cdot 10} = \frac{51}{60} - \frac{50}{60} = \frac{51 - 50}{60} = \frac{1}{60}$$

37. c. Rewrite this as a multiplication problem, cancel factors that are common to the numerator and denominator, and then multiply:

$$\frac{18}{5} \div \frac{9}{20} = \frac{18}{5} \cdot \frac{20}{9} = \frac{9 \cdot 2}{5} \cdot \frac{4 \cdot 5}{9} = 8$$

38. c. A reasonable strategy is to begin with one of the fractions, say $\frac{5}{8}$, and compare it to the next one in the list. Discard whichever is smaller and compare the remaining one with the next in the list. Repeat this until you reach the end of the list. Doing so results in the following three comparisons:

<u>Comparison 1</u>: $\frac{5}{8} \overset{?}{>} \frac{2}{3}$ Cross multiplying yields the false statement $15 > 16$. This implies that $\frac{2}{3}$ is larger.

<u>Comparison 2</u>: $\frac{2}{3} \overset{?}{>} \frac{8}{11}$ Cross multiplying yields the false statement $22 > 24$. This implies that $\frac{8}{11}$ is larger.

<u>Comparison 3</u>: $\frac{8}{11} \overset{?}{>} \frac{4}{10}$ Cross multiplying yields the true statement $80 > 44$. This implies that $\frac{8}{11}$ is larger.

Thus, we conclude that $\frac{8}{11}$ is the largest of the choices.

39. a. The fact that $\frac{1}{4} < \frac{5}{8} < \frac{2}{3}$ is evident from the following two comparisons:

<u>Comparison 1</u>: $\frac{1}{4} \overset{?}{<} \frac{5}{8}$ Cross multiplying yields the true statement $8 < 20$, so the original inequality is true.

<u>Comparison 2</u>: $\frac{5}{8} \overset{?}{<} \frac{2}{3}$ Cross multiplying yields the true statement $15 < 16$, so the original inequality is true.

40. a. Since $\frac{3}{5}(360) = \frac{3}{5}(5 \cdot 72) = 216$, we conclude that Irma has read 216 pages.

41. a. Cancel factors that are common to the numerator and denominator, then multiply:

$$\frac{5}{8} \times \frac{4}{7} = \frac{5}{4 \cdot 2} \times \frac{4}{7} = \frac{5}{14}$$

42. d. The reciprocal of $\frac{21}{42}$ is $\frac{42}{21}$, which is equivalent to 2.

43. d. Note that $1\frac{3}{8} = \frac{11}{8}$. The additive inverse of a real number is its negative. So, the answer is $-\frac{11}{8}$.

44. c. The remaining 28 (of 42) envelopes need to be addressed. Thus, the fraction of envelopes that needs to be addressed is $\frac{28}{42} = \frac{2 \cdot 14}{3 \cdot 14} = \frac{2}{3}$.

45. b. Apply the order of operations:

$$1 + \frac{\left(-\frac{5}{3}\right)(-2)}{\left(-\frac{10}{7}\right)} \div \left(\frac{7}{5} \cdot \frac{10}{3}\right) = 1 + \left(\frac{\frac{10}{3}}{-\frac{10}{7}}\right) \div \left(\frac{7}{5} \cdot \frac{5 \cdot 2}{3}\right)$$

$$= 1 + \left(\frac{10}{3} \div \left(-\frac{10}{7}\right)\right) \div \left(\frac{7}{5} \cdot \frac{5 \cdot 2}{3}\right)$$

$$= 1 + \left(\frac{10}{3} \cdot \left(-\frac{7}{10}\right)\right) \div \left(\frac{14}{3}\right)$$

$$= 1 - \frac{7}{3} \div \frac{14}{3}$$

$$= 1 - \frac{7}{3} \cdot \frac{3}{14}$$

$$= 1 - \frac{1}{2}$$

$$= \frac{1}{2}$$

46. e. Since there are m men in a class of n students, there must be $n - m$ women in the class. So, the ratio of men to women in the class is $\frac{m}{n - m}$.

47. d. Compute the difference between $\frac{1}{2}$ and each of the four choices. Then, compare the absolute values of these differences; the choice that produces the smallest difference is the one closest to $\frac{1}{2}$. The differences are as follows:

$$\frac{2}{3} - \frac{1}{2} = \frac{4}{6} - \frac{3}{6} = \frac{1}{6}$$

$$\frac{3}{10} - \frac{1}{2} = \frac{3}{10} - \frac{5}{10} = -\frac{2}{10} = -\frac{1}{5}$$

$$\frac{5}{6} - \frac{1}{2} = \frac{5}{6} - \frac{3}{6} = \frac{2}{6} = \frac{1}{3}$$

$$\frac{3}{5} - \frac{1}{2} = \frac{6}{10} - \frac{5}{10} = \frac{1}{10}$$

The smallest absolute value of these four differences is $\frac{1}{10}$. So, of the four choices, the one closest to $\frac{1}{2}$ is $\frac{3}{5}$.

48. c. Applying the order of operations, first simplify the exponentiated term. Then, multiply left to right. Finally, compute the resulting difference by first rewriting both fractions using the least common denominator, which is 12:

$$7\left(\frac{5}{6}\right) - 3\left(\frac{1}{2}\right)^2 = 7\left(\frac{5}{6}\right) - 3\left(\frac{1}{4}\right) = \frac{35}{6} - \frac{3}{4} =$$

$$\frac{35 \cdot 2}{6 \cdot 2} - \frac{3 \cdot 3}{4 \cdot 3} = \frac{70}{12} - \frac{9}{12} = \frac{61}{12}$$

Set 4 (Page 7)

49. d. The exponent applies only to 5, not to the –1 multiplied in front. So, $-5^3 = -(5 \times 5 \times 5) = -125$.

50. a. By definition, $(-11)^2 = (-11) \times (-11) = 121$.

51. c. Using the fact that any nonzero base raised to the zero power is 1, we have $5(4^0) = 5(1) = 5$.

52. a. Applying the exponent rules yields:
$$(2^2)^{-3} = 2^{(2 \cdot (3))} = 2^{-6} = \frac{1}{2^6} = \frac{1}{64}$$

53. c. Applying the order of operations and the definition of an exponent yields:
$$\frac{(1-3)^2}{-8} = \frac{(-2)^2}{-8} = \frac{(-2) \times (-2)}{-8} = \frac{4}{-8} = \frac{-1}{2}$$

54. b. Applying the order of operations and the definition of an exponent yields:
$$-5(-1 - 5^{-2}) = -5(-1 - \frac{1}{25}) = -5(-\frac{25}{25} - \frac{1}{25}) = -5(-\frac{26}{25}) = \cancel{5}(\frac{26}{5 \cdot \cancel{5}}) = \frac{26}{5}$$

55. c. First, apply the definition of a negative exponent to simplify the first term within the brackets. Next, rewrite the resulting first term using the fact that "a product raised to a power is the product of the powers." Then, simplify:

$$-\left[\left(-\frac{3}{2}\right)^{-2} - \left(\frac{2}{3}\right)^2\right] = -\left[\left(-\frac{2}{3}\right)^2 - \left(\frac{2}{3}\right)^2\right] =$$

$$-\left[\left(-1\right)^2\left(\frac{2}{3}\right)^2 - \left(\frac{2}{3}\right)^2\right] = -\left[\left(\frac{2}{3}\right)^2 - \left(\frac{2}{3}\right)^2\right] = 0$$

56. b. First, apply the definition of a negative exponent to simplify the two terms to which it applies. Then, apply the order of operations:

$$-\left(-\frac{1}{2}\right)^{-3} - \frac{\left(-\frac{1}{3}\right)^2}{9^{-2}} = -(-2)^3 - \frac{\left(-\frac{1}{3}\right)^2}{\left(\frac{1}{9}\right)^2}$$

$$= -[(-2)(-2)(-2)] - \frac{\left(-\frac{1}{3}\right)\left(-\frac{1}{3}\right)}{\left(\frac{1}{9}\right)\left(\frac{1}{9}\right)}$$

$$= -[-8] - \frac{\frac{1}{9}}{\frac{1}{81}}$$

$$= 8 - \frac{1}{9} \cdot \frac{81}{1}$$

$$= 8 - 9$$

$$= -1$$

57. d. Apply the order of operations and exponent rules:

$$-\left(\frac{2}{5}\right)^0 \cdot (-3^2 + 2^{-3})^{-1} = -1 \cdot (-9 + \frac{1}{2^3})^{-1}$$

$$= -1 \cdot (-9 + \frac{1}{8})^{-1}$$

$$= -1 \cdot (-\frac{72}{8} + \frac{1}{8})^{-1}$$

$$= -1 \cdot (-\frac{71}{8})^{-1}$$

$$= -1 \cdot (-\frac{8}{71})$$

$$= \frac{8}{71}$$

58. c. Apply the order of operations and exponent rules:

$$4^{-2}(1-2(-1)^{-3})^{-2} = \frac{1}{4^2}\left(1-2\left(\frac{1}{-1}\right)^3\right)^{-2} =$$

$$\frac{1}{16}(1-2(-1))^{-2} = \frac{1}{16}(3)^{-2} = \frac{1}{16}\left(\frac{1}{3^2}\right) =$$

$$\frac{1}{144} = \frac{1}{12^2} = 12^{-2}$$

59. d. Apply the order of operations and exponent rules:

$$-2^{-2} + \frac{(-1^3 + (-1)^3)^{-2}}{-2^2} = -\frac{1}{2^2} + \frac{(-1+(-1))^{-2}}{-4}$$

$$= -\frac{1}{4} + \frac{(-2)^{-2}}{-4} = -\frac{1}{4} + \frac{\frac{1}{(-2)^2}}{-4} = -\frac{1}{4} + \frac{\frac{1}{4}}{-4} =$$

$$-\frac{1}{4} - \frac{1}{16} = -\frac{5}{16}$$

60. c. Simplify each expression:

$$\left(-\frac{1}{4}\right)^{-1} = (-4)^1 = -4$$

$$-\frac{3}{8\left(-\frac{1}{4}\right)} = -\frac{3}{-2} = \frac{3}{2}$$

$$4\left(-\frac{1}{4}\right) + 3 = -1 + 3 = 2$$

$$-\left(-\frac{1}{4}\right)^0 = -(1) = -1$$

Hence, the expression with the largest value is $4\left(-\frac{1}{4}\right) + 3$.

61. d. The reciprocal of a fraction p strictly between 0 and 1 is necessarily larger than 1. So, $p^{-1} > 1$. Also, raising a fraction p strictly between 0 and 1 to a positive integer power results in a fraction with a smaller value. (To see this, try it out with $p = \frac{1}{2}$.) As such, both p^2 and p^3 are less than p and are not larger than 1. So, of the four expressions provided, the one with the largest value is p^{-1}.

62. c. The reciprocal of a fraction p strictly between 0 and 1 is necessarily larger than 1. So, $p^{-1} > 1$. Raising a fraction p strictly between 0 and 1 to a positive integer power results in a fraction with a smaller value. (Try this out with $p = \frac{1}{2}$.) We know that $0 < p^3 < p^2 < p < 1$. Therefore, of the four expressions provided, the one with the smallest value is p^3.

63. b. Note that the expressions p, p^3, and p^{-1} are all negative since it is assumed that p is a fraction between -1 and 0. Since squaring a negative fraction results in a positive value, we conclude that p^2 is positive and is, therefore, the largest of the four choices.

64. c. Raising a fraction strictly between 1 and 2 to a positive integer power results in a larger fraction. Thus, we know that $1 < p < p^2$. Moreover, the reciprocals of fractions larger than 1 are necessarily less than 1. In particular, $p^{-1} < 1$, which shows that p^{-1} is smaller than both p and p^2. Finally, multiplying both sides of the inequality, $p^{-1} < 1$ by p^{-1} shows that $p^{-2} = p^{-1} \cdot p^{-1} < p^{-1}$. Therefore, we conclude that the smallest of the four expressions is p^{-2}.

Set 5 (Page 8)

65. a. The quantity $n\%$ means "n parts out of 100." It can be written as $\frac{n}{100}$ or equivalently as $n \times 0.01$. Applying this to 40 yields the equivalent expressions I and II.

66. d. The result of increasing 48 by 55% is given by $48 + 0.55(48) = 74.4$.

67. d. The price resulting from discounting $250 by 25% is given by $\$250 - 0.25(\$250)$. This quantity is equivalent to both $0.75 \times \$250$ and $(1 - 0.25) \times \$250$.

68. c. Note that if $0 < a < b$, then $\frac{1}{a} > \frac{1}{b}$. So, the correct expression would be $\frac{1}{\frac{1}{3}} > \frac{1}{x} > \frac{1}{\frac{1}{2}}$, which is equivalent to $3 > \frac{1}{x} > 2$.

69. c. Since the digit in the thousandths place is 8, we round the digit in the hundredths place up by 1, resulting in 117.33.

70. b. We must determine the value of n for which $\frac{n}{100} \times 300 = 400$. The value of n that satisfies this equation is $133\frac{1}{3}$. So, we conclude that $133\frac{1}{3}\%$ of 300 results in 400.

71. d. Starting with 0.052, moving the decimal place to the left one unit to obtain 0.0052 is equivalent to dividing 0.052 by 10. Therefore, 0.0052 is smaller than 0.052.

72. c. The phrase "400% of 30" is equivalent to the mathematical expression $\frac{400}{100} \times 30$. Simplifying this expression yields 120.

73. c. Note that $x = \frac{3}{8}$ satisfies the condition $\frac{5}{16} < x < \frac{9}{20}$, which is seen by performing the following two comparisons using cross multiplication:

Comparison 1: $\frac{5}{16} \overset{?}{<} \frac{3}{8}$ Cross multiplying yields the true statement $40 < 48$, so the original inequality is true.

Comparison 2: $\frac{3}{8} \overset{?}{<} \frac{9}{20}$ Cross multiplying yields the true statement $60 < 72$, so the original inequality is true.

74. b. 22.5% is equivalent to $\frac{22.5}{100}$, which is equal to 0.225.

75. d. Note that $\frac{2}{5} = 0.40$ and $\frac{3}{7} \approx 0.42857$. So, $\frac{3}{7}$ is not less than $\frac{2}{5}$.

76. b. To see that $-0.01 < -0.005$, first convert both to their equivalent fractional form:

$-0.01 = -\frac{1}{100}$

$-0.005 = -\frac{5}{1000} = -\frac{1}{200}$

Next, we compare these two fractions. To this end, note that $-\frac{1}{100} < -\frac{1}{200}$ is equivalent to $\frac{1}{100} > \frac{1}{200}$. Cross multiplying in the latter inequality yields the true statement $200 > 100$, so the inequality is true. Since -0.005 is clearly less than 1.01, we conclude that -0.005 is between -0.01 and 1.01.

77. b. Observe that $\frac{5}{8} - \frac{2}{5} = \frac{25}{40} - \frac{16}{40} = \frac{9}{40} = 0.225$.

78. c. Observe that $(3.09 \times 10^{12}) \div 3 = \frac{3.09}{3} \times 10^{12} = 1.03 \times 10^{12}$. Alternatively, you could first rewrite 3.09×10^{12} as 3,090,000,000,000 and divide by 3 to obtain 1,030,000,000,000, which is equivalent to 1.03×10^{12}.

79. a. Move the decimal place to the right until just after the first nonzero digit; each place moved contributes an additional -1 power of 10. Doing so in 0.00000321 requires that we move the decimal place 6 units to the right, so that 0.00000321 is equivalent to 3.21×10^{-6}.

80. c. We must determine the value of n for which $\frac{n}{100} \times \frac{8}{9} = \frac{1}{3}$. Solve for n, as follows:

$\frac{n}{100} \times \frac{8}{9} = \frac{1}{3}$

$\frac{n}{100} = \frac{9}{8} \times \frac{1}{3} = \frac{3}{8}$

$n = 100 \times \frac{3}{8} = \frac{300}{8} = 37.5$

Thus, we conclude that 37.5% of $\frac{8}{9}$ is $\frac{1}{3}$.

Set 6 (Page 10)

81. a. Apply the order of operations as follows:
$-2(-3)^2 + 3(-3) - 7 = -2(9) - 9 - 7 = -18 - 9 - 7 = -34$

82. b. Apply the order of operations as follows:
$\frac{7(-2)}{(-2)^2 + (-2)} = \frac{-14}{4 - 2} = -\frac{14}{2} = -7$

83. b. Apply the order of operations as follows:
$2(3)(6) - (-8) = 36 + 8 = 44$

84. c. Apply the order of operations as follows:
$y = -(-3)^3 + 3(-3) - 3 = -(-27) - 9 - 3 = 27 - 9 - 3 = 15$

85. b. Apply the order of operations as follows:
$(-5)(6) + (-8) \div (\frac{1}{2}) = -30 - 8 \div \frac{1}{2} = -30 - 8 \cdot 2 = -30 - 16 = -46$

86. b. Apply the order of operations as follows:
$\frac{6^2}{3} - 4(6) + 10 = \frac{36}{3} - 4(6) + 10 = 12 - 24 + 10 = -2$

87. a. Apply the order of operations as follows:
$4(2^{-2})(2(2)^{-2})(3(-2)^2) = 4\left(\frac{1}{4}\right)\left(2 \cdot \frac{1}{4}\right)(3 \cdot 4) = \frac{4 \cdot 2 \cdot 3 \cdot 4}{4 \cdot 4} = 6$

88. a. Apply the order of operations as follows:
$7(6) + \frac{12}{6} - (-8) = 42 + 2 + 8 = 52$

89. b. Apply the order of operations as follows:
$$(3(2)(5)+2)\left(\tfrac{2}{5}\right)=(32)\left(\tfrac{2}{5}\right)=\tfrac{64}{5}=12.8$$

90. d. Apply the order of operations as follows:
$$\left(\tfrac{7}{5(-2)^2}+\tfrac{3}{10(-2)}\right)^{-2}=\left(\tfrac{7}{5\cdot4}+\tfrac{3}{10(-2)}\right)^{-2}=$$
$$\left(\tfrac{7}{20}-\tfrac{3}{20}\right)^{-2}=\left(\tfrac{4}{20}\right)^{-2}=\left(\tfrac{1}{5}\right)^{-2}=5^2=25$$

91. c. Apply the order of operations as follows:
$$\tfrac{6(2)^2}{2(3)^2}+\tfrac{4(2)}{3(3)}=\tfrac{6\cdot4}{2\cdot9}+\tfrac{4(2)}{3(3)}=\tfrac{24}{18}+\tfrac{8}{9}=$$
$$\tfrac{24}{18}+\tfrac{16}{18}=\tfrac{40}{18}=\tfrac{20}{9}$$

92. c. Apply the order of operations as follows:
$$(1)(-1)+\tfrac{1}{-1}+(1)^2-(-1)^2=-1-1+1-1$$
$$=-2$$

93. b. Note that if $x=2$, then $y=-2$. Now, apply the order of operations as follows:
$$(((2)(-2))^{-2})^2=((-4)^{-2})^2=(-4)^{-2\times2}=$$
$$(-4)^{-4}=\tfrac{1}{(-4)^4}=\tfrac{1}{256}$$

94. b. Apply the order of operations as follows:
$$\tfrac{1}{2}[(\tfrac{6}{2}-3)-4(3)]=\tfrac{1}{2}[(3-3)-12]=\tfrac{1}{2}[-12]$$
$$=-6$$

95. d. Apply the order of operations as follows:
$$(-8)^2-4(3)^2(\tfrac{1}{2})=64-4(9)(\tfrac{1}{2})=64-18$$
$$=46$$

96. a. Apply the order of operations as follows:
$$3(6)^2(-5)(5(3)-3(-5))=3(36)(-5)$$
$$(15+15)=3(36)(-5)(30)=-16,200$$

Set 7 (Page 12)

97. b. $\dfrac{(3x^2)^3}{x^2x^4}=\dfrac{3^3x^6}{x^6}=3^3=27$

98. d. $(4w^9)^3=4^3w^{27}=64w^{27}$

99. b. Note that the power -2 does not apply to the 6 since it is not enclosed in the parentheses to which the exponent applies. Therefore, $6(e^{-2})^{-2}=6e^4$.

100. a. $(-45a^4b^9c^5)\div(9ab^3c^3)=\dfrac{-45a^4b^9c^5}{9ab^3c^3}=-5a^3b^6c^2$

101. a. $(2x^3)^3\cdot(4x^3)^2=2^3\cdot x^9\cdot4^2\cdot x^6$
$$=(2^3\cdot4^2)\cdot(x^9\cdot x^6)$$
$$=64x^{15}$$

102. d. $\left(\dfrac{(ab)^3}{b}\right)^4=\left(\dfrac{a^3b^3}{b}\right)^4=(a^3b^2)^4=a^{12}b^8$

103. b. $\dfrac{(\frac{x}{y})^2(\frac{y}{x})^{-2}}{xy}=\dfrac{\frac{x^2}{y^2}\cdot\frac{y^{-2}}{x^{-2}}}{xy}=\dfrac{\frac{x^2}{y^2}\cdot\frac{x^2}{y^2}}{xy}=\dfrac{\frac{x^4}{y^4}}{xy}=\dfrac{x^4}{y^4(xy)}=\dfrac{x^3}{y^5}$

104. d. $\left(\dfrac{2a}{b}\right)\left(\dfrac{a^{-1}}{(2b)^{-1}}\right)=\dfrac{2a}{b}\cdot\dfrac{2b}{a}=\dfrac{4ab}{ba}=4$

105. c. $3x^2y(2x^3y^2)=6x^5y^3$

106. e. $(\tfrac{a}{b})^2(\tfrac{b}{a})^{-2}(\tfrac{1}{a})^{-1}=\dfrac{a^2}{b^2}\cdot\dfrac{b^{-2}}{a^{-2}}\cdot\dfrac{a}{1}=\dfrac{a^2}{b^2}\cdot\dfrac{a^2}{b^2}\cdot\dfrac{a}{1}=\dfrac{a^5}{b^4}$

107. c. $(3xy^5)^2-11x^2y^2(4y^4)^2=3^2x^2y^{10}-11x^2y^2\cdot4^2y^8=9x^2y^{10}-176x^2y^{10}=-167x^2y^{10}$

108. a. $\dfrac{2(3x^2y)^2(xy)^3}{3(xy)^2}=\dfrac{2(3^2x^4y^2)(x^3y^3)}{3(x^2y^2)}=\dfrac{18x^7y^5}{3x^2y^2}=6x^5y^3$

109. c. $\dfrac{(4b)^2x^{-2}}{(2ab^2x)^2}=\dfrac{4^2b^2x^{-2}}{2^2a^2b^4x^2}=\dfrac{16b^2}{2^2a^2b^4x^2x^2}=\dfrac{16b^2}{4a^2b^4x^4}=\dfrac{4}{a^2b^2x^4}$

110. a. "The product of $6x^2$ and $4xy^2$ is divided by $3x^3y$" can be expressed symbolically as $\dfrac{(6x^2)(4xy^2)}{3x^3y}$, which is simplified as follows:
$$\dfrac{(6x^2)(4xy^2)}{3x^3y}=\dfrac{24x^3y^2}{3x^3y}=8y$$

111. a. The expression described by the phrase "$3x^2$ is multiplied by the quantity $2x^3y$ raised to the fourth power" can be expressed symbolically as $(3x^2)(2x^3y)^4$, which is simplified as follows: $(3x^2)(2x^3y)^4=(3x^2)(2^4x^{12}y^4)=48x^{14}y^4$

112. b. "The product of $-9p^3r$ and the quantity $5p-6r$" can be expressed symbolically as $(-9p^3r)(5p-6r)$, which is simplified using the distributive property as follows: $(-9p^3r)(5p-6r)=-45p^4r+54p^3r^2$.

Set 8 (Page 13)
113. c. $5ab^4-ab^4=4ab^4$
114. a. $5c^2+3c-2c^2+4-7c=(5c^2-2c^2)+(3c-7c)+4=3c^2-4c+4$
115. c. $-5(x-(-3y))+4(2y+x)=-5(x+3y)+4(2y+x)=-5x-15y+8y+4x=-x-7y$

116. b. Gather like terms, as follows: $3x^2 + 4ax - 8a^2 + 7x^2 - 2ax + 7a^2 = (3x^2 + 7x^2) + (4ax - 2ax) + (-8a^2 + 7a^2) = 10x^2 + 2ax - a^2$

117. d. The base expressions of the three terms used to form the sum $9m^3n + 8mn^3 + 2m^3n^3$ are different. So, they cannot be combined.

118. d. $-7g^6 + 9h + 2h - 8g^6 = (-7g^6 - 8g^6) + (9h + 2h) = -15g^6 + 11h$

119. b. $(2x^2)(4y^2) + 6x^2y^2 = 8x^2y^2 + 6x^2y^2 = 14x^2y^2$

120. c. $(5a^2 \cdot 3ab) + 2a^3b = 15a^3b + 2a^3b = 17a^3b$

121. b. $2x^{-3} - \frac{3x^{-1}}{x^4} - (x^3)^{-1} = \frac{2}{x^3} - \frac{3}{x^5} - \frac{1}{x^3} = \frac{1}{x^3} - \frac{3}{x^5} = x^{-3} - 3x^{-5}$

122. a. $(ab^2)^3 + 2b^2 - (4a)^3b^6 = a^3b^6 + 2b^2 - 4^3a^3b^6 = 2b^2 - 63a^3b^6$

123. b. $\frac{(-3x^{-1})^{-2}}{x^{-2}} + \frac{8}{9}(x^2)^2 = \frac{x^2}{(-3x^{-1})^2} + \frac{8x^4}{9} = \frac{x^2}{(-3)^2x^2} + \frac{8x^4}{9} = \frac{x^4}{9} + \frac{8x^4}{9} = x^4$

124. b. $-(-a^{-2}bc^{-3})^{-2} + 5\left(\frac{b}{a^2c^3}\right)^{-2} = -(a^4b^{-2}c^6) + 5\left(\frac{b^{-2}}{a^{-4}c^{-6}}\right) = -\frac{a^4c^6}{b^2} + \frac{5a^4c^6}{b^2} = \frac{4a^4c^6}{b^2}$

125. c. $3(z + 1)^2w^3 - \frac{2w(z + 1)}{((z + 1)w^2)^{-1}}$
$= 3(z + 1)^2w^3 - 2w(z + 1) \cdot ((z + 1)w^2)$
$= 3(z + 1)^2w^3 - 2(z + 1)^2w^3$
$= (z + 1)^2w^3$

126. a. $\left(-2(4x + 1)^5y^{-5} - \frac{2y(4x + 1)^2}{(4x + 1)y^{-2})^{-3}}\right)^{-2}$

$= \left(-\frac{2(4x + 1)^5}{y^5} - \frac{2y(4x + 1)^2}{(4x + 1)^{-3}y^6}\right)^{-2}$

$= \left(-\frac{2(4x + 1)^5}{y^5} - \frac{2(4x + 1)^5}{y^5}\right)^{-2}$

$= \left(-\frac{4(4x + 1)^5}{y^5}\right)^{-2}$

$= \left(-\frac{y^5}{4(4x + 1)^5}\right)^2$

$= \frac{(-1)^2y^{10}}{4^2(4x + 1)^{10}}$

$= \frac{y^{10}}{16(4x + 1)^{10}}$

127. b.

$4z((xy^{-2})^{-3} + (x^{-3}y^6))^{-1} - \left[\frac{1}{z}\left(\frac{2y^6}{x^3}\right)\right]^{-1}$

$= 4z((x^{-3}y^6) + (x^{-3}y^6))^{-1} - \left[\frac{2y^6}{zx^3}\right]^{-1}$

$= 4z(2(x^{-3}y^6))^{-1} - \left[\frac{2y^6}{zx^3}\right]^{-1}$

$= 4z(2^{-1}x^3y^{-6})^{-1} - \frac{zx^3}{2y^6}$

$= \frac{4zx^3}{2y^6} - \frac{zx^3}{2y^6}$

$= \frac{3zx^3}{2y^6}$

128. a. $(0.2x^{-2})^{-1} + \frac{2}{5}x^2 - \frac{5x^4}{(2x)^2}$

$= \left(\frac{2}{10}x^{-2}\right)^{-1} + \frac{2}{5}x^2 - \frac{5x^4}{4x^2}$

$= 5x^2 + \frac{2}{5}x^2 - \frac{5}{4}x^2$

$= \left(\frac{100}{20} + \frac{8}{20} - \frac{25}{20}\right)x^2$

$= \frac{83}{20}x^2$

Set 9 (Page 15)

129. b. According to the order of operations, we perform exponentiation first, then multiplication, and then subtraction. The square of a number x is x^2. Four times this quantity is $4x^2$, and finally, two less is $4x^2 - 2$.

130. d. First, note that 25% of V is equal to $0.25V$, which is equivalent to $\frac{1}{4}V$. Since the original volume of the tank is being *increased* by this quantity, we add it to the original volume V. This results in the equivalent expressions $V + 0.25V$ and $V + \frac{1}{4}V$. Adding the coefficients of V in the first expression yields another equivalent expression $1.25V$. So, they are all correct choices.

131. b. The amount of money paid for the total number hours of tutoring is $40h$. Adding the one-time fee of $30 to this amount results in $x = \$30 + \$40h$.

132. a. According to the order of operations, we interpret parenthetical quantities first, then multiply, and then subtract. The quantity *the sum of a number and 5* is $(x + 5)$. Then, three times this quantity is $3(x + 5)$, and finally nine less results in $3(x + 5) - 9$.

133. a. The total cost for a phone call lasting x minutes is the cost for the first minute plus the cost for the remaining $x - 1$ minutes. The first minute costs $0.35 and the cost for the remaining $x - 1$ minutes is $0.15(x - 1)$. The sum of these results in the total cost $y = 0.15(x - 1) + 0.35$.

134. b. According to the order of operations, we first interpret parenthetical quantities and then multiply. The difference between a number and five is represented by the expression $(x - 5)$. Then, half of this difference is $\frac{1}{2}(x - 5)$.

135. c. The sum of three numbers is represented by $(a + b + c)$. The reciprocals of these numbers are $\frac{1}{a}, \frac{1}{b}$ and $\frac{1}{c}$. The sum of the reciprocals is $(\frac{1}{a} + \frac{1}{b} + \frac{1}{c})$, and so, the product of these two sums is given by $(a + b + c)(\frac{1}{a} + \frac{1}{b} + \frac{1}{c})$.

136. c. The expression $3x$ is described by the phrase "3 times a number." So, the sum of $3x + 15$ is described by the phrase "15 more than 3 times a number." Finally, since the word "is" is interpreted as "equals," we conclude that the given equation is described by choice **c**.

137. c. The cost for x desks, each of which costs D dollars, is xD. Similarly, the costs for the chairs and file cabinets are yE and zF, respectively. Thus, the total cost T is equal to $xD + yE + zF$.

138. a. First, increasing d by 50% is described by the expression $d + 0.50d$, which is equal to $1.50d$. Now, a decrease of this quantity by 50% is described by the expression $1.50d - 0.50(1.50d) = 1.50d - 0.75d = 0.75d$ This value is 75% of the original value d. Hence, it is 25% smaller than d.

139. a. First, since there are w weeks in one month, the number of weeks in m months must be mw. Since we are told that there are m months in one year, the quantity mw represents the number of weeks in one year. Next, since there are d days in one week, and the number of weeks in a year is mw, we conclude that there are mwd days in one year.

140. d. The phrase "40% of j" is expressed symbolically as $0.40j$ and the phrase "50% of k" is expressed symbolically as $0.50k$. Equating these quantities yields $0.40j = 0.50k$. Dividing both sides by 0.40 then results in the equivalent equality $j = \frac{0.50}{0.40}k = 1.25k$. This says that the value of j is 125% of k. So, we conclude that j is 25% larger than k.

141. d. p percent of q can be represented symbolically as $\frac{pq}{100}$. As such, since we are *decreasing* q by this quantity, the resulting quantity is represented by $q - \frac{pq}{100}$.

142. d. The cost of the three meals is $(a + b + c)$ and a 15% tip is represented by $0.15(a + b + c)$. This latter value is added to the cost of the three meals to obtain the total cost of the dinner, namely $(a + b + c) + 0.15(a + b + c) = 1.15(a + b + c)$. Now, splitting this cost evenly between the two brothers amounts to dividing this quantity by 2; this is represented by choices **b** and **c**.

143. d. A 75% increase in enrollment E is represented symbolically as $0.75E$, which is equivalent to $\frac{3}{4}E$. Adding this to the original enrollment E results in the sum $E + \frac{3}{4}E$, which is the new enrollment.

144. a. The total cost of her orders, before the discount is applied, is represented by the sum $W + X + Y + Z$. A 15% discount on this amount is represented symbolically as $0.15(W + X + Y + Z)$. So her cost is $(W + X + Y + Z) - 0.15(W + X + Y + Z)$, which is equivalent to $0.85(W + X + Y + Z)$.

Section 2—Linear Equations and Inequalities

Set 10 (Page 18)

145. a. $z - 7 = -9$

$z - 7 + 7 = -9 + 7$

$z = -2$

146. e. $\frac{k}{8} = 8$

$\frac{k}{8} \cdot 8 = 8 \cdot 8$

$k = 64$

147. a. $-7k - 11 = 10$

$-7k - 11 + 11 = 10 + 11$

$-7k = 21$

$-7k\left(-\frac{1}{7}\right) = 21\left(-\frac{1}{7}\right)$

$k = -3$

148. c. $9a + 5 = -22$

$9a + 5 - 5 = -22 - 5$

$9a = -27$

$9a\left(\frac{1}{9}\right) = -27\left(\frac{1}{9}\right)$

$a = -3$

149. d. $\frac{p}{6} + 13 = p - 2$

$\frac{p}{6} + 13 - 13 = p - 2 - 13$

$\frac{p}{6} = p - 15$

$\frac{p}{6} - p = p - p - 15$

$-\frac{5}{6}p = -15$

$-\frac{5}{6}p\left(-\frac{6}{5}\right) = -15\left(\frac{-6}{5}\right)$

$p = 18$

150. a. $2.5p + 6 = 18.5$

$2.5p + 6 - 6 = 18.5 - 6$

$2.5p = 12.5$

$p = \frac{12.5}{2.5} = 5$

151. a. $\frac{3x}{10} = \frac{15}{25}$

$\frac{3x}{10}\left(\frac{10}{3}\right) = \frac{15}{25}\left(\frac{10}{3}\right)$

$x = \frac{150}{75} = 2$

152. b. $2.3(4 - 3.1x) = 1 - 6.13x$

$9.2 - 7.13x = 1 - 6.13x$

$9.2 - 7.13x + 7.13x = 1 - 6.13x + 7.13x$

$9.2 = 1 + x$

$9.2 - 1 = 1 - 1 + x$

$8.2 = x$

Since $8.2 = \frac{41}{5}$, the correct choice is **b**.

153. d. An application of the distributive property shows that $33c - 21 = 3(11c - 7)$. So, since $11c - 7 = 8$, we conclude that $33c - 21 = 3(8) = 24$.

154. d. $\frac{x}{2} + \frac{1}{6}x = 4$

$\frac{6x}{12} + \frac{2x}{12} = 4$

$\frac{8x}{12} = 4$

$\frac{8x}{12}\left(\frac{12}{8}\right) = 4\left(\frac{12}{8}\right)$

$x = 6$

155. b. $-3b - \frac{2}{3} = \frac{b}{3}$

$-3b - \frac{b}{3} = \frac{2}{3}$

$-\frac{9b}{3} - \frac{b}{3} = \frac{2}{3}$

$-\frac{10b}{3} = \frac{2}{3}$

$b = -\frac{3}{10} \cdot \frac{2}{3} = -\frac{1}{5}$

156. c. $\frac{3c}{4} - 9 = 3$

$\frac{3c}{4} - 9 + 9 = 3 + 9$

$\frac{3c}{4} = 12$

$\frac{3c}{4}\left(\frac{4}{3}\right) = 12\left(\frac{4}{3}\right)$

$c = 16$

157. b. $-\frac{2a}{3} = -54$

$-\frac{2}{3}a\left(-\frac{3}{2}\right) = -54\left(-\frac{3}{2}\right)$

$a = 81$

158. d.
$$1.3 + 5x - 0.1 = -1.2 - 3x$$
$$1.2 + 5x = -1.2 - 3x$$
$$1.2 + 5x - 1.2 = -1.2 - 3x - 1.2$$
$$5x = -2.4 - 3x$$
$$5x + 3x = -2.4 - 3x + 3x$$
$$8x = -2.4$$
$$x = -0.3$$

159. d.
$$4(4v + 3) = 6v - 28$$
$$16v + 12 = 6v - 28$$
$$16v + 12 - 12 = 6v - 28 - 12$$
$$16v = 6v - 40$$
$$16v - 6v = 6v - 6v - 40$$
$$10v = -40$$
$$v = -4$$

160. b.
$$13k + 3(3 - k) = -3(4 + 3k) - 2k$$
$$13k + 9 - 3k = -12 - 9k - 2k$$
$$10k + 9 = -12 - 11k$$
$$10k + 9 - 9 = -12 - 11k - 9$$
$$10k = -21 - 11k$$
$$10k + 11k = -21 - 11k + 11k$$
$$21k = -21$$
$$21k\left(\frac{1}{21}\right) = -21\left(\frac{1}{21}\right)$$
$$k = -1$$

Set 11 (Page 19)

161. a.
$$-2(3v + 5) = 14$$
$$-6v - 10 = 14$$
$$-6v = 14 + 10 = 24$$
$$v = \frac{24}{-6} = -4$$

162. b.
$$\frac{5}{2}(x - 2) + 3x = 3(x + 2) - 10$$
$$\frac{5}{2}x - 5 + 3x = 3x + 6 - 10$$
$$\frac{11}{2}x - 5 = 3x - 4$$
$$\frac{11}{2}x - 3x - 5 = -4$$
$$\frac{5}{2}x = -4 + 5 = 1$$
$$x = \frac{2}{5}$$

163. c. Let x be the unknown number. The sentence "Twice a number increased by 11 is equal to 32 less than three times the number" can be expressed symbolically as $2x + 11 = 3x - 32$. We solve this equation for x, as follows:
$$2x + 11 = 3x - 32$$
$$2x = 3x - 32 - 11$$
$$2x - 3x = -43$$
$$-x = -43$$
$$x = 43$$

164. d.
$$\frac{4a + 4}{7} = -\frac{2 - 3a}{4}$$
$$28 \cdot \left(\frac{4a + 4}{7}\right) = 28 \cdot \left(-\frac{2 - 3a}{4}\right)$$
$$4(4a + 4) = -7(2 - 3a)$$
$$16a + 16 = -14 + 21a$$
$$16 = -14 + 21a - 16a$$
$$16 + 14 = 5a$$
$$30 = 5a$$
$$a = 6$$

165. a. Let x be the smaller of the two unknown integers. The next consecutive even integer is then $x + 2$. The sentence "The sum of two consecutive even integers is 126" can be expressed symbolically as $x + (x + 2) = 126$. We solve this equation for x:
$$x + (x + 2) = 126$$
$$2x + 2 = 126$$
$$2x = 124$$
$$x = 62$$
Thus, the two integers are 62 and 64.

166. b.
$$0.8(x + 20) - 4.5 = 0.7(5 + x) - 0.9x$$
$$8(x + 20) - 45 = 7(5 + x) - 9x$$
$$8x + 160 - 45 = 35 + 7x - 9x$$
$$8x + 115 = 35 - 2x$$
$$10x = -80$$
$$x = -8$$

167. e. First, we solve the equation $4x + 5 = 15$ for x:
$$4x + 5 = 15$$
$$4x = 10$$
$$x = \frac{10}{4} = 2.5$$
Now, substitute $x = 2.5$ into the expression $10x + 5$ to obtain $10(2.5) + 5 = 25 + 5 = 30$.

168. d. Let x be the unknown number. 40% of this number is represented symbolically as $0.40x$. Therefore, the sentence "Ten times 40% of a number is equal to 4 less than six times the number" can be expressed as the equation $10(0.40x) = 6x - 4$. We solve this equation for x:

$10(0.40x) = 6x - 4$

$4x = 6x - 4$

$4x + 4 = 6x$

$4 = 2x$

$x = 2$

169. b. Let x be the unknown number. The sentence "$\frac{7}{8}$ of nine times a number is equal to ten times the number minus 17" can be expressed as the equation $\frac{7}{8}(9x) = 10(x - 17)$. Solve this equation for x:

$\frac{7}{8}(9x) = 10(x - 17)$

$8 \cdot \frac{7}{8}(9x) = 8 \cdot 10(x - 17)$

$63x = 80(x - 17)$

$63x = 80x - 1360$

$-17x = -1360$

$x = 80$

170. d. $a = \frac{7b - 4}{4}$

$4a = 4 \cdot \left(\frac{7b - 4}{4}\right)$

$4a = 7b - 4$

$4a + 4 = 7b$

$b = \frac{4a + 4}{7}$

171. b. $\frac{2x + 8}{5} = \frac{5x - 6}{6}$

$30 \cdot \frac{2x + 8}{5} = 30 \cdot \frac{5x - 6}{6}$

$6(2x + 8) = 5(5x - 6)$

$12x + 48 = 25x - 30$

$12x = 25x - 78$

$-13x = -78$

$x = 6$

172. b. Let x be the unknown number. The sentence "When ten is subtracted from the opposite of a number, the resulting difference is 5" can be expressed symbolically as the equation $-x - 10 = 5$. We solve this equation for x as follows:

$-x - 10 = 5$

$-x = 15$

$x = -15$

173. b. $9x + \frac{8}{3} = \frac{8}{3}x + 9$

$3 \cdot (9x + \frac{8}{3}) = 3 \cdot (\frac{8}{3}x + 9)$

$27x + 8 = 8x + 27$

$27x - 8x + 8 = 27$

$19x = 27 - 8 = 19$

$x = 1$

174. c. Substitute $F = 50°$ into the formula $F = \frac{9}{5}C + 32$ and then solve the resulting equation for C, as follows:

$50 = \frac{9}{5}C + 32$

$5 \cdot 50 = 5 \cdot (\frac{9}{5}C + 32)$

$250 = 9C + 160$

$90 = 9C$

$C = 10$

175. d. Let x be the unknown number. A 22.5% decrease in its value can be expressed symbolically as $x - 0.225x = 0.775x$. We are given that this quantity equals 93, which can be expressed as $0.775x = 93$. We solve this equation for x:

$0.775x = 93$

$x = \frac{93}{0.775} = 120$

176. a. The scenario described in this problem can be expressed as the equation $-4(x + 8) + 6x = 2x + 32$. We solve this equation for x:

$-4(x + 8) + 6x = 2x + 32$

$-4x - 32 + 6x = 2x + 32$

$2x - 32 = 2x + 32$

$0 - 32 = 32$

$0 = 64$

So, there is no such number x.

Set 12 (Page 21)

177. c. $\dfrac{\frac{1}{2}x-4}{3} = \dfrac{x+8}{5}$

$$15 \cdot \left(\dfrac{\frac{1}{2}x-4}{3}\right) = 15 \cdot \dfrac{x+8}{5}$$

$$5 \cdot (\tfrac{1}{2}x - 4) = 3(x+8)$$

$$\tfrac{5}{2}x - 20 = 3x + 24$$

$$2 \cdot (\tfrac{5}{2}x - 20) = 2(3x + 24)$$

$$5x - 40 = 6x + 48$$

$$5x - 88 = 6x$$

$$-88 = x$$

178. a. $5x - 2[x - 3(7-x)] = 3 - 2(x-8)$

$$5x - 2x + 6(7-x) = 3 - 2(x-8)$$

$$5x - 2x + 42 - 6x = 3 - 2x + 16$$

$$-3x + 42 = 19 - 2x$$

$$23 = x$$

179. d. $ax + b = cx + d$

$$ax - cx = d - b$$

$$(a-c)x = (d-b)$$

$$\dfrac{(a-c)x}{(a-c)} = \dfrac{(d-b)}{(a-c)}$$

$$x = \dfrac{d-b}{a-c}$$

180. a. Let x be the smallest of the four whole numbers. The next three consecutive odd whole numbers are then $x + 2$, $x + 4$, and $x + 6$. The sentence "The sum of four consecutive, odd whole numbers is 48" can be expressed as the equation $x + (x + 2) + (x + 4) + (x + 6) = 48$. We solve this equation for x as follows:

$$x + (x + 2) + (x + 4) + (x + 6) = 48$$

$$4x + 12 = 48$$

$$4x = 36$$

$$x = 9$$

Thus, the smallest of the four whole numbers is 9.

181. a. In order to solve for T, we must simply divide both sides of the equation by nR. This results in the equation $T = \dfrac{PV}{nR}$.

182. a. $B = \dfrac{C+A}{D-A}$

$$B \cdot (D - A) = C + A$$

$$BD - BA = C + A$$

$$BD - C = A + BA$$

$$BD - C = A(1 + B)$$

$$A = \dfrac{BD - C}{1 + B}$$

183. b. 30% of r is represented symbolically as $0.30r$, and 75% of s is represented symbolically as $0.75s$. The fact that these two quantities are equal is represented by the equation $0.30r = 0.75s$. We are interested in 50% of s when $r = 30$. So, we substitute $r = 30$ into this equation, solve for s, and then multiply the result by 0.50:

$$0.30(30) = 0.75s$$

$$9 = 0.75s$$

$$s = \dfrac{9}{0.75} = 12$$

So, 50% of s is equal to $0.50(12) = 6$.

184. e. We must solve the given equation for g:

$$fg + 2f - g = 2 - (f + g)$$

$$fg + 2f - g = 2 - f - g$$

$$fg = 2 - f - g - 2f + g$$

$$fg = 2 - 3f$$

$$g = \dfrac{2 - 3f}{f}$$

185. b. Let x be the width of the room. Then, the length of the room is equal to $2x + 3$. The perimeter of the room is given by $2x + 2(2x + 3)$. Since this quantity is known to be 66, we must solve the equation $2x + 2(2x + 3) = 66$ as follows:

$$2x + 2(2x + 3) = 66$$

$$2x + 4x + 6 = 66$$

$$6x + 6 + 66$$

$$6x = 60$$

$$x = 10$$

Thus, the length of the room is $2(10) + 3 = 23$ feet.

186. b. $\frac{4-2x}{3} - 1 = \frac{1-y}{2}$

$6 \cdot \frac{4-2x}{3} - 6 \cdot 1 = 6 \cdot \frac{1-y}{2}$

$2(4-2x) - 6 = 3(1-y)$

$8 - 4x - 6 = 3 - 3y$

$2 - 4x = 3 - 3y$

$3y = 1 + 4x$

$y = \frac{1+4x}{3}$

187. e. Let x be the smallest of five consecutive odd integers. The next four consecutive odd integers are given by $x + 2, x + 4, x + 6$, and $x + 8$. The average of these five integers is equal to their sum divided by 5, which is expressed symbolically by $\frac{x+(x+2)+(x+4)+(x+6)+(x+8)}{5}$. Since this quantity is -21, we must solve the equation $\frac{x+(x+2)+(x+4)+(x+6)+(x+8)}{5} = -21$ as follows:

$\frac{x+(x+2)+(x+4)+(x+6)+(x+8)}{5} = -21$

$\frac{5x+20}{5} = -21$

$x + 4 = -21$

$x - -25$

Thus, the least of the five integers is -25.

188. a. First, we solve $\frac{a}{b} + 6 = 4$ for a:

$\frac{a}{b} + 6 = 4$

$\frac{a}{b} = -2$

$a = -2b$

Next, substitute this expression for a into the equation $-6b + 2a - 25 = 5$ and solve for b:

$-6b + 2a - 25 = 5$

$-6b + 2(-2b) - 25 = 5$

$-6b - 4b - 25 = 5$

$-10b - 25 = 5$

$-10b = 30$

$b = -3$

Plugging this in for b in the expression $a = -2b$ yields $a = -2(-3) = 6$. Finally, we substitute these numerical values for a and b into $\left(\frac{b}{a}\right)^2$ to obtain $\left(\frac{-3}{6}\right)^2 = \left(-\frac{1}{2}\right)^2 = \frac{1}{4}$.

189. d. Let x be the unknown number. The sentence "Three more than one-fourth of a number is three less than the number" can be expressed as the equation $\frac{1}{4}x + 3 = x - 3$. We must solve this equation for x as follows:

$\frac{1}{4}x + 3 = x - 3$

$4 \cdot (\frac{1}{4}x + 3) = 4 \cdot (x - 3)$

$x + 12 = 4x - 12$

$x + 24 = 4x$

$24 = 3x$

$x = 8$

190. c. $\frac{5x-2}{2-x} = y$

$(2-x) \cdot \frac{5x-2}{2-x} = (2-x) \cdot y$

$5x - 2 = y(2-x)$

$5x - 2 = 2y - xy$

$5x + xy = 2 + 2y$

$x(5+y) = 2 + 2y$

$x = \frac{2+2y}{5+y}$

191. b. Solve this problem by determining the weight of each portion. The sum of the weights of the initial batches of corn is equal to the weight of the final mixture. Therefore,

$(20 \text{ bushels})\left(\frac{56 \text{ pounds}}{\text{bushel}}\right) +$

$(x \text{ bushels})\left(\frac{50 \text{ pounds}}{\text{bushel}}\right) =$

$[(20 + x) \text{ bushels}]\left(\frac{54 \text{ pounds}}{\text{bushel}}\right)$

Suppressing units yields the equation $20 \times 56 + 50x = (20 + x) \times 54$.

192. d. $-5[x - (3 - 4x - 5) - 5x] - 2^2 = 4[2 - (x - 3)]$

$-5[x - 3 + 4x + 5 - 5x] - 4 = 4[2 - x + 3]$

$-5[2] - 4 = 4[5 - x]$

$-10 - 4 = 20 - 4x$

$-14 = 20 - 4x$

$-34 = -4x$

$x = \frac{-34}{-4} = \frac{17}{2} = 8.5$

Set 13 (Page 22)

193. c. $3x + 2 < 11$
$3x < 9$
$x < 3$

194. a. $-7x \geq -35$
$(-\frac{1}{7})-7x \leq (-\frac{1}{7}) \cdot (-35)$
$x \leq 5$
So, the solution set is $\{x : x \leq 5\}$.

195. a. $1 - 2x > -5$
$-2x > -6$
$x < 3$
(Note: Remember to reverse the inequality sign when dividing both sides of an inequality by a negative number.)

196. d. All values to the right of and including −4 are shaded. Thus, the inequality that depicts this situation is $x \geq -4$.

197. b. $4x + 4 > 24$
$4x > 20$
$x > 5$

198. a. $-8x + 11 < 83$
$-8x < 72$
$x > -9$
(Note: Remember to reverse the inequality sign when dividing both sides of an inequality by a negative number.)

199. c. $-4(x - 1) \leq 2(x + 1)$
$-4x + 4 \leq 2x + 2$
$4 \leq 6x + 2$
$2 \leq 6x$
$\frac{1}{3} = \frac{2}{6} \leq x$
The answer can be written equivalently as $x \geq \frac{1}{3}$.

200. c. $x + 5 \geq 3x + 9$
$5 \geq 2x + 9$
$-4 \geq 2x$
$-2 \geq x$
The answer can be written equivalently as $x \leq -2$.

201. d. $-6(x + 1) \geq 60$
$-6x - 6 \geq 60$
$-6x \geq 66$
$x \leq -11$
(Note: Remember to reverse the inequality sign when dividing both sides of an inequality by a negative number.)

202. b. The right side of the inequality $2x - 4 < 7$ $(x - 2)$ can be described as "the product of seven and the quantity two less than a number," and the left side can be described as "four less than two times the number." Reading from right to left, the quantity on the right side is *greater than* the one on the left. Hence, the correct choice is **b**.

203. a. $\frac{-x}{0.3} \leq 20$
$x \geq (-0.3)(20) = -6$
(Note: Remember to reverse the inequality sign when dividing both sides of an inequality by a negative number.)

204. d. $-8(x + 3) \leq 2(-2x + 10)$
$-8x - 24 \leq -4x + 20$
$-24 \leq 4x + 20$
$-44 \leq 4x$
$-11 \leq x$
The answer can be written equivalently as $x \geq -11$.

205. b. $3(x - 16) - 2 < 9(x - 2) - 7x$
$3x - 48 - 2 < 9x - 18 - 7x$
$3x - 50 < 2x - 18$
$x - 50 < -18$
$x < 32$

206. b. $-5[9 + (x - 4)] \geq 2(13 - x)$
$-5[5 + x] \geq 2(13 - x)$
$-25 - 5x \geq 26 - 2x$
$-51 - 5x \geq -2x$
$-51 \geq 3x$
$-17 \geq x$
The answer can be written equivalently as $x \leq -17$.

207. a. When solving a compound inequality for which the only expression involving the variable is located between the two inequality signs and is linear, the goal is to simplify the inequality by adding/subtracting the constant term in the middle portion of the inequality to/from all three parts of the inequality, and then to divide all three parts of the inequality by the coefficient of x. The caveat in the latter step is that when the coefficient of x is negative, *both* inequality signs are switched. We proceed as follows:

$-4 < 3x - 1 \le 11$
$-3 < 3x \le 12$
$-1 < x \le 4$

208. b. Using the same steps as in question 207, proceed as follows:

$10 \le 3(4 - 2x) - 2 < 70$
$10 \le 12 - 6x - 2 < 70$
$10 \le 10 - 6x < 70$
$0 \le -6x < 60$
$0 \ge x > -10$

The last compound inequality above can be written equivalently as $-10 < x \le 0$.

Set 14 (Page 24)

209. c. Using the fact that $|a| = b$ if and only if $a = \pm b$, we see that solving the equation $|-x| - 8 = 0$, or equivalently $|-x| = 8$, is equivalent to solving $-x = \pm 8$. We solve these two equations separately:

$-x = 8$	$-x = -8$
$x = -8$	$x = 8$

So, both -8 and 8 are solutions of this equation.

210. a. We rewrite the given equation as an equivalent one solved for $|x|$, as follows:

$2|x| + 4 = 0$
$2|x| = -4$
$|x| = -2$

The left side must be nonnegative for any value of x (since it is the absolute value of an expression) while the right side is negative. So, there can be no solution to this equation.

211. c. We rewrite the given equation as an equivalent one solved for $|x|$:

$-3|x| + 2 = 5|x| - 14$
$-3|x| + 16 = 5|x|$
$16 = 8|x|$
$2 = |x|$

Using the fact that $|a| = b$ if and only if $a = \pm b$, it follows that the two solutions of the equation $2 = |x|$ are $x = \pm 2$. Thus, there are two distinct values of x that satisfy the given equation.

212. a. Using the fact that $|a| = b$ if and only if $a = \pm b$, we see that solving the equation $|3x - \frac{2}{3}| - \frac{1}{9} = 0$, or equivalently $|3x - \frac{2}{3}| = \frac{1}{9}$, is equivalent to solving $3x - \frac{2}{3} = \pm \frac{1}{9}$. We solve these two equations separately:

$3x - \frac{2}{3} = -\frac{1}{9}$	$3x - \frac{2}{3} = \frac{1}{9}$
$3x = \frac{2}{3} - \frac{1}{9} = \frac{5}{9}$	$3x = \frac{2}{3} + \frac{1}{9} = \frac{7}{9}$
$x = \frac{5}{27}$	$x = \frac{7}{27}$

So, both $\frac{5}{27}$ and $\frac{7}{27}$ are solutions to this equation.

213. b. Using the fact that $|a| = b$ if and only if $a = \pm b$, we see that solving the equation $|3x + 5| = 8$ is equivalent to solving $3x + 5 = \pm 8$. We solve these two equations separately, as follows:

$3x + 5 = -8$	$3x + 5 = 8$
$3x = -13$	$3x = 3$
$x = -\frac{13}{3}$	$x = 1$

Thus, the solutions to the equation are $x = -\frac{13}{3}$ and $x = 1$.

214. b. First, we rewrite the equation in an equivalent form:
$$-6(4 - |2x + 3|) = -24$$
$$-24 + 6|2x + 3| = -24$$
$$6|2x + 3| = 0$$
$$|2x + 3| = 0$$
Now, using the fact that $|a| = b$ if and only if $a = \pm b$, we see that solving the equation $|2x + 3| = 0$ is equivalent to solving $2x + 3 = 0$. The solution of this equation is $x = -\frac{3}{2}$. So, we conclude that there is only one value of x that satisfies this equation.

215. a. First, we rewrite the equation in an equivalent form:
$$1 - (1 - (2 - |1 - 3x|)) = 5$$
$$1 - (1 - 2 + |1 - 3x|) = 5$$
$$1 - (-1 + |1 - 3x|) = 5$$
$$1 + 1 - |1 - 3x| = 5$$
$$2 - |1 - 3x| = 5$$
$$-|1 - 3x| = 3$$
$$|1 - 3x| = -3$$
Since the left side is non-negative (being the absolute value of a quantity) and the right side is negative, there can be no value of x that satisfies this equation.

216. c. Note that $|a| = |b|$ if and only if $a = \pm b$. Using this fact, solving the equation $|2x + 1| = |4x - 5|$ is equivalent to solving $2x + 1 = \pm(4x - 5)$. We solve these two equations separately:

$2x + 1 = (4x - 5)$	$2x + 1 = -(4x - 5)$
$6 = 2x$	$2x + 1 = -4x + 5$
$3 = x$	$6x = 4$
	$x = \frac{2}{3}$

Thus, there are two solutions to the original equation.

217. d. Note that $|a| > c$ if and only if ($a > c$ or $a < -c$). Using this fact, we see that the values of x that satisfy the inequality $|x| > 3$ are precisely those values of x that satisfy either $x > 3$ or $x < -3$. So, the solution set is $(-\infty, -3) \cup (3, \infty)$.

218. a. First, note that $|-2x| = |-1| \cdot |2x| = |2x|$. Also, $|a| > c$ if and only if ($a > c$ or $a < -c$). The values of x that satisfy the inequality $|2x| > 0$ are those that satisfy either $2x > 0$ or $2x < 0$. Dividing both of these inequalities by 2 yields $x > 0$ or $x < 0$. So, the solution set is $(-\infty, 0) \cup (0, \infty)$.

219. c. First, note that the inequality $-|-x - 1| \geq 0$ is equivalent to $|-x - 1| \leq 0$. Moreover, since $|-x - 1| = |-(x + 1)| = |-1| \cdot |x + 1| = |x + 1|$, this inequality is also equivalent to $|x + 1| \leq 0$. The left side must be nonnegative since it is the absolute value of a quantity. The only way that it can be less than or equal to zero is if it actually *equals* zero. This happens only when $x + 1 = 0$, which occurs when $x = -1$.

220. c. Note that $|a| \geq c$ if and only if ($a \geq c$ or $a \leq -c$). Using this fact, we see that the values of x that satisfy the inequality $|8x + 3| \geq 3$ are precisely those values of x that satisfy either $8x + 3 \geq 3$ or $8x + 3 \leq -3$. We solve these two inequalities separately:

$8x + 3 \geq 3$	$8x + 3 \leq -3$
$8x \geq 0$	$8x \leq -6$
$x \geq 0$	$x \leq -\frac{6}{8} = -\frac{3}{4}$

Thus, the solution set is $[0, \infty) \cup (-\infty, -\frac{3}{4}]$

221. d. Note that $|a| < c$ if and only if $-c < a < c$. Using this fact, we see that the values of x that satisfy the inequality $|2x - 3| < 5$ are precisely those values of x that satisfy $-5 < 2x - 3 < 5$. We solve this compound inequality as follows:
$$-5 < 2x - 3 < 5$$
$$-2 < 2x < 8$$
$$-1 < x < 4$$
Thus, the solution set is $(-1, 4)$.

222. a. First, we rewrite the given inequality in an equivalent form:

$$2 - (1 - (2 - |1 - 2x|)) > -6$$
$$2 - (1 - 2 + |1 - 2x|) > -6$$
$$2 - (-1 + |1 - 2x|) > -6$$
$$2 + 1 - |1 - 2x| > -6$$
$$3 - |1 - 2x| > -6$$
$$-|1 - 2x| > -9$$
$$|1 - 2x| < 9$$

Now, note that $|a| < c$ if and only if if $-c < a < c$. Using this fact, we see that the values of x that satisfy the inequality $|1 - 2x| < 9$ are precisely those values of x that satisfy $-9 < 1 - 2x < 9$. We solve this compound inequality:

$$-9 < 1 - 2x < 9$$
$$-10 < -2x < 8$$
$$5 > x > -4$$

So, the solution set is $(-4, 5)$.

223. c. First, we rewrite the given inequality in an equivalent form:

$$-7|1 - 4x| + 20 \leq -2|1 - 4x| - 15$$
$$-7|1 - 4x| + 35 \leq -2|1 - 4x|$$
$$35 \leq 5|1 - 4x|$$
$$7 \leq |1 - 4x|$$

The last inequality is equivalent to $|1 - 4x| \geq 7$. Now, $|a| \geq c$ if and only if ($a \geq c$ or $a \leq -c$). Using this fact, we see that the values of x that satisfy the inequality $|1 - 4x| \geq 7$ are precisely those values of x that satisfy either $1 - 4x \geq 7$ or $1 - 4x \leq -7$. We solve these two inequalities separately:

$$1 - 4x \geq 7 \qquad 1 - 4x \leq -7$$
$$-4x \geq 6 \qquad -4x \leq -8$$
$$x \leq -\frac{6}{4} = -\frac{3}{2} \qquad x \geq 2$$

So, the solution set is $(-\infty, -\frac{3}{2}] \cup [2, \infty)$.

224. d. First, we rewrite the given inequality in an equivalent form:

$$|1 - (-2^2 + x) - 2x| \geq |3x - 5|$$
$$|1 - (-4 + x) - 2x| \geq |3x - 5|$$
$$|1 + 4 - x - 2x| \geq |3x - 5|$$
$$|5 - 3x| \geq |3x - 5|$$

Now, note that the left side of the last inequality is equivalent to

$$|5 - 3x| = |-1(3x - 5)| = |-1| \cdot |3x - 5|$$
$$= |3x - 5|$$

Thus, the original inequality is actually equivalent to $|3x - 5| \geq |3x - 5|$. Since the left and right sides of the inequality are identical, every real number x satisfies the inequality. So, the solution set is the set of all real numbers.

Set 15 (Page 26)

225. c. The coordinates of points in the third quadrant are both negative.

226. e. The x-coordinate of J is -3 and the y-coordinate is 4. So, J is identified as the point $(-3, 4)$.

227. b. Since $ABCD$ is a square, the x-coordinate of B will be the same as the x-coordinate of A, namely -1, and the y-coordinate of B will be the same as the y-coordinate of C, namely 4. So, the coordinates of B are $(-1, 4)$.

228. e. Since $ABCD$ is a square, the x-coordinate of D is the same as the x-coordinate of C, which is 6, and the y-coordinate of D is the same as the y-coordinate of A, which is -3. So, the coordinates of D are $(6, -3)$.

229. d. Points in Quadrant IV have positive x-coordinates and negative y-coordinates. Therefore, $(2, -5)$ lies in Quadrant IV.

230. a. For all nonzero real numbers, both x^2 and $(-y)^2$ are positive, so points of the form $(x^2, (-y)^2)$ must lie in Quadrant I.

231. d. First, note that $|-x-2| = 0$ only when $x = -2$ and $|-x-1| = 0$ when $x = -1$. So, for all other values of x, these expressions are positive. For all real numbers $x < -2$, we conclude that $|-x-2| > 0$ and $-|-x-1| < 0$. Therefore, points whose coordinates are given by $(|-x-2|, -|-x-1|)$ must lie in Quadrant IV.

232. b. The fact that x is a positive real number requires that the point (x, y) lie to the right of the y-axis, so it cannot lie in Quadrants II or III, or be on the y-axis. It can, however, lie in Quadrants I or IV, or be on the x-axis. The fact that y can be any real number does not further restrict the location of (x, y). Hence, the correct choice is **b**.

233. c. Because y is a nonnegative real number, the point (x, y) must lie on or above the x-axis, so it cannot lie in Quadrants III or IV. The point can also be on the y-axis. The fact that x can be any real number does not further restrict the location of (x, y), so the correct choice is **c**.

234. a. We need to choose the point that has a positive x-coordinate and negative y-coordinate. Since $a < 0$, it follows that $-a > 0$. Thus, the point that lies in Quadrant IV is $(-a, a)$.

235. b. The correct point will have a negative x-coordinate and negative y-coordinate. Note that for any nonzero real number, that $-a^2 < 0$ and $(-a)^2 > 0$. So, the point that lies in Quadrant III is $(a, -a^2)$.

236. a. Look for the point that has a negative x-coordinate and positive y-coordinate. Since $a > 0$, it follows that $-a < 0$. So, the point that lies in Quadrant II is $(-a, a)$.

237. d. Note that if x is a negative integer, then $-x^3 = -(x)(x)(x)$ must be positive (because it is a product of an even number of negative integers). Likewise, since x and y are both negative integers, xy^2 is negative (because it is a product of an odd number of negative integers). Hence, the x-coordinate of $(-x^3, xy^2)$ is positive and its y-coordinate is negative. So, the point lies in Quadrant IV.

238. b. Using the fact that x and y are both assumed to be negative integers, we must determine the signs of the coordinates of the point $(\frac{-x^2}{(-y)^3}, \frac{1}{xy})$. To this end, note that $-x^2$ is negative, $(-y)^3$ is positive (since $-y$ is a positive integer and the cubes of positive integers are positive integers), and xy is positive (since it is a product of an even number of negative integers). The x-coordinate of the given point is therefore negative (since the numerator is negative and denominator is positive, thereby creating a quotient involving an odd number of negative integers) and the y-coordinate is positive. So, the point lies in Quadrant II.

239. c. Since the y-coordinate of the point $(-x, -2)$ is -2, it follows that for any real number x, the point must lie somewhere strictly below the x-axis. If $x > 0$, the x-coordinate of the point is negative, so that it lies in Quadrant III, while it lies in Quadrant IV if $x < 0$ and on the y-axis if $x = 0$. So, the correct choice is **c**.

240. d. The phrase "y is nonpositive" can be expressed symbolically as $y \le 0$. As such, $-y \ge 0$. Since the x-coordinate of the point $(1, -y)$ is positive and the y-coordinate is nonnegative, the point must be in Quadrant I or on the x-axis. It can be *on* the x-axis if $y = 0$. So, neither **a** nor **b** is true.

Set 16 (Page 28)

241. a. Convert the given equation $3y - x = 9$ into slope-intercept form by solving for y, as follows:

$3y - x = 9$

$3y = x + 9$

$y = \frac{1}{3}x + 3$

The slope of this line is the coefficient of x, namely $\frac{1}{3}$.

242. b. The line whose equation is $y = -3$ is horizontal. Any two distinct points on the line share the same y-value, but have different x-values. So, computing the slope as "change in y over change in x" results in 0, no matter which two points are used.

243. a. Convert the given equation $8y = 16x - 4$ into slope-intercept form by solving for y:

$8y = 16x - 4$

$y = 2x - \frac{1}{2}$

So, the y-intercept is $(0, -\frac{1}{2})$.

244. d. Substituting $x = 3$ and $y = 1$ into the equation $y = \frac{2}{3}x - 1$ yields the true statement $1 = \frac{2}{3}(3) - 1$, which implies that the point $(3,1)$ is on this line.

245. b. The slope-intercept form of a line with slope m and y-intercept $(0,b)$ is $y = mx + b$. So, the equation of the line with slope -3 and y-intercept of $(0,2)$ is $y = -3x + 2$.

246. a. First, choose two of the five points listed and compute the slope. We will use the first two listed, $(1,7)$ and $(2,10)$. The slope is $m = \frac{10-7}{2-1} = 3$. Next, use one of the points, such as $(1,7)$, and the slope $m = 3$ to write the equation of the line using the point-slope formula $y - y_1 = m(x - x_1)$, where (x_1,y_1) is the point on the line. Applying this yields the equation $y - 7 = 3(x - 1)$, which simplifies to $y - 7 = 3x - 3$, or equivalently $y = 3x + 4$.

247. b. Transforming the equation $3x + y = 5$ into slope-intercept form simply requires that we solve for y to obtain the equation $y = -3x + 5$.

248. b. First, the slope of the line containing the points $(2,3)$ and $(-2,5)$ is $m = \frac{5-3}{(-2)-2} = \frac{2}{-4} = -\frac{1}{2}$. Next, use one of the points, such as $(2,3)$, and the slope $m = -\frac{1}{2}$ to write the equation of the line using the point-slope formula $y - y_1 = m(x - x_1)$, where, (x_1,y_1) is the point on the line. This yields the equation $y - 3 = -\frac{1}{2}(x - 2)$, which simplifies to $y - 3 = -\frac{1}{2}x + 1$, or equivalently $y = -\frac{1}{2}x + 4$.

249. c. We transform the equation $y = -\frac{2}{15}x - \frac{3}{5}$ into standard form $Ax + By = C$, as follows:

$y = -\frac{2}{15}x - \frac{3}{5}$

$-15y = 2x + 9$

$0 = 2x + 15y + 9$

$2x + 15y = -9$

250. a. We must solve the equation $-3y = 12x - 3$ for y, which can be done by simply dividing both sides by -3. This yields $y = -4x + 1$. The slope of this line is -4.

251. d. Solving the equation $6y + x = 7$ for y yields the equivalent equation $y = -\frac{1}{6}x + \frac{7}{6}$. The slope of this line is $-\frac{1}{6}$.

252. c. We must first determine the equation of the line. To do so, choose two of the five points listed and compute the slope. Using $(-4,15)$ and $(-2,11)$, the slope is $m = \frac{15-11}{(-4)-(-2)} = \frac{4}{-2} = -2$. Next, use one of the points, say $(-4,15)$, and the slope $m = -2$ to write the equation of the line using the point-slope formula $y - y_1 = m(x - x_1)$, where (x_1, y_1) is the point on the line. Applying this yields the equation $y - 15 = -2(x - (-4))$, which simplifies to $y - 15 = -2x - 8$, or equivalently $y = -2x + 7$. Now, to determine the missing value z, we simply substitute $x = 2$ into this equation; the resulting value of y is equal to the missing value of z. The substitution yields $y = -2(2) + 7 = 3$.

253. c. First, the slope of the line containing the points $(0,-1)$ and $(2,3)$ is $m = \frac{3-(-1)}{2-0} = \frac{4}{2}$ $= 2$. Next, use one of the points, such as $(2,3)$, and the slope $m = 2$ to write the equation of the line using the point-slope formula $y - y_1 = m(x - x_1)$, where (x_1, y_1) is the point on the line. This yields the equation $y - 3 = 2(x - 2)$, which simplifies to $y - 3 = 2x - 4$, or equivalently $y = 2x - 1$.

254. a. Consider the line whose equation is $x = 2$. All points on this line are of the form $(2,y)$, where y can be any real number. However, in order for this line to have a y-intercept, at least one of the points on it would need to have an x-coordinate of 0, which is not the case. So, a vertical line need not have a y-intercept.

255. a. First, we must determine the equation of the line. The slope of the line is given by $m = \frac{0-(-6)}{9-0} = \frac{6}{9} = \frac{2}{3}$. Since the y-intercept of the line is given to be $(0, -6)$, we conclude that the equation of the line is $y = \frac{2}{3}x - 6$. Now, observe that substituting the point $(-6,-10)$ into the equation yields the true statement $-10 = \frac{2}{3}(-6) - 6$. Therefore, the point $(-6,-10)$ lies on this line.

256. d. The slope of a line containing the points $(-3,-1)$, $(0,y)$, and $(3,-9)$ can be computed using any two pairs of these points. Specifically, using $(-3,-1)$ and $(3,-9)$, we see that the slope is $m = \frac{(-1)-(-9)}{(-3)-3} = -\frac{8}{6} = -\frac{4}{3}$. Now, we equate the expression obtained by computing the slope of this line using the points $(-3,-1)$ and $(0,y)$ to $-\frac{4}{3}$, and solve for y:

$$\frac{y-(-1)}{0-(-3)} = -\frac{4}{3}$$

$$\frac{y+1}{3} = -\frac{4}{3}$$

$$y + 1 = -4$$

$$y = -5$$

Set 17 (Page 31)

257. d. The points on the line $y = -3$ are of the form $(x,-3)$, for all real numbers x. This set of points forms a horizontal line containing the point $(0,-3)$. The correct graph is given by choice **d.**

258. b. The slope of this line segment is $m = \frac{0-(-5)}{-3-0}$ $= -\frac{5}{3}$.

259. b. The y-axis is a vertical line and hence, its slope is undefined.

260. d. The slope of this line segment is $m = \frac{2-(-6)}{10-(-2)} = \frac{8}{12} = \frac{2}{3}$.

261. c. The slope is 2 (so that the graph of the line rises from left to right at a rate of two vertical units up per one horizontal unit right) and the y-intercept is $(0,3)$. The correct graph is shown in choice **c.**

262. a. The slope is -2 (so that the graph of the line falls from left to right at the rate of two vertical units down per one horizontal unit right) and the y-intercept is $(0,9)$. The correct graph is shown in choice **a.**

263. d. The slope is $-\frac{5}{2}$ (so that the graph of the line falls from left to right at a rate of five vertical units down per two horizontal units right) and the y-intercept is $(0,-5)$. The correct graph is in choice **d.**

264. d. The line falls from left to right at a rate of one vertical unit down per one horizontal unit right, and it crosses the y-axis at the point $(0,7)$. So, the slope of the line is -1 and its y-intercept is $(0,7)$. Its equation is therefore $y = -x + 7$.

265. b. Using the two points $(0,5)$ and $(-9,-1)$ on the line, we observe that the line rises from left to right at a rate of six vertical units up per nine horizontal units right. Hence, its slope is $\frac{6}{9} = \frac{2}{3}$. Also, it crosses the y-axis at $(0,5)$. Thus, the equation of this line is $y = \frac{2}{3}x + 5$.

266. c. First, convert the equation $\frac{2}{3}y - \frac{1}{2}x = 0$ into slope-intercept form:

$$\frac{2}{3}y - \frac{1}{2}x = 0$$
$$\frac{2}{3}y = \frac{1}{2}x$$
$$y = \left(\frac{3}{2}\right) \cdot \left(\frac{1}{2}x\right) = \frac{3}{4}x$$

From this, we observe that since the slope is $\frac{3}{4}$, the graph of the line rises from left to right at a rate of 3 vertical units up per 4 horizontal units right. The y-intercept is the origin. So, the correct graph is shown in choice **c**.

267. b. A line with a positive slope rises from left to right. The only line that rises from left to right is the one in choice **b**.

268. a. A line with an undefined slope must be vertical. The only graph that satisfies this criterion is choice **a**.

269. d. First, convert the equation into slope-intercept form as follows:

$$0.1x - 0.7y = 1.4$$
$$0.1x = 0.7y + 1.4$$
$$0.1x - 1.4 = 0.7y$$
$$y = \frac{0.1}{0.7}x - \frac{1.4}{0.7} = \frac{1}{7}x - 2$$

Since the slope is $\frac{1}{7}$, the graph of the line rises from left to right at a rate of one vertical unit up per seven horizontal units right. The y-intercept is $(0,-2)$, and the correct graph is given by choice **d**.

270. c. The graph of $y = c$, where $c \neq 0$ is a horizontal line that is either above or below the x-axis. If it lies above the x-axis, the graph crosses into only Quadrant I and Quadrant II, while if it lies below the x-axis, it crosses into only Quadrant III and Quadrant IV.

271. b. The graph of $y = c$, where $c \neq 0$, is a horizontal line that lies either above or below the x-axis, and must cross the y-axis.

272. a. For instance, consider the line whose equation is $y = -x - 1$. Its graph is shown here:

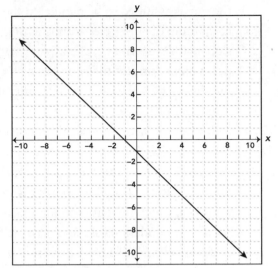

Observe that the graph does indeed cross into three of the four quadrants.

Set 18 (Page 42)

273. e. A line perpendicular to the given line must have a slope $m = \frac{-3}{2}$. So, the line given in choice **e** is the correct choice.

274. e. Two lines are parallel if and only if they have the same slope. This is true for the lines provided in choice **e** since the slopes of both lines are 6.

275. b. The line provided in choice **b** is equivalent to $y = -2x + 6$. Since this has the same slope as the given line, namely -2, we conclude that the correct answer is **b**.

276. d. Since we want a line perpendicular to a line with slope $m_1 = \frac{3}{4}$, we must use $m_2 = -\frac{1}{\frac{3}{4}} = -\frac{4}{3}$ as the slope. Since the point $(-6,4)$ must be on the line, the point-slope formula for the line is $y - 4 = -\frac{4}{3}(x + 6)$. This is equivalent to the equation $y = -\frac{4}{3}x - 4$.

277. b. A line parallel to $y = 3x + 8$ must have slope 3. Using the point-slope formula for a line with the point $(4,4)$, we see that the equation of the line we seek is $y - 4 = 3(x - 4)$, which simplifies to $y = 3x - 8$.

278. b. The slope of the line passing through the two given points is $m = \frac{6-2}{-5-4} = -\frac{4}{9}$. This is actually the slope of the line we seek because the line is assumed to be parallel to the one containing the two given points. Using this slope with the point $(0,12)$, we see that the point-slope form of the equation of the lines is $y - 12 = -\frac{4}{9}(x - 0)$, which simplifies to $y = -\frac{4}{9}x + 12$.

279. c. A line perpendicular to the given line must have slope $\frac{18}{13}$. Using this slope with the point $(0,0)$, we see that the point-slope form of the equation of the line we seek is $y - 0 = \frac{18}{13}(x - 0)$, which simplifies to $y = \frac{18}{13}x$.

280. b. A line with undefined slope is vertical. Only horizontal lines are perpendicular to vertical ones, and $y = -2$ is a horizontal line.

281. b. Only horizontal lines have zero slopes. The only horizontal line among the choices provided is given by choice **b**.

282. c. Let $x = $ the length of the first piece. Then, $2x - 1 = $ the length of the second piece and $3(2x - 1) + 10 = $ the length of the third piece. The sum of the lengths of these three smaller pieces will be the length of the original piece of rope. This is represented as the equation $x + (2x - 1) + 3(2x - 1) + 10 = 60$. To solve this equation, we first simplify the left side to obtain $9x + 6 = 60$. Solving this equation gives us $x = 6$. Therefore, the length of the first piece is 6 feet, the second piece is 11 feet long, and finally, the third piece is 43 feet long. So, we conclude that the longest piece of rope is 43 feet long.

283. b. Let $x = $ number of canisters of Ace balls. Then, $x + 1 = $ number of canisters of Longline balls. The important observation is that multiplying the price of one canister of Ace balls by the number of canisters of Ace balls results in the portion of the total amount spent on Ace balls. The same reasoning is true for the Longline balls. So, we must solve an equation of the form:

amount spent on Ace balls + amount spent on Longline balls = total amount spent

Using the information provided, this equation becomes $3.50x + 2.75(x + 1) = 40.25$. Simplifying the left side of the equation yields $6.25x + 2.75 = 40.25$. Subtracting 2.75 from both sides and then dividing by 6.25 yields the solution $x = 6$. So, we conclude that he bought 6 canisters of Ace balls and 7 canisters of Longline balls.

284. d. Let $x = $ the number of gallons needed of the 30% nitrogen. Then, since we are supposed to end up with 10 gallons, it must be the case that $10 - x = $ the number of gallons needed of the 90% nitrogen. Multiplying the number of gallons of 30% nitrogen by its concentration yields the amount of nitrogen contained within the 30% solution. A similar situation holds for the 90% nitrogen, as well as for the final 70% solution. We must solve an equation of the following form:

$$\left(\begin{array}{c}\text{amount of nitrogen contributed}\\\text{from the 30\% solution}\end{array}\right) +$$

$$\left(\begin{array}{c}\text{amount of nitrogen contributed}\\\text{from the 90\% solution}\end{array}\right) =$$

$$\left(\begin{array}{c}\text{total amount of nitrogen}\\\text{in the entire 10 gallons}\end{array}\right)$$

Using the information provided, this equation becomes $0.30x + 0.90(10 - x) = 0.70(10)$, which is solved as follows:

$$0.30x + 0.90(10 - x) = 0.70(10)$$
$$30x + 90(10 - x) = 70(10)$$
$$30x + 900 - 90x = 700$$
$$-60x + 900 = 700$$
$$-60x = -200$$
$$x = \frac{-200}{-60} = \frac{10}{3}$$

Thus, rounding to two decimal places, we conclude that she should mix approximately 3.33 gallons of the 30% nitrogen solution with 6.67 gallons of the 90% nitrogen solution to obtain the desired mixture.

285. a. The important concept in this problem is how rate, time, and distance interrelate. It is known that distance = rate × time. We need to determine the amount of time that the girl is riding, and at precisely what time the girl and the instructor meet and have therefore traveled the exact same distance from the starting point. So, we must determine expressions for the distances traveled by both the girl and her instructor, and then equate them. To this end, let x = number of hours the girl has been riding when she intercepts her instructor. Then, since the instructor had a 3-hour head start, the amount of time that he has been riding when the girl catches him must be $3 + x$ hours.

Now, write an equation for the girl, and one for the instructor that relates their respective times, rates, and distance traveled.

Let R_g = rate of the girl = 17 mph
T_g = time the girl is riding when she meets her instructor = x hours
D_g = distance the girl has ridden when she finally intercepts the instructor = $17x$
R_I = rate of the instructor = 7 mph
T_I = time the instructor is riding when he meets the girl = $3 + x$ hours
D_I = distance the instructor has ridden when he is intercepted by the girl = $7(3 + x)$

Using the information provided, we must solve the equation $17x = 7(3 + x)$, as follows:
$17x = 7(3 + x)$
$17x = 21 + 7x$
$10x = 21$
$x = 2.1$
Thus, it takes the girl 2.1 hours (or 2 hours 6 minutes) to overtake her instructor.

286. d. Let x = the amount invested at 10% interest. Then, she invested $1,500 + x$ dollars at 11% interest. The amount of interest she earns in one year from the 10% investment is $0.10x$, and the amount of interest earned in one year from the 11% investment is $0.11(1,500 + x)$. Since her total yearly interest earned is 795 dollars, the following equation describes this scenario:
$0.10x + 0.11(1,500 + x) = 795$
This equation is solved as follows:
$0.10x + 0.11(1,500 + x) = 795$
$0.10x + 165 + 0.11x = 795$
$0.21x = 630$
$x = \frac{630}{0.21} = 3,000$
Hence, she invested $3,000 at 10% interest and $4,500 at 11% interest.

287. b. Let x = the number of nickels in the piggy bank. Then there are $65 - x$ dimes in the bank. The amount contributed to the total by the nickels is $0.05x$ and the amount contributed by the dimes is $0.10(65 - x)$. Since the total in the bank is $5.00, we must solve the following equation.
$0.05x + 0.10(65 - x) = 5.00$
The equation is solved as follows:
$0.05x + 0.10(65 - x) = 5.00$
$0.05x + 6.5 - 0.10x = 5.00$
$-0.05x = -1.5$
$x = \frac{-1.5}{-0.05} = 30$
Thus, there are 30 nickels and 35 dimes in the piggy bank.

288. b. Let x = Lisa's current age (in years). Lori's age is $2x$. The statement, "In 5 years, Lisa will be the same age as her sister was 10 years ago" can be expressed symbolically as the equation $x + 5 = 2x - 10$, which is solved as follows:
$x + 5 = 2x - 10$
$x = 15$
Thus, Lisa is currently 15 years old and Lori is 30 years old.

Set 19 (Page 45)

289. b. The fact that the graph of the line is solid means that it is included in the solution set, so the inequality describing the shaded region must include equality (that is, it must be either $\geq$ or $\leq$). Next, since the shaded region is below the horizontal line $y = -2$, all points in the solution set have a y-value that is less than or equal to -2. Hence, the inequality illustrated by this graph is $y \leq -2$.

290. a. The graph of the line is solid, so it is included in the solution set and the inequality describing the shaded region must include equality (either $\geq$ or $\leq$). Next, since the graph of the line rises from left to right at the rate of two vertical units up per one horizontal unit right, its slope is 2. And, since it crosses the y-axis at $(0,7)$, we conclude that the equation of the line is $y = 2x + 7$. Finally, since the shaded region is below the line $y = 2x + 7$, the inequality illustrated by this graph is $y \leq 2x + 7$. This can be verified by choosing a point in the shaded region, say $(0,0)$, and observing that substituting it into the inequality results in the true statement $0 \leq 7$.

291. b. The graph of the line is dashed, so it is not included in the solution set and the inequality describing the shaded region must not include equality (it must be either $>$ or $<$). Next, since the graph of the line falls from left to right at the rate of four vertical units down per one horizontal unit right, its slope is -4. Since it crosses the y-axis at $(0,-3)$, we conclude that the equation of the line is $y = -4x - 3$. Finally, the shaded region is below the line $y = -4x - 3$, so we conclude that the inequality illustrated by this graph is $y < -4x - 3$. This can be verified by choosing any point in the shaded region, say $(0,-4)$, and observing that substituting it into

the inequality results in the true statement $-4 < -3$.

292. c. The graph of the line is dashed, so it is not included in the solution set and the inequality describing the shaded region must not include equality (it must be either $>$ or $<$). Next, since the shaded region is to the left of the vertical line $x = 8$, all points in the solution set have an x-value that is less than 8. Hence, the inequality illustrated by this graph is $x < 8$.

293. c. The graph of the line is solid, so it is included in the solution set and the inequality describing the shaded region must include equality (it must be either $\geq$ or $\leq$). Next, since the graph of the line rises from left to right at the rate of one vertical unit up per one horizontal unit right, its slope is 1. Since it crosses the y-axis at $(0,2)$, the equation of the line is $y = x + 2$. Finally, since the shaded region is above the line $y = x + 2$, the inequality illustrated by this graph is $y \geq x + 2$, which is equivalent to $x - y \leq -2$. We can verify this by choosing any point in the shaded region, such as $(0,3)$, and observing that substituting it into the inequality results in the true statement $-3 \leq -2$.

294. c. The fact that the graph of the line is solid means that it is included in the solution set, so the inequality describing the shaded region must include equality (it must be either $\geq$ or $\leq$). Next, since the graph of the line rises from left to right at the rate of one vertical unit up per one horizontal unit right, its slope is 1. Since it crosses the y-axis at $(0,0)$, the equation of the line is $y = x$. The shaded region is above the line $y = x$, so we conclude that the inequality illustrated by this graph is $y \geq x$, which is equivalent to $y - x \geq 0$. Substituting a point from the shaded region, such as $(0,3)$ into the inequality results in the true statement $3 \geq 0$.

295. a. The graph of the line is dashed, so it is not included in the solution set, and the inequality describing the shaded region must not include equality (it must be either $>$ or $<$). Next, since the graph of the line falls from left to right at the rate of one vertical unit down per six horizontal units right, its slope is $-\frac{1}{6}$. It crosses the y-axis at $(0,-\frac{1}{2})$, so the equation of the line is $y = -\frac{1}{6}x - \frac{1}{2}$. Finally, since the shaded region is above the line $y = -\frac{1}{6}x - \frac{1}{2}$, the inequality illustrated by this graph is $y > -\frac{1}{6}x - \frac{1}{2}$. Multiplying both sides of this inequality by 2 and moving the x-term to the left results in the equivalent inequality $\frac{1}{3}x + 2y > -1$. We can verify this by substituting any point from the shaded region, such as $(0,1)$, into the inequality, which results in the true statement $2 > -1$.

296. a. Because the graph of the line is solid, we know that it is included in the solution set, so the inequality describing the shaded region must include equality (it must be either $\geq$ or $\leq$). Next, the graph of the line falls from left to right at the rate of three vertical unit down per one horizontal unit right, so its slope is -3. And, since it crosses the y-axis at $(0, 4)$, we conclude that the equation of the line is $y = -3x + 4$. The shaded region is below the line $y = -3x + 4$, so the inequality illustrated by this graph is $y \leq -3x + 4$ Multiplying both sides of the inequality by 2 and moving the x-term to the left results in the equivalent inequality $2y + 6x \leq 8$. This can be further verified by choosing a point from the shaded region, such as $(0,0)$, and observing that substituting it into the inequality results in the true statement $0 \leq 8$.

297. c. The graph of the line is dashed, so it is not included in the solution set and the inequality describing the shaded region must not include equality (it must be either $>$ or $<$). Next, since the graph of the line rises from left to right at the rate of three vertical units up per one horizontal unit right, its slope is 3. And, since it crosses the y-axis at $(0,-2)$, the equation of the line is $y = 3x - 2$. Finally, since the shaded region is above the line $y = 3x - 2$, we conclude that the inequality illustrated by this graph is $y > 3x - 2$. Moving the y-term to the right results in the equivalent inequality $3x - y - 2 < 0$. This can be verified by choosing an arbitrary point in the shaded region, say $(0,0)$, and observing that we can verify this by substituting a point from the shaded region, such as $(0,0)$ into the inequality, resulting in the true statement $-2 < 0$.

298. a. The graph of the line is solid, so we know that it is included in the solution set, and that the inequality describing the shaded region must include equality (either $\geq$ or $\leq$). Next, since the graph of the line rises from left to right at the rate of three vertical unit up per one horizontal unit right, its slope is 3. It crosses the y-axis at $(0,1)$, so the equation of the line is $y = 3x + 1$. Finally, since the shaded region is above the line $y = 3x + 1$, the inequality illustrated by this graph is $y \geq 3x + 1$. This can be verified by choosing any point in the shaded region, such as $(0,2)$, and substituting it into the inequality, which results in the true statement $2 \geq 1$.

299. d. The fact that the graph of the line is solid means that it is included in the solution set, so the inequality describing the shaded region must include equality (that is, it must be either ≥ or ≤). Next, since the graph of the line falls from left to right at the rate of two vertical unit down per one horizontal unit right, its slope is –2. And, since it crosses the y-axis at $(0,4)$, we conclude that the equation of the line is $y = -2x + 4$. Finally, since the shaded region is above the line $y = -2x + 4$, we conclude that the inequality illustrated by this graph is $y \geq -2x + 4$. This can be verified by choosing a point in the shaded region, such as $(0,5)$, substituting it into the inequality to produce the true statement $5 \geq 4$. Observe that simplifying $3x - y \leq 7x + y - 8$ results in this inequality.

300. d. Substituting $x = 3$ and $y = -2$ into the inequality $9x - 1 > y$ yields the true statement $26 > -2$. We can therefore conclude that $(3,-2)$ satisfies this inequality.

301. d. First, since the given inequality does not include equality, the horizontal line $y = 4$ is not included in the solution set and should be dashed. Because $y > 4$, any point in the solution set (the shaded region) must have a y-coordinate that is larger than 4. Such points occur only above the line $y = 4$. So, the correct graph is given by choice **d**.

302. c. Since the given inequality does not include equality, the vertical line $x = 4$ is not included in the solution set and should be dashed. Also, since $x > 4$, any point in the solution set (the shaded region) must have an x-coordinate that is larger than 4. Such points occur to the right of the line $x = 4$. So, the correct graph is shown in choice **c**.

303. b. The graph of the line is solid, so it is included in the solution set, and the inequality describing the shaded region must include equality (it must be either ≥ or ≤). Next, since the graph of the line falls from left to right at the rate of one vertical unit down per seven horizontal units right, its slope is $-\frac{1}{7}$. It crosses the y-axis at $(0,10)$, so the equation of the line is $y = -\frac{1}{7}x + 10$. Finally, since the shaded region is below the line $y = -\frac{1}{7}x + 10$, the inequality illustrated by this graph is $y \leq -\frac{1}{7}x + 10$. This can be verified by choosing any point in the shaded region, such as $(0,5)$, and substituting it into the inequality to produce the true statement $5 \leq 10$. Observe that simplifying $-28y \geq 2x - 14(y + 10)$ results in this inequality.

304. b. The following graph illustrates the inequality $y \leq 2x + 7$, whose solution set intersects all four quadrants.

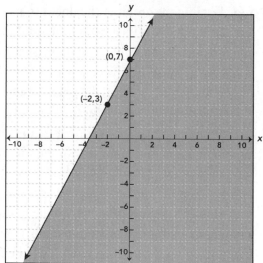

Set 20 (Page 50)
305. c. Adding the two equations together yields the equation $10a = -40$, the solution of which is $a = -4$. Now, substitute –4 in for a in the first equation and solve for b:
$5(-4) + 3b = -2$
$-20 + 3b = -2$
$3b = 18$
$b = 6$

306. d. Add the two equations together to get the equation $-2y = 8$, which simplifies to $y = -4$. Next, substitute -4 for y in the second equation and solve for x:

$x - 5(-4) = -3$
$x + 20 = -3$
$x = -23$

307. b. In the first equation, multiply the $(x + 4)$ term by 3 to obtain $3(x + 4) = 3x + 12$. Then, subtract 12 from both sides of the equation, so that the first equation becomes $3x - 2y = -7$. Now, add the two equations together to obtain $-x = 1$, or $x = -1$.

308. d. First, multiply the second equation by 2 to obtain $y + 8x = 24$. Then, subtract the first equation from this one to obtain $6x = 18$, which simplifies to $x = 3$.

309. b. First, simplify the second equation by subtracting 9 from both sides of the equation. The second equation becomes $-2x - 6 = y$. Then, multiply the equation by 2 and add it to the first equation to obtain $-6 = -y$, the solution of which is $y = 6$. Now, substitute the value of y into the first equation and solve for x:

$4x + 6 = -3(6)$
$4x + 6 = -18$
$4x = -24$
$x = -6$

Since $x = -6$ and $y = 6$, the value of $\frac{x}{y} = \frac{-6}{6} = -1$.

310. e. First, multiply the first equation by -4 to obtain $28a - b = -100$. Then, add this to the second equation to obtain $29a = -87$ so that $a = -3$. To find b, substitute this value into the second equation to obtain $b + (-3) = 13$, so $b = 16$.

311. b. Simplify the first equation as follows:

$2(m + n) + m = 2m + 2n + m = 3m + 2n = 9$.

Now, subtract the second equation from the first equation to obtain the equation $5n = -15$, which simplifies to $n = -3$.

312. a. Simplifying the left side of the first equation results in $14a + 21b = 56$. Multiplying the second equation by -7 yields $-7b - 14a = 28$. Now, adding these two equations together yields $14b = 84$, the solution of which is $b = 6$.
Finally, substitute this into the second equation and solve for a:

$6 + 2a = -4$
$2a = -10$
$a = -5$

313. c. Multiply the first equation by 8 to obtain the equivalent equation $4x + 48y = 56$. Adding this to the second equation in the system results in the equation $33y = 66$, which simplifies to $y = 2$.

314. d. Divide the second equation by 2 and add it to the first equation to obtain the equation $7a = 21$, the solution of which is $a = 3$. Now, substitute the value of a into the first equation and solve for b:

$4(3) + 6b = 24$
$12 + 6b = 24$
$6b = 12$
$b = 2$

Since $a = 3$ and $b = 2$, the value of $a + b = 3 + 2 = 5$.

315. b. First, simplify the first equation by multiplying $(a + 3)$ by $\frac{1}{2}$ to obtain the equivalent equation $\frac{1}{2}a + \frac{3}{2} - b = -6$. Then, subtract $\frac{3}{2}$ from both sides to further obtain $\frac{1}{2}a - b = -\frac{15}{2}$. Next, multiply the equation by -6 and add it to the second equation to obtain the equation $4b = 40$, or $b = 10$. Now, substitute the value of b into the second equation and solve for a:

$3a - 2(10) = -5$
$3a - 20 = -5$
$3a = 15$
$a = 5$

Since $a = 5$ and $b = 10$, the value of $a + b = 5 + 10 = 15$.

316. b. Multiply the first equation to 5 and simplify to obtain $c - d = 10$. Then, subtract the second equation from this to obtain $5d = 10$, or $d = 2$. Now, substitute this value into the second equation and solve for c:

$c - 6(2) = 0$

$c = 12$

So, the value of $\frac{c}{d}$ is $\frac{12}{2} = 6$.

317. a. Multiply the second equation by 2, then add to the first equation to obtain $-7x = -63$, which simplifies $x = 9$. Now, substitute this value into the second equation and solve for y:

$-9 - y = -6$

$-y = 3$

$y = -3$

So, the value of xy is $(9)(-3) = -27$.

318. e. First, simplify the first equation by multiplying $(x - 1)$ by 9 to obtain $9(x - 1) = 9x - 9$. Then, add 9 and $4y$ to both sides of the equation. The first equation becomes $9x + 4y = 11$. Multiply the second equation by -2 and add it to the first equation to obtain $-5x = 5$ or $x = -1$. Now, substitute the value of x into the second equation and solve for y:

$2y + 7(-1) = 3$

$2y - 7 = 3$

$2y = 10$

$y = 5$

Since $y = 5$ and $x = -1$, the value of $(y - x)^2 = (5 - (-1))^2 = 6^2 = 36$.

319. c. Multiply the second equation by 3 and add it to the first equation to obtain $14q = 98$, which simplifies to $q = 7$. Now, substitute the value of q into the second equation and solve for p:

$-5p + 2(7) = 24$

$-5p + 14 = 24$

$-5p = 10$

$p = -2$

Since $p = -2$ and $q = 7$, the value of $(p + q)^2 = (-2 + 7)^2 = 5^2 = 25$.

320. b. Multiply the first equation by 2 to obtain $8x - 6y = 20$ and the second equation by 3 to obtain $15x + 6y = 3$. Then, add these equations to obtain $23x = 23$, which simplifies to $x = 1$. Now, substitute this into the first equation and solve for y:

$4(1) - 3y = 10$

$-3y = 6$

$y = -2$

So, the solution of the system is $x = 1$, $y = -2$.

Set 21 (Page 53)

321. a. Since the first equation is already solved for x, substitute it directly into the second equation and solve for y:

$2(-5y) + 2y = 16$

$-10y + 2y = 16$

$-8y = 16$

$y = -2$

Now, substitute this value for y into the first equation to find the corresponding value of x: $x = -5(-2) = 10$. Hence, the solution of the system is $x = 10$, $y = -2$.

322. d. Solve the first equation for y in terms of x:

$2x + y = 6$

$y = 6 - 2x$

Substitute this expression for y in the second equation and solve for x:

$\frac{6 - 2x}{2} + 4x = 12$

$3 - x + 4x = 12$

$3x + 3 = 12$

$3x = 9$

$x = 3$

323. d. Solve the first equation for a in terms of b by multiplying both sides of the equation by 2 to obtain $a = 2b + 2$. Now, substitute this expression for a in the second equation and solve for b:

$3(2b + 2 - b) = -21$

$3(b + 2) = -21$

$3b + 6 = -21$

$3b = -27$

$b = -9$

Substitute the value of b into the first equation and solve for a:

$\frac{a}{2} = -9 + 1$

$\frac{a}{2} = -8$

$a = -16$

Since $a = -16$ and $b = -9$, the value of

$\sqrt{\frac{a}{b}} = \sqrt{\frac{-16}{-9}} = \sqrt{\frac{16}{9}} = \frac{4}{3}$.

324. e. Solve the second equation for a in terms of b:

$b + a = 13$

$a = 13 - b$

Substitute this expression for a in the first equation and solve for b:

$-7a + \frac{b}{4} = 25$

$-7(13 - b) + \frac{b}{4} = 25$

$7b - 91 + \frac{b}{4} = 25$

$\frac{29b}{4} = 116$

$29b = 464$

$b = 16$

325. a. Solve the second equation for b in terms of a:

$b + 2a = -4$

$b - -2a - 4$

Substitute this expression for b in the first equation and solve for a:

$7(2a + 3(-2a - 4)) = 56$

$7(2a - 6a - 12) = 56$

$7(-4a - 12) = 56$

$-28a - 84 = 56$

$-28a = 140$

$a = -5$

326. b. Solve the second equation for c in terms of d:

$c - 6d = 0$

$c = 6d$

Substitute this expression for c in the first equation and solve for d:

$\frac{c - d}{5} - 2 = 0$

$\frac{6d - d}{5} - 2 = 0$

$\frac{5d}{5} - 2 = 0$

$d - 2 = 0$

$d = 2$

Substitute the value of d into the second equation and solve for c:

$c - 6(2) = 0$

$c - 12 = 0$

$c = 12$

Since $c = 12$ and $d = 2$, the value of

$\frac{c}{d} = \frac{12}{2} = 6$.

327. a. Solve the second equation for y in terms of x:

$-x - y = -6$

$-y = x - 6$

$y = -x + 6$

Substitute this expression for y in the first equation and solve for x:

$-5x + 2(-x + 6) = -51$

$-5x - 2x + 12 = -51$

$-7x + 12 = -51$

$-7x = -63$

$x = 9$

Substitute the value of x into the second equation and solve for y:

$-9 - y = -6$

$-y = 3$

$y = -3$

Since $x = 9$ and $y = -3$, the value of

$xy = (9)(-3) = -27$.

328. e. Solve the second equation for b in terms of a:

$b - a = 1$

$b = a + 1$

Substitute this expression for b in the first equation and solve for a:

$10(a + 1) - 9a = 6$

$10a + 10 - 9a = 6$

$a + 10 = 6$

$a = -4$

Substitute the value of a into the second equation and solve for b:

$b - (-4) = 1$

$b + 4 = 1$

$b = -3$

Since $a = -4$ and $b = -3$, the value of $ab = (-4)(-3) = 12$.

329. b. Solve the second equation for y in terms of x:

$2x - y = 9$

$-y = -2x + 9$

$y = 2x - 9$

Substitute this expression for y in the first equation and solve for x:

$\frac{x + 2x - 9}{3} = 8$

$\frac{3x - 9}{3} = 8$

$x - 3 = 8$

$x = 11$

Substitute the value of x into the second equation and solve for y:

$2(11) - y = 9$

$22 - y = 9$

$-y = -13$

$y = 13$

Since $x = 11$ and $y = 13$, the value of $x - y = 11 - 13 = -2$.

330. c. Two lines are parallel if and only if they have the same slope. The slope-intercept form of the line $x - y = 7$ is $y = x - 7$, and the slope-intercept form of the line $2 - y = -x$ is $y = x + 2$. The slope of each of these lines is 1, so, they are parallel.

331. b. Since the two lines can intersect in exactly one point, we conclude that the system of equations represented by the graph has one solution.

332. b. The slope-intercept form of the line $y - 3x = -2$ is $y = 3x - 2$. As such, since the slope of this line, namely 3, is the same as the slope of the line given by the first equation, we conclude that the lines are parallel. Their graphs never intersect, so the system has no solution.

333. b. Solve the first equation for y to obtain $y = 3x - 2$. Now, substitute this into the second equation and solve for x, as follows:

$2(3x - 2) - 3x = 8$

$6x - 4 - 3x = 8$

$3x - 4 = 8$

$3x = 12$

$x = 4$

Next, substitute this value of x into the first equation to determine that the corresponding value of y is $y = 3(4) - 2 = 10$. Thus, the value of $\frac{2x}{y}$ is $\frac{2(4)}{10} = \frac{4}{5}$.

334. c. Since the graph consists of a single line, we conclude that the two equations that make up the system are exactly the same, so every point on the line is a solution of the system. There are infinitely many such points.

335. b. Since the two lines are parallel, they never intersect. There are no solutions of this system.

336. c. Observe that dividing both sides of the second equation $-3y + 9x = -6$ by -3 and rearranging terms results in the first equation. This means that the equations are identical, so any point that satisfies the first equation automatically satisfies the second. Since there are infinitely many such points, the system has infinitely many solutions.

Set 22 (Page 56)

337. c. The graphs of the lines $y = 4$ and $y = x + 2$ are dashed, so that the inequality signs used in both of the inequalities comprising the system are either $<$ or $>$. Next, note that points in the shaded region lie above the line $y = 4$ and below the line $y = x + 2$. This means that the system of linear inequalities for which the shaded region is the solution set is given by $y > 4$, $y < x + 2$.

338. a. The graphs of the lines $y = 5$ and $x = 2$ are solid, which means that the inequality signs used in both of the inequalities are either $\leq$ or $\geq$. Next, note that points in the shaded region lie above (or on) the line $y = 5$ and to the left of (or on) the line $x = 2$. Therefore, the system of linear inequalities for which the shaded region is the solution set is given by $y \geq 5$, $x \leq 2$.

339. a. First, note that the graphs of the lines $y = -x + 4$ and $y = x + 2$ are dashed, which means that the inequality signs used in both of the inequalities in the system are either $<$ or $>$. Next, note that points in the shaded region lie below the line $y = x + 2$ and below the line $y = -x + 4$. This implies that the system of linear inequalities for which the shaded region is the solution set is given by $y < x + 2$, $y < -x + 4$.

340. a. First, the graph of the line $y = \frac{1}{4}x$ is dashed, so the corresponding inequality should involve one of the signs $<$ or $>$. The graph of $y = -4x$ is solid (so the corresponding inequality should involve one of the signs $\leq$ or $\geq$). Points in the shaded region lie above the line $y = \frac{1}{4}x$ and below the line $y = -4x$. Therefore, the system of linear inequalities for which the shaded region is the solution set is given by $y > \frac{1}{4}x$, $y \leq -4x$.

341. d. The slope-intercept form of the line $2y - 3x = -6$ is $y = \frac{3}{2}x - 3$. The graphs of this line and $y = 5 - \frac{5}{2}x$ are solid, so the inequality signs used in both of the inequalities are either $\leq$ or $\geq$. Points in the shaded region lie above (or on) the line $2y - 3x = -6$ and above (or on) the line $y = 5 - \frac{5}{2}x$. This means that the system of linear inequalities for which the shaded region is the solution set is given by $2y - 3x \geq -6$, $y \geq 5 - \frac{5}{2}x$.

342. a. Given that the first inequality does not include equality, but the second inequality does, we know that the graph of the line $y = 2$ is dashed and the graph of the line $y = 2x + 1$ is solid. Points that satisfy the inequality $y > 2$ must be above the line $y = 2$, and those satisfying $y \leq 2x + 1$ must lie below the line $y = 2x + 1$. The intersection of these two regions is given by the illustration in choice **a.**

343. b. The slope-intercept forms of the lines $5y = 8(x + 5)$ and $12(5 - x) = 5y$ are, respectively, $y = \frac{8}{5}x + 8$ and $y = -\frac{12}{5}x + 12$. The graph of the line $y = \frac{8}{5}x + 8$ is solid (so that the corresponding inequality should involve one of the signs $\leq$ or $\geq$). The graph of $y = -\frac{12}{5}x + 12$ is dashed, so that the corresponding inequality should involve one of the signs $<$ or $>$. Points in the shaded region lie below (or on) the line $5y = 8(x + 5)$ and below the line $12(5 - x) = 5y$. This implies that the system of linear inequalities for which the shaded region is the solution set is given by $5y \leq 8(x + 5)$, $12(5 - x) > 5y$.

344. d. The graph of the line $y = 3x$ is dashed, so that the corresponding inequality should involve one of the signs $<$ or $>$. The graph of $y = -5$ is solid, so that the corresponding inequality should involve one of the signs $\leq$ or $\geq$. Note that points in the shaded region lie above the line $y = 3x$ and below (or on) the line $y = -5$. The system of linear inequalities for which the shaded region is the solution set is given by $y > 3x$, $y \leq -5$.

345. b. The slope-intercept form of the lines $9(y - 4) = 4x$ and $-9y = 2(x + 9)$ are, respectively, $y = \frac{4}{9}x + 4$ and $y = -\frac{2}{9}x - 2$. The graphs of both lines are dashed, so the inequality signs used in both inequalities are either $<$ or $>$. Next, note that points in the shaded region lie below the line $y = \frac{4}{9}x + 4$ and above the line $y = -\frac{2}{9}x - 2$. This tells us that the system of linear inequalities for which the shaded region is the solution set is given by $y < \frac{4}{9}x + 4$, $y > -\frac{2}{9}x - 2$. This system is equivalent to $9(y - 4) < 4x$, $-9y < 2(x + 9)$, which can be seen by reversing the simplification process used to obtain the slope-intercept forms of the lines in the first step. In doing so, remember that multiplying both sides of an inequality results in a switching of the inequality sign.

346. c. The slope-intercept forms of the lines $y - x = 6$ and $11y = -2(x + 11)$ are $y = x + 6$ and $y = -\frac{2}{11}x - 2$, respectively. The graphs of both lines are solid, so the inequality signs used in both inequalities are either $\leq$ or $\geq$. Next, points in the shaded region lie above (or on) the line $y = x + 6$ and above (or on) the line $y = -\frac{2}{11}x - 2$. The system of linear inequalities for which the shaded region is the solution set is given by $y \geq x + 6$, $y \geq -\frac{2}{11}x - 2$. This system is equivalent to $y - x \geq 6$, $11y \geq -2(x + 11)$, which can be seen by reversing the simplification process used to obtain the slope-intercept forms of the lines in the first step.

347. c. The slope-intercept forms of the lines $5x - 2(y + 10) = 0$ and $2x + y = 3$ are, respectively, $y = \frac{5}{2}x - 10$ and $y = -2x - 3$. The graphs of both lines are solid, which means that both inequality signs are either $\leq$ or $\geq$. Points in the shaded region lie above (or on) the line $y = \frac{5}{2}x - 10$ and below (or on) the line $y = -2x - 3$, so the system of linear inequalities for which the shaded region is the solution set is given by $y \geq \frac{5}{2}x - 10$, $y \leq -2x - 3$ This system is equivalent to $5x - 2(y + 10) \leq 0$, $2x + y \leq -3$, which can be seen by reversing the simplification process used to obtain the slope-intercept forms of the lines in the first step. In doing so, remember that multiplying both sides of an inequality results in a reversing of the inequality sign.

348. b. The slope-intercept forms of the lines $7(y - 5) = -5x$ and $-3 = \frac{1}{4}(2x - 3y)$ are, respectively, $y = -\frac{5}{7}x + 5$ and $y = \frac{2}{3}x + 4$. The graphs of both lines are dashed, so the inequality signs used in both of the inequalities comprising the system are either $<$ or $>$. Points in the shaded region lie below the line $y = -\frac{5}{7}x + 5$ and below the line $y = \frac{2}{3}x + 4$. This tells us that the system of linear inequalities for which the shaded region is the solution set is given by $y < -\frac{5}{7}x + 5$, $y < \frac{2}{3}x + 4$. This system is equivalent to $7(y - 5) < -5x$, $-3 < \frac{1}{4}(2x - 3y)$, which can be seen by reversing the simplification process used to obtain the slope-intercept forms of the lines in the first step. In doing so, remember that multiplying both sides of an inequality results in a reversing of the inequality sign.

349. d. The solution set for the system in choice **a** is the empty set. The solution set for the system in choice **b** consists of only the points that lie on the line $y = 3x + 2$, and the solution set of the system in choice **c** consists of only the points that lie on the line $y = x$. So, the solution sets of none of these systems span the entire Cartesian plane. In fact, it is impossible for such a system of linear inequalities to have a solution set that spans the entire Cartesian plane.

350. b. Note that the graphs of the lines $y = x + 3$ and $y = x - 1$ are parallel, and the graph of $y = x + 3$ lies strictly above the graph of $y = x - 1$. Using the first inequality specified in the system, any point that it is in the solution set of the system to $y > x + 3$, $y < x - 1$ would necessarily be above the line $y = x + 3$, and therefore, by the previous observation, also above the line $y = x - 1$. However, the second inequality in the system requires that the point be *below* the line $y = x - 1$, which is not possible. Hence, the solution set of this system is the empty set.

351. d. The boundaries of Quadrant III are the x-axis and y-axis; the equations of these axes are $y = 0$ and $x = 0$, respectively. Since points in the solution set are not to be on either axis, both inequalities comprising the system must involve one of the signs $<$ or $>$. Next, note that the sign of both the x- and y-coordinate of a point in Quadrant

III is negative. We conclude that the system with this solution set is given by $x < 0$, $y < 0$.

352. b. A system of linear inequalities whose solution set consists of the points on a single line must be of the form $y \geq mx + b$, $y \leq mx + b$, assuming that the lines are not vertical. Observe that the first inequality in the system $2y - 6x \leq 4$, is equivalent to $y \leq 2 + 3x$. Since the second inequality in the system of choice **b** is $y \geq 2 + 3x$, we conclude that the solution set consists of those points on the line $y = 3x + 2$.

Section 3—
Polynomial Expressions

Set 23 (Page 66)

353. d. $(x^2 - 3x + 2) + (x^3 - 2x^2 + 11) =$
$x^3 + x^2 - 2x^2 - 3x + 2 + 11 =$
$x^3 - x^2 - 3x + 13$

354. a. $(3x^2 - 5x + 4) - (-\frac{2}{3}x + 5)$
$= 3x^2 - 5x + 4 + \frac{2}{3}x - 5$
$= 3x^2 - 5x + \frac{2}{3}x + 4 - 5$
$= 3x^2 - \frac{15}{3}x + \frac{2}{3}x - 1$
$= 3x^2 - \frac{13}{3}x - 1$

355. b. $(\frac{1}{3}x^2 - \frac{1}{5}x - \frac{2}{3}) - (\frac{2}{3}x^2 - \frac{7}{10}x + \frac{1}{2})$
$= \frac{1}{3}x^2 - \frac{1}{5}x - \frac{2}{3} - \frac{2}{3}x^2 + \frac{7}{10}x - \frac{1}{2}$
$= \frac{1}{3}x^2 - \frac{2}{3}x^2 - \frac{1}{5}x + \frac{7}{10}x - \frac{2}{3} - \frac{1}{2}$
$= -\frac{1}{3}x^2 - \frac{2}{10}x + \frac{7}{10}x - \frac{4}{6} - \frac{3}{6}$
$= -\frac{1}{3}x^2 + \frac{5}{10}x - \frac{7}{6}$
$= -\frac{1}{3}x^2 + \frac{1}{2}x - \frac{7}{6}$

356. c. $(9a^2b + 2ab - 5a^2) - (-2ab - 3a^2 + 4a^2b)$
$= 9a^2b + 2ab - 5a^2 + 2ab + 3a^2 - 4a^2b$
$= 9a^2b - 4a^2b + 2ab + 2ab - 5a^2 + 3a^2$
$= 5a^2b + 4ab - 2a^2$

357. a. $(\frac{1}{6}x^2 + \frac{2}{3}x + 1) + (2x - \frac{2}{3}x^2 + 4) - (\frac{7}{2} + 3x + \frac{1}{2}x^2)$
$= \frac{1}{6}x^2 + \frac{2}{3}x + 1 + 2x - \frac{2}{3}x^2 + 4 - \frac{7}{2} - 3x - \frac{1}{2}x^2$
$= \frac{1}{6}x^2 - \frac{2}{3}x^2 - \frac{1}{2}x^2 + \frac{2}{3}x + 2x - 3x + 1 + 4 - \frac{7}{2}$
$= \frac{1}{6}x^2 - \frac{4}{6}x^2 - \frac{3}{6}x^2 + \frac{2}{3}x - x + 5 - \frac{7}{2}$
$= -x^2 - \frac{1}{3}x + \frac{3}{2}$

358. d. $(2 - 3x^3) - [(3x^3 + 1) - (1 - 2x^3)]$
$= 2 - 3x^3 - [3x^3 + 1 - 1 + 2x^3]$
$= 2 - 3x^3 - [5x^3]$
$= 2 - 3x^3 - 5x^3$
$= 2 - 8x^3$

359. b. The degree of a polynomial is the highest power to which the variable x is raised. For the polynomial $-5x^8 + 9x^4 - 7x^3 - x^2$, the term involving the highest power of x is $-5x^8$, so the degree of the polynomial is 8.

360. c. For the polynomial $-\frac{3}{2}x + 5x^4 - 2x^2 + 12$, the term involving the highest power of x is $5x^4$, so the degree of the polynomial is 4.

361. a. A constant polynomial is of the form $cx^0 = c$, where c is a constant. By this definition, the degree of the constant polynomial 4 is zero.

362. c. By definition, a polynomial is an expression of the form $a_nx^n + a_{n-1}x^{n-1} + \ldots + a_1x + a_0$ where $a_0, a_1, \ldots, a_n$ are real numbers and n is a nonnegative integer. Put simply, once the expression has been simplified, it cannot contain negative powers of the variable x. Therefore, the expression $x - 3x^{-2}$ is not a polynomial.

363. c. A polynomial is an expression of the form $a_nx^n + a_{n-1}x^{n-1} + \ldots + a_1x + a_0$, where $a_0, a_1, \ldots, a_n$ are real numbers and n is a nonnegative integer. That is, once the expression has been simplified, it cannot contain negative powers of the variable x. If we simplify the expression $(-2x)^{-1} - 2$ using the exponent rules, we obtain $-\frac{1}{2}x^{-1} - 2$, which cannot be a polynomial because of the term $-\frac{1}{2}x^{-1}$. Note that the expression given in choice **a** *is* a polynomial; the coefficients, not the variable, involve negative exponents. The expression in choice **b** is a polynomial for similar reasons; note that the first term is really just a constant since $x^0 = 1$.

364. d. The statements in choices **a**, **b**, and **c** are all true, and follow from the fact that simplifying such arithmetic combinations of polynomials simply involves adding and subtracting the coefficients of like terms. Note also that, by definition, a *trinomial* is a polynomial with three terms and a *binomial* is a polynomial with two terms.

365. a. In general, dividing one polynomial by another will result in an expression involving a term in which the variable is raised to a negative power. For instance, the quotient of even the very simple polynomials 3 and x^2 is $\frac{3}{x^2} = 3x^{-2}$, which is not a polynomial.

366. b. $-(-2x^0)^{-3} + 4^{-2}x^2 - 3^{-1}x - 2$

$= -(-2)^{-3} + \frac{1}{4^2}x^2 - \frac{1}{3}x - 2$

$= -\frac{1}{(-2)^3} + \frac{1}{4^2}x^2 - \frac{1}{3}x - 2$

$= -\frac{1}{-8} + \frac{1}{16}x^2 - \frac{1}{3}x - 2$

$= \frac{1}{16}x^2 - \frac{1}{3}x - \frac{15}{8}$

367. d. $-(2 - (1 - 2x^2 - (2x^2 - 1))) - (3x^2 - (1 - 2x^2))$

$= -(2 - (1 - 2x^2 - 2x^2 + 1)) - (3x^2 - 1 + 2x^2)$

$= -(2 - (2 - 4x^2)) - (5x^2 - 1)$

$= -(2 - 2 + 4x^2) - (5x^2 - 1)$

$= -4x^2 - 5x^2 + 1$

$= -9x^2 + 1$

368. b. $-2^2(2^{-3} - 2^{-2}x^2) + 3^3(3^{-2} - 3^{-3}x^3)$

$= -4\left(\frac{1}{2^3} - \frac{1}{2^2}x^2\right) + 27\left(\frac{1}{3^2} - \frac{1}{3^3}x^3\right)$

$= -4\left(\frac{1}{8} - \frac{1}{4}x^2\right) + 27\left(\frac{1}{9} - \frac{1}{27}x^3\right)$

$= -\frac{1}{2} + x^2 + 3 - x^3$

$= -x^3 + x^2 + \frac{5}{2}$

Set 24 (Page 67)

369. a. $(3x^3)(7x^2) = (3 \cdot 7)(x^3x^2) = 21(x^{3+2}) = 21x^5$

370. c. $2x(5x^2 + 3y) = 2x(5x^2) + 2x(3y) = 10x^3 + 6xy$

371. a. $x^3 + 6x = x \cdot x^2 + 6 \cdot x = x(x^2 + 6)$

372. b. $2x^2(3x + 4xy - 2xy^3) = 2x^2(3x) + 2x^2(4xy) - 2x^2(2xy^3) = 6x^3 + 8x^3y - 4x^3y^3$

373. d. $7x^5(x^8 + 2x^4 - 7x - 9)$

$= 7x^5(x^8) + 7x^5(2x^4) - 7x^5(7x) - 7x^5(9)$

$= (7)(x^5x^8) + (7 \cdot 2)(x^5x^4) - (7 \cdot 7)(x^5x) - (7 \cdot 9)(x^5)$

$= 7x^{13} + 14x^9 - 49x^6 - 63x^5$

374. c. $4x^2z(3xz^3 - 4z^2 + 7x^5)$

$= 4x^2z(3xz^3) + 4x^2z(-4z^2) + 4x^2z(7x^5)$

$= 12x^3z^4 - 16x^2z^3 + 28x^7z$

375. c. To find the product of two binomials, multiply the first term of each binomial, the outside terms, the inside terms, and the last terms (FOIL). Then, add the products:

$(x - 3)(x + 7) = x^2 + 7x - 3x - 21$

$= x^2 + 4x - 21$

376. d. Use FOIL to find the product of two binomials. Then, add the products:

$(x - 6)(x - 6) = x^2 - 6x - 6x + 36$

$= x^2 - 12x + 36$

377. a. Use FOIL to find the product of two binomials. Then, add the products:

$(x - 1)(x + 1) = x^2 + x - x - 1 = x^2 - 1$

378. e. First, note that $(x + c)^2 = (x + c)(x + c)$. Then, use FOIL to find the product of the two binomials. Finally, add the products:

$(x + c)(x + c) = x^2 + cx + cx + c^2$

$= x^2 + 2cx + c^2$

379. b. Use FOIL to find the product of two binomials. Then, add the products:

$(2x + 6)(3x - 9) = 6x^2 - 18x + 18x - 54$

$= 6x^2 - 54$

380. e. Begin by multiplying the first two terms: $-3x(x + 6) = -3x^2 - 18x$. Then, multiply the two binomials, $-3x^2 - 18x$ and $x - 9$:

$(-3x^2 - 18x)(x - 9) = -3x^3 + 27x^2 - 18x^2 + 162x = -3x^3 + 9x^2 + 162x$

381. c. $(x-4)(3x^2+7x-2)$
$= x(3x^2+7x-2)-4(3x^2+7x-2)$
$= x(3x^2)+x(7x)-x(2)-4(3x^2)-4(7x)$
$\quad -4(-2)$
$= 3x^3+7x^2-2x-12x^2-28x+8$
$= 3x^3-5x^2-30x+8$

382. e. Begin by multiplying the first two terms:
$(x-6)(x-3)=x^2-3x-6x+18$
$= x^2-9x+18$
Then, multiply $(x^2-9x+18)$ by $(x-1)$:
$(x^2-9x+18)(x-1)=x^3-9x^2+18x-x^2+$
$9x-18=x^3-10x^2+27x-18$

383. c. First, simplify the left side of the equation:
$(5x+1)(2y+2)=10xy+2y+10x+2$
Now, simplify the equation by rearranging
and combining like terms:
$(5x+1)(2y+2)=10xy+12$
$10xy+2y+10x+2=10xy+12$
$2y+10x+2=12$
$10x+2y=10$
$5x+y=5$

384. a. $(2x^3-2x^2+1)(6x^3+7x^2-5x-9)$
$= 2x^3(6x^3+7x^2-5x-9)-2x^2$
$\quad (6x^3+7x^2-5x-9)+(6x^3+7x^2-5x-9)$
$= 12x^6+14x^5-10x^4-18x^3-12x^5-14x^4+$
$\quad 10x^3+18x^2+6x^3+7x^2-5x-9$
$= 12x^6+(14x^5-12x^5)+(-10x^4-14x^4)+$
$\quad (-18x^3+10x^3+6x^3)+(18x^2+7x^2)$
$\quad -5x-9$
$= 12x^6+2x^5-24x^4-2x^3+25x^2-5x-9$

Set 25 (Page 69)

385. b. $15x-10=5(3x)-5(2)=5(3x-2)$

386. b. $9x^5+24x^2-6x$
$= 3x(3x^4)+3x(8x)-3x(2)$
$= 3x(3x^4+8x-2)$

387. c. $36x^4-90x^3-18x$
$= 18x(2x^3)+18x(-5x^2)+18x(-1)$
$= 18x(2x^3-5x^2-1)$

388. a. $x^3-x=x(x^2)+x(-1)=x(x^2-1)$

389. d. $5x^2+49$ cannot be factored further.

390. a. $36-81x^2=9(4)-9(9x^2)=9(4-9x^2)$

391. c. $125x^3-405x^2=5x^2(25x)+5x^2(-81)$
$= 5x^2(25x-81)$

392. c. $7^3x^3-7^2x^2+7x-49=7(7^2x^3-7x^2+x-7)$
$= 7(49x^3-7x^2+x-7)$

393. b. $5x(2x+3)-7(2x+3)=(2x+3)(5x-7)$

394. c. $5x(6x-5)+7(5-6x)=5x(6x-5)-7(6x-5)$
$= (5x-7)(6x-5)$

395. a. $6(4x+1)-3y(1+4x)+7z(4x+1)$
$= 6(4x+1)-3y(4x+1)+7z(4x+1)$
$= (6-3y+7z)(4x+1)$

396. b. $5x(\frac{2}{3}x+7)-(\frac{2}{3}x+7)=(5x-1)(\frac{2}{3}x+7)$

397. c. $3x(x+5)^2-8y(x+5)^3+7z(x+5)^2$
$= (x+5)^2(3x)+(x+5)^2(-8y(x+5))+$
$\quad (x+5)^2(7z)$
$= (x+5)^2(3x-8y(x+5)+7z)$
$= (x+5)^2(3x-8yx-40y+7z)$

398. a. $8x^4y^2(x-9)^2-16x^3y^5(x-9)^3+12x^5y^3(9-x)$
$= 8x^4y^2(x-9)^2-16x^3y^5(x-9)^3$
$\quad -12x^5y^3(x-9)$
$= 4x^3y^2(x-9)[2x(x-9)]+4x^3y^2(x-9)$
$\quad [-4y^3(x-9)^2]+4x^3y^2(x-9)[-3x^2y]$
$= 4x^3y^2(x-9)[2x(x-9)-4y^3(x-9)-3x^2y]$
$= 4x^3y^2(x-9)[2x^2-18x-4y^3(x^2-18x+81)$
$\quad -3x^2y]$
$= 4x^3y^2(x-9)[2x^2-18x-4y^3x^2+72y^3$
$\quad -324y^3-3x^2y]$

399. c. $8x^4y^2z(2w-1)^3-16x^2y^4z^3(2w-1)^3$
$\quad +12x^4y^4z(2w-1)^4$
$= 4x^2y^2z(2w-1)^3[2x^2]+4x^2y^2z(2w-1)^3$
$\quad [-4y^2z^2]+4x^2y^2z(2w-1)^3[3x^2y^2(2w-1)]$
$= 4x^2y^2z(2w-1)^3[2x^2-4y^2z^2+3x^2y^2$
$\quad (2w-1)]$
$= 4x^2y^2z(2w-1)^3[2x^2-4y^2z^2+6x^2y^2w$
$\quad -3x^2y^2]$

400. b. $-22a^3bc^2(d-2)^3(1-e)^2+55a^2b^2c^2(d-2)^2$
$\quad (1-e)-44a^2bc^4(d-2)(1-e)$
$= 11a^2bc^2(d-2)(1-e)[-2a(d-2)^2(1-e)]$
$\quad +11a^2bc^2(d-2)(1-e)[5b(d-2)]+$
$\quad 11a^2bc^2(d-2)(1-e)[-4c^2]$
$= 11a^2bc^2(d-2)(1-e)[-2a(d-2)^2(1-e)$
$\quad +5b(d-2)-4c^2]$

Set 26 (Page 70)

401. b. $x^2 - 36 = x^2 - 6^2 = (x - 6)(x + 6)$

402. a. $144 - y^2 = 12^2 - y^2 = (12 - y)(12 + y)$

403. d. $4x^2 + 1$ cannot be factored further.

404. b. $9x^2 - 25 = (3x)^2 - (5)^2 = (3x - 5)(3x + 5)$

405. a. $121x^4 - 49z^2 = (11x^2)^2 - (7z)^2 =$
$(11x^2 - 7z)(11x^2 + 7z)$

406. c. $6x^2 - 24 = 6(x^2) - 6(4) = 6(x^2 - 4) = 6((x)^2 - (2)^2) = 6(x - 2)(x + 2)$

407. c. $32x^5 - 162x = 2x(16x^4 - 81)$
$= 2x[(4x^2)^2 - 9^2] = 2x(4x^2 - 9)(4x^2 + 9)$
$= 2x[(2x)^2 - 3^2](4x^2 + 9) = 2x(2x - 3)$
$(2x + 3)(4x^2 + 9)$

408. b. $28x(5 - x) - 7x^3(5 - x) = (28x - 7x^3)(5 - x)$
$= 7x(4 - x^2)(5 - x)$
$= 7x(2^2 - x^2)(5 - x) = 7x(2 - x)(2 + x)(5 - x)$

409. a. $x^2(3x - 5) + 9(5 - 3x) = x^2(3x - 5) - 9(3x - 5) = (x^2 - 9)(3x - 5) = (x - 3)(x + 3)(3x - 5)$

410. a. $x(x^2 + 7x) - 9x^3(x^2 + 7x) = (x - 9x^3)(x^2 + 7x)$
$= [x(1 - 9x^2)][x(x + 7)]$
$= x(1 - (3x)^2) \cdot x(x + 7) = x(1 - 3x)(1 + 3x)$
$x(x + 7) = x^2(1 - 3x)(1 + 3x)(x + 7)$

411. b. $1 + 2x + x^2 = x^2 + 2x + 1 = x^2 + x + x + 1 = (x^2 + x) + (x + 1)$
$= x(x + 1) + (x + 1) = (x + 1)(x + 1)$
$= (x + 1)^2$

412. c. $4x^2 - 12x + 9 = 4x^2 - 6x - 6x + 9 = (4x^2 - 6x) - (6x - 9)$
$= 2x(2x - 3) - 3(2x - 3) = (2x - 3)(2x - 3)$
$= (2x - 3)^2$

413. c. $75x^4 + 30x^3 + 3x^2 = 3x^2[25x^2 + 10x + 1] = 3x^2[25x^2 + 5x + 5x + 1]$
$= 3x^2[5x(5x + 1) + (5x + 1)] = 3x^2[(5x + 1)(5x + 1)] = 3x^2(5x + 1)^2$

414. a. $9x^2(3 + 10x) - 24x(10x + 3) + 16(3 + 10x)$
$= 9x^2(3 + 10x) - 24x(3 + 10x) + 16(3 + 10x)$
$= (3 + 10x)(9x^2 - 24x + 16)$
$= (3 + 10x)(9x^2 - 12x - 12x + 16)$
$= (3 + 10x)(3x(3x - 4) - 4(3x - 4)$
$= (3 + 10x)(3x - 4)(3x - 4)$
$= (3 + 10x)(3x - 4)^2$

415. b. $1 - 6x^2 + 9x^4 = 1 - 3x^2 - 3x^2 + 9x^4$
$= (1 - 3x^2) - 3x^2(1 - 3x^2)$
$= (1 - 3x^2)(1 - 3x^2) = (1 - 3x^2)^2$

416. b. $8x^7 - 24x^4 + 18x = 2x(4x^6 - 12x^3 + 9) =$
$2x[4x^6 - 6x^3 - 6x^3 + 9]$
$= 2x[2x^3(2x^3 - 3) - 3(2x^3 - 3)]$
$= 2x[(2x^3 - 3)(2x^3 - 3)] = 2x(2x^3 - 3)^2$

Set 27 (Page 71)

417. a. $x^2 + 2x - 8 = x^2 + 4x - 2x - 8$
$= (x^2 + 4x) - (2x + 8) = x(x + 4) - 2(x + 4)$
$= (x + 4)(x - 2)$

418. a. $x^2 - 9x + 20 = x^2 - 5x - 4x + 20$
$= (x^2 - 5x) - (4x - 20) = x(x - 5) - 4(x - 5)$
$= (x - 4)(x - 5)$

419. c. $6x^2 + 11x - 2 = 6x^2 - x + 12x - 2$
$= (6x^2 - x) + (12x - 2) = x(6x - 1) + 2(6x - 1)$
$= (x + 2)(6x - 1)$

420. b. $12x^2 - 37x - 10 = 12x^2 + 3x - 40x - 10$
$= 3x(4x + 1) - 10(4x + 1)$
$= (3x - 10)(4x + 1)$

421. c. $7x^2 - 12x + 5 = 7x^2 - 5x - 7x + 5$
$= x(7x - 5) - (7x - 5) = (7x - 5)(x - 1)$

422. a. $9 - 7x - 2x^2 = 9 + 2x - 9x - 2x^2$
$= 1(9 + 2x) - x(9 + 2x) = (9 + 2x)(1 - x)$

423. c. $2x^3 + 6x^2 + 4x = 2x(x^2 + 3x + 2)$
$= 2x(x^2 + x + 2x + 2)$
$= 2x[x(x + 1) + 2(x + 1)] = 2x(x + 2)(x + 1)$

424. c. $-4x^5 + 24x^4 - 20x^3 = -4x^3(x^2 - 6x + 5)$
$= -4x^3(x^2 - x - 5x + 5)$
$= -4x^3[x(x - 1) - 5(x - 1)]$
$= -4x^3(x - 5)(x - 1)$

425. b. $-27x^4 + 27x^3 - 6x^2 = -3x^2(9x^2 - 9x + 2) = -3x^2(9x^2 - 6x - 3x + 2)$
$= -3x^2[(9x^2 - 6x) - (3x - 2)]$
$= -3x^2[3x(3x - 2) - (3x - 2)]$
$= -3x^2(3x - 1)(3x - 2)$

426. b. $x^2(x + 1) - 5x(x + 1) + 6(x + 1)$
$= (x + 1)(x^2 - 5x + 6)$
$= (x + 1)(x^2 - 2x - 3x + 6)$
$= (x + 1)[x(x - 2) - 3(x - 2)]$
$= (x + 1)(x - 3)(x - 2)$

427. a. $2x^2(x^2-4) - x(x^2-4) + (4-x^2)$
$= 2x^2(x^2-4) - x(x^2-4) - (x^2-4)$
$= (x^2-4)[2x^2-x-1]$
$= (x^2-4)[2x^2-2x+x-1]$
$= (x^2-4)[2x(x-1)+(x-1)]$
$= (x^2-4)(2x+1)(x-1)$
$= (x-2)(x+2)(2x+1)(x-1)$

428. b. $27(x-3) + 6x(x-3) - x^2(x-3)$
$= -(x-3)(x^2-6x-27)$
$= -(x-3)(x^2-9x+3x-27)$
$= -(x-3)[x(x-9)+3(x-9)]$
$= -(x-3)(x+3)(x-9)$

429. c. $(x^2+4x+3)x^2 + (x^2+4x+3)3x + 2(x^2+4x+3) = (x^2+4x+3)[x^2+3x+2]$
$= (x^2+3x+x+3)[x^2+x+2x+2]$
$= [x(x+3)+(x+3)][x(x+1)+2(x+1)]$
$= [(x+1)(x+3)][(x+2)(x+1)]$
$= (x+1)^2(x+2)(x+3)$

430. a. $18(x^2+6x+8) - 2x^2(x^2+6x+8)$
$= (x^2+6x+8)[18-2x^2]$
$= (x^2+6x+8)[2(9-x^2)]$
$= (x^2+4x+2x+8)[2(3^2-x^2)]$
$= (x(x+4)+2(x+4))[2(3-x)(3+x)]$
$= 2(x+2)(x+4)(3-x)(3+x)$

431. a. $2x^2(16+x^4) + 3x(16+x^4) + (16+x^4)$
$= (16+x^4)[2x^2+3x+1]$
$= (16+x^4)[2x^2+2x+x+1]$
$= (16+x^4)[2x(x+1)+(x+1)]$
$= (16+x^4)(2x+1)(x+1)$

432. c. $6x^2(1-x^4) + 13x(1-x^4) + 6(1-x^4) = (1-x^4)$
$[6x^2+13x+6] = (1-x^4)[6x^2+4x+9x+6]$
$= (1^2-(x^2)^2)[2x(3x+2)+3(3x+2)]$
$= (1-x^2)(1+x^2)(2x+3)(3x+2)$
$= (1-x)(1+x)(1+x^2)(2x+3)(3x+2)$

Set 28 (Page 73)

433. c. First, factor the polynomial:
$x^2-36 = x^2-6^2 = (x-6)(x+6)$
Now, set each of the factors equal to zero and solve for x to conclude that the zeros of the polynomial are -6 and 6.

434. a. First, factor the polynomial:
$9x^2-25 = (3x)^2-(5)^2 = (3x-5)(3x+5)$
The factors are $3x-5$ and $3x+5$. Now, set each factor equal to zero and solve for x. The zeros of the polynomial are $-\frac{5}{3}$ and $\frac{5}{3}$.

435. d. Note that $5x^2+49$ cannot be factored further. Since both terms are positive, the sum is positive, so there is no x-value that makes the expression equal to zero.

436. b. First, factor the polynomial:
$6x^2-24 = 6(x^2)-6(4) = 6(x^2-4)$
$= 6(x-2)(x+2)$
The factors are 6, $x-2$, and $x+2$. Set each factor equal to zero and solve for x to conclude that the zeros of the polynomial are -2 and 2.

437. d. Begin by factoring, the polynomial:
$5x(2x+3) - 7(2x+3) = (2x+3)(5x-7)$
The factors are $2x+3$ and $5x-7$. Now, set each factor equal to zero and solve for x to find that the zeros of the polynomial are $-\frac{3}{2}$ and $\frac{7}{5}$.

438. d. First, factor the polynomial:
$5x(\frac{2}{3}x+7) - (\frac{2}{3}x+7) = (5x-1)(\frac{2}{3}x+7)$
The factors are $5x-1$ and $\frac{2}{3}x+7$. Now, set each factor equal to zero and solve for x. The zeros of the polynomial are $\frac{1}{5}$ and $-\frac{21}{2}$.

439. a. Begin by factoring the polynomial:
$28x(5-x) - 7x^3(5-x) = (28x-7x^3)(5-x)$
$= 7x(4-x^2)(5-x)$
$= 7x(2^2-x^2)(5-x)$
$= 7x(2-x)(2+x)(5-x)$
There are four factors: $7x$, $2-x$, $2+x$, and $5-x$. Now, set each of these factors equal to zero and solve for x. The zeros of the polynomial are 0, -2, 2, and 5.

440. c. First, factor the polynomial:
$75x^4 + 30x^3 + 3x^2 = 3x^2[25x^2 + 10x + 1]$
$= 3x^2[25x^2 + 5x + 5x + 1]$
$= 3x^2[5x(5x + 1) + (5x + 1)]$
$= 3x^2[(5x + 1)(5x + 1)] = 3x^2(5x + 1)^2$
The factors are $3x^2$ and $(5x + 1)^2$. Now, set each factor equal to zero and solve for x to conclude that the zeros of the polynomial are 0 and $-\frac{1}{5}$.

441. a. First, factor the polynomial:
$x^2 - 9x + 20 = x^2 - 5x - 4x + 20$
$= (x^2 - 5x) - (4x - 20) = x(x - 5) - 4(x - 5)$
$= (x - 4)(x - 5)$
The new factors are $x - 4$ and $x - 5$. Now, set each factor equal to zero and solve for x. The zeros of the polynomial are 4 and 5.

442. c. Begin by factoring the polynomial:
$12x^2 - 37x - 10 = 12x^2 + 3x - 40x - 10$
$= 3x(4x + 1) - 10(4x + 1)$
$= (3x - 10)(4x + 1)$
The factors are $3x - 10$ and $4x + 1$. Now, set each factor equal to zero and solve for x to find that the zeros of the polynomial are $\frac{10}{3}$ and $-\frac{1}{4}$.

443. d. First, factor the polynomial:
$9 - 7x - 2x^2 = 9 + 2x - 9x - 2x^2 = 1(9 + 2x) -$
$x(9 + 2x) = (9 + 2x)(1 - x)$
The factors are $9 + 2x$ and $1 - x$. Set each factor equal to zero and solve for x. The zeros of the polynomial are $-\frac{9}{2}$ and 1.

444. b. Begin by factoring the polynomial:
$2x^3 + 6x^2 + 4x = 2x(x^2 + 3x + 2)$
$= 2x(x^2 + x + 2x + 2)$
$2x[x(x + 1) + 2(x + 1)] = 2x(x + 2)(x + 1)$
There are three factors: $2x$, $x + 2$, and $x + 1$. Set each factor equal to zero and solve for x to conclude that the zeros of the polynomial are -2, -1, and 0.

445. c. First, factor the polynomial:
$-4x^5 + 24x^4 - 20x^3 = -4x^3(x^2 - 6x + 5)$
$= -4x^3(x^2 - x - 5x + 5)$
$= -4x^3[x(x - 1) - 5(x - 1)]$
$= -4x^3(x - 5)(x - 1)$
The three factors are $-4x^3$, $x - 5$, and $x - 1$. Now, set each factor equal to zero and solve for x to find that zeros of the polynomial: 0, 1, and 5.

446. a. First, factor the polynomial:
$2x^2(x^2 - 4) - x(x^2 - 4) + (4 - x^2)$
$= 2x^2(x^2 - 4) - x(x^2 - 4) - (x^2 - 4)$
$= (x^2 - 4)[2x^2 - x - 1]$
$= (x^2 - 4)[2x^2 - 2x + x - 1]$
$= (x^2 - 4)[2x(x - 1) + (x - 1)]$
$= (x^2 - 4)(2x + 1)(x - 1)$
$= (x - 2)(x + 2)(2x + 1)(x - 1)$
Now, set each of the four factors equal to zero and solve for x. The zeros of the polynomial are 1, 2, -2, and $-\frac{1}{2}$.

447. d. Begin by factoring the polynomial:
$2x^2(16 + x^4) + 3x(16 + x^4) + (16 + x^4)$
$= (16 + x^4)[2x^2 + 3x + 1]$
$= (16 + x^4)[2x^2 + 2x + x + 1]$
$= (16 + x^4)[2x(x + 1) + (x + 1)]$
$= (16 + x^4)(2x + 1)(x + 1)$
The three factors are $16 + x^4$, $2x + 1$, and $x + 1$. Now, set each factor equal to zero. Solve for x to find that zeros of the polynomial: -1 and $-\frac{1}{2}$.

448. b. First, factor the polynomial:
$18(x^2 + 6x + 8) - 2x^2(x^2 + 6x + 8)$
$= (x^2 + 6x + 8)[18 - 2x^2]$
$= (x^2 + 6x + 8)[2(9 - x^2)]$
$= (x^2 + 4x + 2x + 8)[2(3^2 - x^2)]$
$= (x(x + 4) + 2(x + 4))[2(3 - x)(3 + x)]$
$= 2(x + 2)(x + 4)(3 - x)(3 + x)$
Set each of the four factors equal to zero and solve for x. The zeros of the polynomial are -4, -2, -3, and 3.

Set 29 (Page 74)

449. b. The strategy is to determine the x-values that make the expression on the left side equal to zero. Doing so requires that we first factor the polynomial:

$$x^2 - 36 = x^2 - 6^2 = (x - 6)(x + 6)$$

Next, set each factor equal to zero and solve for x to conclude that the zeros of the polynomial are –6 and 6. Now, we assess the sign of the expression on the left side on each subinterval formed using these values. To this end, we form a number line, choose a real number in each of the subintervals, and record the sign of the expression above each:

Since the inequality does not include "equals," we do not include those values from the number line that make the polynomial equal to zero. Therefore, the solution set is $(-^\circ, -6) \cup (6, ^\circ)$.

450. b. Determine the x-values that make the expression on the left side equal to zero. First, factor the polynomials:

$$9x^2 - 25 = (3x)^2 - (5)^2 = (3x - 5)(3x + 5)$$

Next, set each factor equal to zero and solve for x to conclude that the zeros of the polynomial are $-\frac{5}{3}$ and $\frac{5}{3}$. Now, assess the sign of the expression on the left side on each subinterval formed using these values. To this end, form a number line, choose a real number in each of the subintervals, and record the sign of the expression above each:

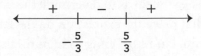

Since the inequality includes "equals," include those values from the number line that make the polynomial equal to zero. The solution set is $[-\frac{5}{3}, \frac{5}{3}]$.

451. c. Determine the x-values that make the expression on the left side equal to zero. Begin by factoring the polynomial, if possible. However, note that $5x^2 + 49$ cannot be factored further. Moreover, since both terms are positive for any value of x, the sum is positive for every value of x. Therefore, the solution set is the empty set.

452. a. Find the x-values that make the expression on the left side equal to zero. First, factor the polynomial:

$$6x^2 - 24 = 6(x^2) - 6(4) = 6(x^2 - 4) =$$
$$6((x)^2 - (2)^2) = 6(x - 2)(x + 2)$$

Next, set each factor equal to zero. Solve for x to find that the zeros of the polynomial: –2 and 2. Now, assess the sign of the expression on the left side on each subinterval formed using these values. Form a number line, choose a real number in each of the subintervals, and record the sign of the expression above each:

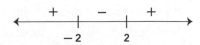

Since the inequality includes "equals," we include those values from the number line that make the polynomial equal to zero. The solution set is $(-^\circ, -2] \cup [2, ^\circ)$.

453. a. The strategy is to determine the x-values that make the expression on the left side equal to zero. Doing so requires that we first factor the polynomial:

$5x(2x + 3) - 7(2x + 3) = (2x + 3)(5x - 7)$

Set each factor equal to zero and solve for x. The zeros of the polynomial: $-\frac{3}{2}$ and $\frac{7}{5}$. Next assess the sign of the expression on the left side on each subinterval formed using these values. Form a number line, choose a real number in each of the duly formed subintervals, and record the sign of the expression above each:

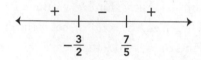

Since the inequality does not include "equals," do not include those values from the number line that make the polynomial equal to zero. As such, the solution set is $(-\infty, -\frac{3}{2}) \cup (\frac{7}{5}, \infty)$.

454. c. Find the x-values that make the expression on the left side equal to zero. First, factor the polynomial:

$5x(\frac{2}{3}x + 7) - (\frac{2}{3}x + 7) = (5x - 1)(\frac{2}{3}x + 7)$

Next, set each factor equal to zero and solve for x. The zeros of the polynomial are $\frac{1}{5}$ and $-\frac{21}{2}$. Assess the sign of the expression on the left side on each subinterval formed using these values. Form a number line, choose a real number in each of the subintervals, and record the sign of the expression above each:

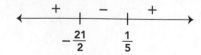

The inequality includes "equals," so we include those values from the number line that make the polynomial equal to zero. The solution set is $[-\frac{21}{2}, \frac{1}{5}]$.

455. a. Determine the x-values that make the expression on the left side equal to zero. To do this, factor the polynomial:

$28x(5 - x) - 7x^3(5 - x) = (28x - 7x^3)(5 - x)$
$= 7x(4 - x^2)(5 - x) = 7x(2^2 - x^2)(5 - x)$
$= 7x(2 - x)(2 + x)(5 - x)$

Next, set each factor equal to zero. Solve for x to find the zeros of the polynomial, which are 0, −2, 2, and 5. Now, assess the sign of the expression on the left side on each subinterval formed using these values. To this end, we form a number line, choose a real number in each of the subintervals, and record the sign of the expression above each:

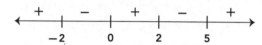

Since the inequality includes "equals," include the values from the number line that make the polynomial equal to zero. The solution set is $(-\infty, -2] \cup [0, 2] \cup [5, \infty)$.

456. c. First, determine the x-values that make the expression on the left side equal to zero. Doing so requires that we factor the polynomial:

$75x^4 + 30x^3 + 3x^2 = 3x^2[25x^2 + 10x + 1]$
$= 3x^2[25x^2 + 5x + 5x + 1]$

$= 3x^2[5x(5x + 1) + (5x + 1)]$
$= 3x^2[(5x + 1) + (5x + 1)] = 3x^2(5x + 1)^2$

Set each factor equal to zero, then solve for x to find the zeros of the polynomial: 0 and $-\frac{1}{5}$. Assess the sign of the expression on the left side on each subinterval formed using these values: Form a number line, choose a real number in each subinterval, and record the sign of the expression above each:

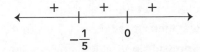

The inequality includes "equals," so we include those values from the number line that make the polynomial equal to zero. Since every x-value that is not a zero of the polynomial results in a positive quantity, the solution set consists of only the zeros of the polynomial, namely $\{-\frac{1}{5}, 0\}$.

457. d. Find the x-values that make the expression on the left side equal to zero. Begin by factoring the polynomial:
$x^2 - 9x + 20 = x^2 - 5x - 4x + 20$
$= (x^2 - 5x) - (4x - 20) = x(x - 5) - 4(x - 5)$
$= (x - 4)(x - 5)$
Set each factor equal to zero, then solve for x to find the zeros of the polynomial, which are 4 and 5. Now, assess the sign of the expression on the left side on each subinterval formed using these values. To this end, form a number line, choose a real number in each subinterval, and record the sign of the expression above each:

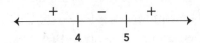

The inequality does not include "equals," so we exclude those values from the number line that make the polynomial equal to zero. Therefore, the solution set is (4,5).

458. d. First, find the x-values that make the expression on the left side equal to zero. Doing so requires that we factor the polynomial:
$12x^2 - 37x - 10 = 12x^2 + 3x - 40x - 10 =$
$3x(4x + 1) - 10(4x + 1) = (3x - 10)(4x + 1)$

Next, set each factor equal to zero and solve for x to conclude that the zeros of the polynomial are $\frac{10}{3}$ and $-\frac{1}{4}$. Now, we assess the sign of the expression on the left side on each subinterval formed using these values. To this end, we form a number line, choose a real number in each subinterval, and record the sign of the expression above each, as follows:

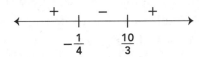

Because the inequality does not include "equals," we exclude those values from the number line that make the polynomial equal to zero. The solution set is $(-\frac{1}{4}, \frac{10}{3})$.

459. d. Determine the x-values that make the expression on the left side equal to zero. To do this, we first factor the polynomial:
$9 - 7x - 2x^2 = 9 + 2x - 9x - 2x^2$
$= 1(9 + 2x) - x(9 + 2x) = (9 + 2x)(1 - x)$
Next, set each factor equal to zero and solve for x to find the zeros of the polynomial which are $-\frac{9}{2}$ and 1. Now, we assess the sign of the expression on the left side on each subinterval formed using these values. Form a number line, choose a real number in each subinterval, and record the sign of the expression above each:

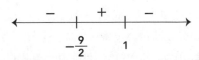

The inequality does not include "equals," so we do not include those values from the number line that make the polynomial equal to zero. The solution set is $(-\frac{9}{2}, 1)$.

460. b. The strategy is to determine the x-values that make the expression on the left side equal to zero. First, factor the polynomial:
$2x^3 + 6x^2 + 4x = 2x(x^2 + 3x + 2) =$
$2x(x^2 + x + 2x + 2)$
$= 2x[x(x + 1) + 2(x + 1)] = 2x(x + 2)(x + 1)$
Next, set each factor equal to zero and solve for x to conclude that the zeros of the polynomial are –2, –1, and 0. Now, we assess the sign of the expression on the left side on each subinterval formed using these values. To this end, we form a number line, choose a real number in each subinterval, and record the sign of the expression above each:

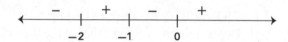

Since the inequality includes "equals," we include those values from the number line that make the polynomial equal to zero. The solution set is $[-2, -1] \cup [0, °)$.

461. a. Find the x-values that make the expression on the left side equal to zero. First, factor the polynomial:
$-4x^5 + 24x^4 - 20x^3 = -4x^3(x^2 - 6x + 5)$
$= -4x^3(x^2 - x - 5x + 5)$
$= -4x^3[x(x - 1) - 5(x - 1)]$
$= -4x^3(x - 5)(x - 1)$
Next, set each factor equal to zero and solve for x. The zeros of the polynomial are 0, 1, and 5. Now, we assess the sign of the expression on the left side on each subinterval formed using these values: We form a number line, choose a real number in each subinterval, and record the sign of the expression above each:

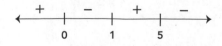

The inequality includes "equals," so we include those values from the number line that make the polynomial equal to zero. The solution set is $(-°, 0] \cup [1, 5)$.

462. a. First, determine the x-values that make the expression on the left side equal to zero. This requires that we factor the polynomial:
$2x^2(x^2 - 4) - x(x^2 - 4) + (4 - x^2)$
$= 2x^2(x^2 - 4) - x(x^2 - 4) - (x^2 - 4)$
$= (x^2 - 4)[2x^2 - x - 1]$
$= (x^2 - 4)[2x^2 - 2x + x - 1]$
$= (x^2 - 4)[2x(x - 1) + (x - 1)]$
$= (x^2 - 4)(2x + 1)(x - 1)$
$= (x - 2)(x + 2)(2x + 1)(x - 1)$
Set each factor equal to zero and solve for x to find the zeros of the polynomial are 1, 2, –2, and $-\frac{1}{2}$. Assess the sign of the expression on the left side on each subinterval formed using these values. To this end, form a number line, choose a real number in each subinterval, and record the sign of the expression above each:

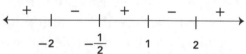

The inequality does not include "equals," so we exclude those values from the number line that make the polynomial equal to zero. The solution set is $(-2, -\frac{1}{2}) \cup (1, 2)$.

463. c. Determine the x-values that make the expression on the left side equal to zero. First, factor the polynomial:
$2x^2(16 + x^4) + 3x)(16 + x^4) + (16 + x^4)$
$= (16 + x^4)[2x^2 + 3x + 1]$
$= (16 + x^4)[2x^2 + 2x + x + 1]$
$= (16 + x^4)[2x(x + 1) + (x + 1)]$
$= (16 + x^4)(2x + 1)(x + 1)$
Set each factor equal to zero and solve for x. The zeros of the polynomial are –1 and $-\frac{1}{2}$. Assess the sign of the expression on the left side on each subinterval formed using these values. To this end, form a number line,

choose a real number in each of the subinterval, and record the sign of the expression above each, as follows:

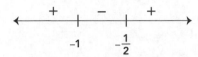

The inequality includes "equals," so we include those values from the number line that make the polynomial equal to zero. Therefore, the solution set is $[-1, -\frac{1}{2}]$.

464. b. Find the x-values that make the expression on the left side equal to zero. First, factor the polynomial:

$18(x^2 + 6x + 8) - 2x^2(x^2 + 6x + 8)$
$= (x^2 + 6x + 8)[18 - 2x^2]$
$= (x^2 + 6x + 8)[2(9 - x^2)]$
$= (x^2 + 4x + 2x + 8)[2(3^2 - x^2)]$
$= [x(x + 4) + 2(x + 4)][2(3 - x)(3 + x)]$
$= 2(x + 2)(x + 4)(3 - x)(3 + x)$

Set each factor equal to zero and solve for x to find the zeros of the polynomial, which are $-4, -2, -3$, and 3. Assess the sign of the expression on the left side on each subinterval formed using these values. To this end, form a number line, choose a real number in each subinterval, and record the sign of the expression above each:

Because the inequality does not include "equals," we exclude those values from the number line that make the polynomial equal to zero. The solution set is $(-4,-3)\cup(-2,3)$.

Section 4—Rational Expressions
Set 30 (Page 78)

465. d. $\frac{2z^2 - z - 15}{z^2 + 2z - 15} = \frac{(2z + 5)(z - 3)}{(z + 5)(z - 3)} = \frac{2z + 5}{z + 5}$

466. d. $\frac{25(-x)^4}{x(5x^2)^2} = \frac{25x^4}{x \cdot 25x^4} = \frac{1}{x}$

467. a. $\frac{z^3 - 16z}{8z - 32} = \frac{z(z^2 - 16)}{8(z - 4)} = \frac{z(z-4)(z+4)}{8(z-4)} = \frac{z(z + 4)}{8}$

468. b. $\frac{y^2 - 64}{8 - y} = \frac{(y - 8)(y + 8)}{8 - y} = \frac{(y - 8)(y + 8)}{-(y - 8)} = -(y + 8)$

469. a. $\frac{x^2 + 8x}{x^3 - 64x} = \frac{x(x + 8)}{x(x^2 - 64)} = \frac{x(x + 8)}{x(x + 8)(x - 8)} = \frac{1}{x - 8}$

470. c. $\frac{2x^2 + 4x}{4x^3 - 16x^2 - 48x} = \frac{2x(x + 2)}{4x(x^2 - 4x - 12)} =$

$\frac{2x(x + 2)}{4x(x - 6)(x + 2)} = \frac{1}{2(x - 6)} = \frac{1}{2x - 12}$

471. a. A rational expression is undefined at any value of x that makes the denominator equal to zero even if the corresponding factor cancels with one in the numerator. Observe that the denominator factors as
$4x^3 + 44x^2 + 120x = 4x(x^2 + 11x + 30)$
$= 4x(x + 5)(x + 6)$.
Setting each factor equal to zero shows that the rational expression is undefined at $x = 0$, -5, and -6.

472. c. The domain of a rational expression is the set of all real numbers that do not make the denominator equal to zero. For this expression, the values of x that must be excluded from the domain are the solutions of the equation $x^3 - 4x = 0$. Factoring the left side yields the equivalent equation $x^3 - 4x = x(x^2 - 4) = x(x - 2)(x + 2) = 0$ The solutions are $x = -2, 0$, and 2. Hence, the expression is defined for any x in the set $(-\infty,-2)\cup(-2,0)\cup(0,2)\cup(2,\infty)$.

473. d. $\dfrac{x^2-16}{x^3+x^2-20x} = \dfrac{(x-4)(x+4)}{x(x^2+x-20)} = \dfrac{(x-4)(x+4)}{x(x+5)(x-4)}$

$= \dfrac{x+4}{x^2+5x}$

474. e. $\dfrac{x}{4x} = \dfrac{1}{4} = \dfrac{5}{20}$, provided that x is not zero.

475. b. Any value of x that makes the denominator equal to zero, even if it also makes the numerator equal to zero, renders a rational expression undefined. For the given expression, both 4 and -4 make the denominator equal to zero.

476. b. Any value of x that makes the denominator equal to zero, even if it also makes the numerator equal to zero, renders a rational expression undefined. To determine these values for the given expression, we factor the denominator as $x^3+3x^2-4x = x(x^2+3x-4) = x(x+4)(x-1)$. Note that the values -4, 0, and 1 all make the given expression undefined.

477. b. $\dfrac{5x^2(x-1)-3x(x-1)-2(x-1)}{10x^2(x-1)+9x(x-1)+2(x-1)} =$

$\dfrac{(x-1)(5x^2-3x-2)}{(x-1)(10x^2+9x+2)} = \dfrac{(x-1)(5x+2)(x-1)}{(x-1)(5x+2)(2x+1)} =$

$\dfrac{x-1}{2x+1}$

478. b. $\dfrac{6x^3-12x}{24x^2} = \dfrac{6x(x^2-2)}{4\cdot 6\cdot x\cdot x} = \dfrac{x^2-2}{4x}$

479. a. $\dfrac{4ab^2-b^2}{8a^2+2a-1} = \dfrac{b^2(4a-1)}{(4a-1)(2a+1)} = \dfrac{b^2}{2a+1}$

480. c. $\dfrac{(2x-5)(x+4)-(2x-5)(x+1)}{9(2x-5)}$

$= \dfrac{(2x-5)[(x+4)-(x+1)]}{9(2x-5)} = \dfrac{3}{9} = \dfrac{1}{3}$

Set 31 (Page 79)

481. a. $\dfrac{4x-45}{x-9} + \dfrac{2x-9}{x-9} - \dfrac{3x+1}{x-9} =$

$\dfrac{(4x-45)+(2x-9)-(3x+1)}{x-9} =$

$\dfrac{4x-45+2x-9-3x-1}{x-9} = \dfrac{3x-55}{x-9}$

482. a. $\dfrac{5a}{ab^3} + \dfrac{2a}{ab^3} = \dfrac{5a+2a}{ab^3} = \dfrac{7a}{ab^3} = \dfrac{7}{b^3}$

483. d. $\dfrac{3-2x}{(x+2)(x-1)} - \dfrac{2-x}{(x-1)(x+2)} = \dfrac{3-2x-(2-x)}{(x-1)(x+2)} =$

$\dfrac{1-x}{(x-1)(x+2)} = \dfrac{-(x-1)}{(x-1)(x+2)} = -\dfrac{1}{x+2}$

484. d. $\dfrac{4}{sr^3} + \dfrac{2}{rs^2} = \dfrac{4s}{s^2r^3} + \dfrac{2r^2}{s^2r^3} = \dfrac{4s+2r^2}{s^2r^3} = \dfrac{2(2s+r^2)}{s^2r^3}$

485. c. $\dfrac{2}{x(x-2)} - \dfrac{5-2x}{(x-2)(x-1)} = \dfrac{2}{x(x-2)} \cdot \dfrac{(x-1)}{(x-1)} -$

$\dfrac{5-2x}{(x-2)(x-1)} \cdot \dfrac{x}{x} = \dfrac{2(x-1)-x(5-2x)}{x(x-1)(x-2)} =$

$\dfrac{2x-2-5x+2x^2}{x(x-1)(x-2)} = \dfrac{2x^2-3x-2}{x(x-1)(x-2)}$

486. b. $\dfrac{4}{t(t+2)} - \dfrac{2}{t} = \dfrac{4}{t(t+2)} - \dfrac{2(t+2)}{t(t+2)} = \dfrac{4-2(t+2)}{t(t+2)}$

$= \dfrac{-2t}{t(t+2)} = \dfrac{-2}{t+2}$

487. b. $\dfrac{1}{x(x+1)} - \dfrac{2x}{(x+1)(x+2)} + \dfrac{3}{x} = \dfrac{1}{x(x+1)} \cdot \dfrac{(x+2)}{(x+2)}$

$- \dfrac{2x}{(x+1)(x+2)} \cdot \dfrac{x}{x} + \dfrac{3}{x} \cdot \dfrac{(x+1)(x+2)}{(x+1)(x+2)} =$

$\dfrac{x+2-2x^2+3(x+1)(x+2)}{x(x+1)(x+2)} =$

$\dfrac{x+2-2x^2+3x^2+3x+6x+6}{x(x+1)(x+2)} = \dfrac{x^2+10x+8}{x(x+1)(x+2)}$

488. b. $\dfrac{x}{2x+1} - \dfrac{1}{2x-1} + \dfrac{2x^2}{4x^2-1} = \dfrac{x}{2x+1} \cdot \dfrac{2x-1}{2x-1} -$

$\dfrac{1}{2x-1} \cdot \dfrac{2x+1}{2x+1} + \dfrac{2x^2}{4x^2-1} =$

$\dfrac{x(2x-1)-1(2x+1)+2x^2}{(2x-1)(2x+1)} = \dfrac{2x^2-x-2x-1+2x^2}{(2x-1)(2x+1)} =$

$\dfrac{4x^2-3x-1}{(2x-1)(2x+1)} = \dfrac{(4x+1)(x-1)}{(2x-1)(2x+1)}$

489. c. $\dfrac{3y+2}{(y-1)^2} - \dfrac{7y-3}{(y-1)(y+1)} + \dfrac{5}{y+1} = \dfrac{3y+2}{(y-1)^2} \cdot \dfrac{y+1}{y+1}$

$-\dfrac{7y-3}{(y-1)(y+1)} \cdot \dfrac{(y-1)}{(y-1)} + \dfrac{5}{y+1} \cdot \dfrac{(y-1)^2}{(y-1)^2} =$

$\dfrac{(3y+2)(y+1) - (7y-3)(y-1) + 5(y-1)^2}{(y-1)^2(y+1)} =$

$\dfrac{3y^2 + 5y + 2 - (7y^2 - 10y + 3) + 5(y^2 - 2y + 1)}{(y-1)^2(y+1)} =$

$\dfrac{3y^2 + 5y + 2 - 7y^2 + 10y - 3 + 5y^2 - 10y + 5}{(y-1)^2(y+1)} =$

$\dfrac{y^2 + 5y + 4}{(y-1)^2(y+1)} = \dfrac{(y+4)(y+1)}{(y-1)^2(y+1)} = \dfrac{y+4}{(y-1)^2}$

490. a. $\left(\dfrac{6z+12}{4z+3} + \dfrac{2z-6}{4z+3}\right)^{-1} = \left(\dfrac{6z+12+2z-6}{4z+3}\right)^{-1} =$

$\left(\dfrac{8z+6}{4z+3}\right)^{-1} = \left(\dfrac{2(4z+3)}{4z+3}\right)^{-1} = 2^{-1} = \dfrac{1}{2}$

491. a. $\dfrac{4}{x-3} + \dfrac{x+5}{3-x} = \dfrac{4}{x-3} + \dfrac{x+5}{-(x-3)} = \dfrac{4}{x-3} - \dfrac{x+5}{x-3}$

$= \dfrac{4 - (x+5)}{x-3} = -\dfrac{x+1}{x-3}$

492. a. $\dfrac{x}{x^2 - 10x + 24} - \dfrac{3}{x-6} + 1 = \dfrac{x}{(x-6)(x-4)} - \dfrac{3}{x-6}$

$+1 = \dfrac{x}{(x-6)(x-4)} - \dfrac{3(x-4)}{(x-6)(x-4)} + \dfrac{(x-6)(x-4)}{(x-6)(x-4)} =$

$= \dfrac{x - 3(x-4) + (x-6)(x-4)}{(x-6)(x-4)}$

$= \dfrac{x - 3x + 12 + x^2 - 10x + 24}{(x-6)(x-4)}$

$= \dfrac{x^2 - 12x + 36}{(x-6)(x-4)} = \dfrac{(x-6)(x-6)}{(x-6)(x+4)} = \dfrac{x-6}{x+4}$

493. d. $\dfrac{-x^2 + 5x}{(x-5)^2} + \dfrac{x+1}{x+5} = \dfrac{-x(x-5)}{(x-5)^2} + \dfrac{x+1}{x+5} =$

$\dfrac{-x}{x-5} + \dfrac{x+1}{x+5} = \dfrac{-x(x+5)}{(x-5)(x+5)} + \dfrac{(x+1)(x-5)}{(x+5)(x-5)} =$

$\dfrac{-x(x+5) + (x+1)(x-5)}{(x-5)(x+5)} = \dfrac{-x^2 - 5x + x^2 - 4x - 5}{(x-5)(x+5)} =$

$\dfrac{5 - 9x}{(x-5)(x+5)}$

494. b. $\dfrac{2x^2}{x^4 - 1} - \dfrac{1}{x^2 - 1} + \dfrac{1}{x^2 + 1} = \dfrac{2x^2}{(x^2-1)(x^2+1)} - \dfrac{1}{x^2-1}$

$+ \dfrac{1}{x^2 + 1} = \dfrac{2x^2}{(x^2-1)(x^2+1)} - \dfrac{x^2+1}{(x^2-1)(x^2+1)} +$

$\dfrac{x^2-1}{(x^2-1)(x^2+1)} = \dfrac{2x^2 - x^2 - 1 + x^2 - 1}{(x^2-1)(x^2+1)} =$

$\dfrac{2x^2 - 2}{(x^2-1)(x^2+1)} = \dfrac{2(x^2-1)}{(x^2-1)(x^2+1)} = \dfrac{2}{x^2+1}$

495. c. $\dfrac{x-1}{x-2} - \dfrac{3x-4}{x^2-2x} = \dfrac{x-1}{x-2} - \dfrac{3x-4}{x(x-2)} = \dfrac{x(x-1)}{x(x-2)} -$

$\dfrac{3x-4}{x(x-2)} = \dfrac{x(x-1) - (3x-4)}{x(x-2)} = \dfrac{x^2 - x - 3x + 4}{x(x-2)} =$

$\dfrac{x^2 - 4x + 4}{x(x-2)} = \dfrac{(x-2)^2}{x(x-2)} = \dfrac{x-2}{x}$

496. c. $1 + \dfrac{x-1}{x} - \dfrac{3x-3}{x^2+3x} = 1 + \dfrac{x-1}{x} - \dfrac{3x-3}{x(x+3)} =$

$\dfrac{x(x+3)}{x(x+3)} + \dfrac{(x-1)(x+3)}{x(x+3)} - \dfrac{3x-3}{x(x+3)} =$

$\dfrac{x(x+3) + (x-1)(x+3) - (3x-3)}{x(x+3)} =$

$\dfrac{x^2 + 3x + x^2 + 2x - 3 - 3x + 3}{x(x+3)} =$

$\dfrac{2x^2 + 2x}{x(x+3)} = \dfrac{2x(x+1)}{x(x+3)} = \dfrac{2(x+1)}{x+3}$

Set 32 (Page 81)

497. c. $\dfrac{4x^3y^2}{z^3} \cdot \dfrac{y^3z^4}{2x^5} = \dfrac{4x^3y^5z^4}{2x^5z^3} = \dfrac{2y^5z}{x^2}$

498. d. $\dfrac{8a^4}{9 - a^2} \cdot \dfrac{5a^2 + 13a - 6}{24a - 60a^2} = \dfrac{8a^4}{(3-a)(3+a)} \cdot$

$\dfrac{(5a-2)(a+3)}{12a(2-5a)} = \dfrac{8a^4}{(3-a)(3+a)} \cdot \dfrac{(5a-2)(a+3)}{12a(5a-2)} =$

$\dfrac{2a^3}{-3(3-a)}$

499. b. $\dfrac{9x-2}{8-4x} \cdot \dfrac{10-5x}{2-9x} = \dfrac{9x-2}{4(2-x)} \cdot \dfrac{5(2-x)}{-(9x-2)} = \dfrac{5}{-4}$

500. a. $\dfrac{12x^2y}{-18xy} \cdot \dfrac{-24xy^2}{56y^3} = \dfrac{(12)(-24)x^3y^3}{(-18)(56)xy^4} = \dfrac{2x^2}{7y}$

501. a. $\dfrac{x^2 - x - 12}{3x^2 - x - 2} \div (3x^2 - 10x - 8) = \dfrac{x^2 - x - 12}{3x^2 - x - 2} \cdot$

$\dfrac{1}{3x^2 - 10x - 8} = \dfrac{(x-4)(x+3)}{(3x+2)(x-1)} \cdot \dfrac{1}{(3x+2)(x-4)} =$

$\dfrac{x+3}{(3x+2)^2(x-1)}$

502. c. $\dfrac{x-3}{2x^3} \div \dfrac{x^2 - 3x}{4x} = \dfrac{x-3}{2x^3} \cdot \dfrac{4x}{x(x-3)} = \dfrac{2}{x^3}$

503. d. $\dfrac{x^2 - 64}{x^2 - 9} \div \dfrac{6x^2 + 48x}{2x - 6} = \dfrac{x^2 - 64}{x^2 - 9} \cdot \dfrac{2x - 6}{6x^2 + 48x} =$

$\dfrac{(x-8)(x+8) \cdot 2(x-3)}{(x-3)(x+3) \cdot 6x(x+8)} = \dfrac{x-8}{3x(x+3)}$

504. a. $\dfrac{2(x-6)^2}{x+5} \cdot \dfrac{-(5+x)}{4(x-6)} = \dfrac{-(x-6)}{2}$

505. b. $\dfrac{9x^2y^3}{14x} \cdot \dfrac{21y}{15xy^2} \cdot \dfrac{10x}{12y^3} = \dfrac{(9)(21)(10)x^3y^4}{(14)(15)(12)x^2y^5} = \dfrac{3x}{4y}$

506. b. $\dfrac{4x^2 + 4x + 1}{4x^2 - 4x} \div \dfrac{2x^2 + 3x + 1}{2x^2 - 2x} = \dfrac{4x^2 + 4x + 1}{4x^2 - 4x} \cdot$

$\dfrac{2x^2 - 2x}{2x^2 + 3x + 1} = \dfrac{(2x+1)^2}{4x(x-1)} \cdot \dfrac{2x(x-1)}{(2x+1)(x+1)} =$

$\dfrac{(2x+1)}{2(x-1)(x+1)}$

507. b. $\dfrac{x^2 - 1}{x^2 + x} \cdot \dfrac{2x + 2}{1 - x^2} \cdot \dfrac{x^2 + x - 2}{x^2 - x} = \dfrac{(x-1)(x+1)}{x(x+1)} \cdot$

$\dfrac{2(x+1)}{(1-x)(1+x)} \cdot \dfrac{(x+2)(x-1)}{x(x-1)} = \dfrac{(x-1)}{x} \cdot \dfrac{2}{-(x-1)} \cdot$

$\dfrac{(x+2)}{x} = \dfrac{-2(x+2)}{x^2}$

508. c. $(4x^2 - 8x - 5) \div \left[\dfrac{-(x-3)}{x+1} \cdot \dfrac{2x^2 - 3x - 5}{x - 3} \right] =$

$(2x + 1)(2x - 5) \div \left[\dfrac{-(x-3)}{x+1} \cdot \dfrac{(2x-5)(x+1)}{x-3} \right] =$

$(2x + 1)(2x - 5) \div [-(2x - 5)] =$

$(2x + 1)(2x - 5) \cdot \dfrac{1}{-(2x-5)} = -(2x + 1)$

509. a. $\dfrac{a^2 - b^2}{2a^2 - 3ab + b^2} \cdot \left[\dfrac{2a^2 - 7ab + 3b^2}{a^2 + ab} \div \dfrac{ab - 3b^2}{a^2 + 2ab + b^2} \right] =$

$\dfrac{a^2 - b^2}{2a^2 - 3ab + b^2} \cdot \left[\dfrac{2a^2 - 7ab + 3b^2}{a^2 + ab} \cdot \dfrac{a^2 + 2ab^2 + b^2}{ab - 3b^2} \right] =$

$\dfrac{(a-b)(a+b)}{(2a-b)(a-b)} \cdot \dfrac{(2a-b)(a-3b)}{a(a+b)} \cdot \dfrac{(a+b)(a+b)}{b(a-3b)} =$

$\dfrac{(a+b)}{(2a-b)} \cdot \dfrac{(2a-b)(a-3b)}{a(a+b)} \cdot \dfrac{(a+b)(a+b)}{b(a-3b)} =$

$\dfrac{(a+b)^2}{ab}$

510. a. $(x - 3) \div \dfrac{x^2 + 3x - 18}{x} = (x - 3) \cdot \dfrac{x}{x^2 + 3x - 18} =$

$(x - 3) \cdot \dfrac{x}{(x-3)(x+6)} = \dfrac{x}{x + 6}$

511. b. $\left[\dfrac{x^2 - x}{4y} \cdot \dfrac{10xy^2}{2x - 2} \right] \div \dfrac{3x^2 + 3x}{15x^2 y^2} = \dfrac{x^2 - x}{4y} \cdot \dfrac{10xy^2}{2x - 2} \cdot$

$\dfrac{15x^2 y^2}{3x^2 + 3x} = \dfrac{x(x-1)}{4y} \cdot \dfrac{10xy^2}{2(x-1)} \cdot \dfrac{15x^2 y^2}{3x(x+1)} =$

$\dfrac{25x^4 y^4}{4xy(x+1)} = \dfrac{25 x^3 y^3}{4(x+1)}$

512. d. $\dfrac{x+2}{x^2 + 5x + 6} \cdot \dfrac{2x^2 + 7x + 3}{4x^2 + 4x + 1} \cdot \dfrac{6x^2 + 5x + 1}{3x^2 + x} \cdot \dfrac{x^2 - 4}{x^2 + 2x} =$

$\dfrac{x+2}{(x+2)(x+3)} \cdot \dfrac{(2x+1)(x+3)}{(2x+1)(2x+1)} \cdot \dfrac{(3x+1)(2x+1)}{x(3x+1)} \cdot$

$\dfrac{(x-2)(x+2)}{x(x+2)} = \dfrac{x-2}{x^2}$

Set 33 (Page 82)

513. b. $1 - \dfrac{\frac{3}{4}}{\frac{9}{16}} \cdot \dfrac{1}{4} + \left[\dfrac{5}{2} - \dfrac{1}{4} \right]^2 = 1 - \left(\dfrac{3}{4} \div \dfrac{9}{16} \right) \cdot$

$\dfrac{1}{4} + \left[\dfrac{10 - 1}{4} \right]^2 = 1 - \left(\dfrac{3}{4} \div \dfrac{9}{16} \right) \cdot \dfrac{1}{4} + \dfrac{81}{16} =$

$1 - \left(\dfrac{3}{4} \cdot \dfrac{16}{9} \right) \cdot \dfrac{1}{4} + \dfrac{81}{16} = 1 - \left(1 \cdot \dfrac{4}{3} \right) \cdot \dfrac{1}{4} + \dfrac{81}{16} =$

$1 - \dfrac{1}{3} + \dfrac{81}{16} = \dfrac{48 - 16 + 243}{48} = \dfrac{275}{48}$

514. b. $\dfrac{\frac{2}{3} + \frac{3}{4}}{\frac{3}{4} - \frac{1}{2}} = \dfrac{\frac{8}{12} + \frac{9}{12}}{\frac{3}{4} - \frac{2}{4}} = \dfrac{\frac{17}{12}}{\frac{1}{4}} = \dfrac{17}{12} \cdot 4 = \dfrac{17}{3}$

515. b. $\left[\dfrac{3x^2 + 6x}{x - 5} + \dfrac{2 + x}{5 - x} \right] \div \dfrac{3x - 1}{25 - x^2} =$

$\left[\dfrac{3x(x+2)}{x - 5} - \dfrac{x + 2}{x - 5} \right] \cdot \dfrac{(5-x)(5+x)}{3x - 1} =$

$\left[\dfrac{3x(x+2) - (x+2)}{x - 5} \right] \cdot \dfrac{-(x-5)(x+5)}{3x - 1} =$

$\dfrac{(3x-1)(x+2)}{x - 5} \cdot \dfrac{-(x-5)(x+5)}{3x - 1}$

$= -(x + 2)(x + 5)$

516. d. $\left[\dfrac{1}{(x+h)^2} - \dfrac{1}{x^2} \right] \div h =$

$\left[\dfrac{1}{(x+h)^2} \cdot \dfrac{x^2}{x^2} - \dfrac{1}{x^2} \cdot \dfrac{(x+h)^2}{(x+h)^2} \right] \div h =$

$\left[\dfrac{x^2 - (x+h)^2}{x^2 (x+h)^2} \right] \div h = \dfrac{x^2 - x^2 - 2hx - h^2}{x^2 (x+h)^2} \div h =$

$\dfrac{-h(2x + h)}{x^2 (x+h)^2} \cdot \dfrac{1}{h} = \dfrac{-(2x + h)}{x^2 (x+h)^2}$

517. d. $\dfrac{a + \frac{1}{b}}{b + \frac{1}{a}} = \dfrac{\frac{ab}{b} + \frac{1}{b}}{\frac{ba}{a} + \frac{1}{a}} = \dfrac{ab + 1}{b} \cdot \dfrac{a}{ba + 1} = \dfrac{a}{b}$

518. a. $\dfrac{\frac{3}{x}-\frac{1}{2}}{\frac{5}{4x}-\frac{1}{2x}} = \dfrac{4x\left[\frac{3}{x}-\frac{1}{2}\right]}{4x\left[\frac{5}{4x}-\frac{1}{2x}\right]} =$

$\dfrac{4x\left(\frac{3}{x}\right)-4x\left(\frac{1}{2}\right)}{4x\left(\frac{5}{4x}\right)-4x\left(\frac{1}{2x}\right)} = \dfrac{12-2x}{5-2} = \dfrac{2(6-x)}{3}$

519. c. $\dfrac{\frac{5}{(x-1)^3}-\frac{2}{(x-1)^2}}{\frac{2}{(x-1)^3}-\frac{5}{(x-1)^4}} \cdot \dfrac{(x-1)^4}{(x-1)^4} =$

$\dfrac{\frac{5}{(x-1)^3}\cdot(x-1)^4-\frac{2}{(x-1)^2}\cdot(x-1)^4}{\frac{2}{(x-1^3)^3}\cdot(x-1)^4-\frac{5}{(x-1)^4}\cdot(x-1)^4} =$

$\dfrac{5(x-1)-2(x-1)^2}{2(x-1)-5} = \dfrac{5x-5-2x^2+4x-2}{2x-7} =$

$\dfrac{-2x^2+9x-7}{2x-7} = \dfrac{-(2x^2-9x+7)}{2x-7} = \dfrac{-(2x-7)(x-1)}{2x-7} =$

$-(x-1)$

520. a. $1-\dfrac{\frac{x}{5}}{1+\frac{x}{5}} = 1-\dfrac{\frac{x}{5}}{\frac{5+x}{5}} = 1-\left(\frac{x}{5}\cdot\frac{5}{x+5}\right) =$

$1-\dfrac{x}{x+5} = \dfrac{x+5}{x+5}-\dfrac{x}{x+5} = \dfrac{5}{x+5}$

521. d. $\dfrac{\frac{a-2}{a+2}-\frac{a+2}{a-2}}{\frac{a-2}{a+2}+\frac{a+2}{a-2}} =$

$\dfrac{\frac{(a-2)(a-2)}{(a+2)(a-2)}-\frac{(a+2)(a+2)}{(a+2)(a-2)}}{\frac{(a-2)(a-2)}{(a+2)(a-2)}+\frac{(a+2)(a+2)}{(a+2)(a-2)}} =$

$\dfrac{\frac{(a-2)(a-2)-(a+2)(a+2)}{(a+2)(a-2)}}{\frac{(a-2)(a-2)+(a+2)(a+2)}{(a+2)(a-2)}} =$

$\dfrac{\frac{a^2-4a+4-(a^2+4a+4)}{(a+2)(a-2)}}{\frac{a^2-4a+4+(a^2+4a+4)}{(a+2)(a-2)}} =$

$\dfrac{a^2-4a+4-(a^2+4a+4)}{(a+2)(a-2)} \cdot \dfrac{(a+2)(a-2)}{a^2-4a+4+(a^2+4a+4)}$

$= -\dfrac{8a}{2a^2+8} = -\dfrac{4a}{a^2+4}$

522. a. $\dfrac{\frac{4}{4-x^2}-1}{\frac{1}{x+2}+\frac{1}{x-2}} =$

$\dfrac{\frac{4}{4-x^2}-\frac{4-x^2}{4-x^2}}{\frac{x-2}{(x+2)(x-2)}+\frac{x+2}{(x+2)(x-2)}} =$

$\dfrac{\frac{4-4+x^2}{4-x^2}}{\frac{x-2+x+2}{(x+2)(x-2)}} =$

$\dfrac{x^2}{4-x^2} \cdot \dfrac{(x+2)(x-2)}{2x} =$

$\dfrac{x^2}{4-x^2} \cdot \dfrac{x^2-4}{2x} = \dfrac{x}{-(x^2-4)} \cdot \dfrac{x^2-4}{2} = -\dfrac{x}{2}$

523. a. $(a^{-1}+b^{-1})^{-1}=\left(\frac{1}{a}+\frac{1}{b}\right)^{-1} = \left(\frac{b+a}{ab}\right)^{-1} = \dfrac{ab}{b+a}$

524. b. $\dfrac{x^{-1}-y^{-1}}{x^{-1}+y^{-1}} = \dfrac{\frac{1}{x}-\frac{1}{y}}{\frac{1}{x}+\frac{1}{y}} = \dfrac{\frac{y-x}{xy}}{\frac{y+x}{xy}} = \dfrac{y-x}{xy}\cdot\dfrac{xy}{y+x} = \dfrac{y-x}{y+x}$

525. b. $\left[\dfrac{x^2+4x-5}{2x^2+x-3}\cdot\dfrac{2x+3}{x+1}\right]-\dfrac{2}{x+2} =$

$\left[\dfrac{(x+5)(x-1)}{(2x+3)(x-1)}\cdot\dfrac{2x+3}{x+1}\right]-\dfrac{2}{x+2} = \dfrac{x+5}{x+1}-\dfrac{2}{x+2} =$

$\dfrac{(x+5)(x+2)}{(x+1)(x+2)}-\dfrac{2(x+1)}{(x+2)(x+1)} =$

$\dfrac{(x+5)(x+2)-2(x+1)}{(x+1)(x+2)} = \dfrac{x^2+7x+10-2x-2}{(x+1)(x+2)} =$

$\dfrac{x^2+5x+8}{(x+1)(x+2)}$

526. c. $\left[\dfrac{x+5}{x-3}-x\right]\div\dfrac{1}{x-3}-\left[\dfrac{x+5}{x-3}-\dfrac{x(x-3)}{x-3}\right]\div\dfrac{1}{x-3}$

$= \dfrac{x+5-x^2+3x}{x-3}\cdot(x-3)$

$= -(x^2-4x-5) = -(x-5)(x+1)$

527. b. $\left[3+\dfrac{1}{x+3}\right]\cdot\dfrac{x+3}{x-2} = \left[\dfrac{3(x+3)}{x+3}+\dfrac{1}{x+3}\right]\cdot\dfrac{x+3}{x-2} =$

$\dfrac{3x+10}{x+3}\cdot\dfrac{x+3}{x-2} = \dfrac{3x+10}{x-2}$

528. d. $1-\left(\dfrac{2}{x}-\left(\dfrac{3}{2x}-\dfrac{1}{6x}\right)\right) =$

$1-\left(\dfrac{2}{x}-\dfrac{3}{2x}+\dfrac{1}{6x}\right) =$

$1-\left(\dfrac{2(6)}{6x}-\dfrac{3(3)}{6x}+\dfrac{1}{6x}\right) =$

$1-\dfrac{12-9+1}{6x} = \dfrac{6x-4}{6x} = \dfrac{3x-2}{3x}$

Set 34 (Page 84)

529. a. First, clear the fractions from all terms in the equation by multiplying both sides by the least common denominator. Then, solve the resulting equation using factoring techniques:

$$\frac{3}{x} = 2 + x$$
$$x \cdot \frac{3}{x} = x \cdot (2 + x)$$
$$3 = 2x + x^2$$
$$x^2 + 2x - 3 = 0$$
$$(x + 3)(x - 1) = 0$$
$$x + 3 = 0 \text{ or } x - 1 = 0$$
$$x = -3 \text{ or } x = 1$$

Since neither of these values makes any of the expressions in the original equation, or any subsequent step of the solution, undefined, we conclude that both are solutions to the original equation.

530. d. First, clear the fractions from all terms in the equation by multiplying both sides by the least common denominator. Then, solve the resulting equation using factoring techniques:

$$\frac{2}{3} - \frac{3}{x} = \frac{1}{2}$$
$$\frac{2}{3} \cdot 6x - \frac{3}{x} \cdot 6x = \frac{1}{2} \cdot 6x$$
$$4x - 18 = 3x$$
$$x = 18$$

This value does not make any of the expressions in the original equation, or any subsequent step of the solution, undefined, so we conclude that it is indeed a solution of the original equation.

531. c. First, clear the fractions from all terms in the equation by multiplying both sides by the least common denominator. Then, solve the resulting equation using factoring techniques:

$$(t-7)(t-1) \cdot \left[\frac{2t}{t-7} + \frac{1}{t-1} \right] = 2 \cdot (t-7)(t-1)$$
$$\frac{2t}{t-7} \cdot (t-7)(t-1) + \frac{1}{(t-1)} \cdot (t-7)(t-1) =$$
$$2(t-7)(t-1) = 2t(t-1) + (t-7) = 2(t-7)(t-1)$$
$$2t^2 - 2t + t - 7 = 2t^2 - 16t + 14$$
$$-t - 7 = -16t + 14$$
$$15t = 21$$
$$t = \frac{21}{15} = \frac{7}{5}$$

Since this value does not make any of the expressions in the original equation, or any subsequent step of the solution, undefined, it is indeed a solution of the original equation.

532. a. First, clear the fractions from all terms in the equation by multiplying both sides by the least common denominator. Then, solve the resulting equation using factoring techniques:

$$x(x+2) \cdot \frac{x+8}{x+2} + x(x+2) \cdot \frac{12}{x^2+2x} = x(x+2) \cdot \frac{2}{x}$$
$$x(x+8) + 12 = 2(x+2)$$
$$x^2 + 8x + 12 = 2x + 4$$
$$x^2 + 6x + 8 = 0$$
$$(x+4)(x+2) = 0$$
$$x = -4 \text{ or } x = -2$$

Note that $x = -2$ makes some of the terms in the original equation undefined, so it cannot be a solution of the equation. Thus, we conclude that the only solution of the equation is $x = -4$.

533. c. First, clear the fractions from all terms in the equation by multiplying both sides by the least common denominator. Then, solve the resulting equation using factoring techniques:

$$\frac{x}{x-3} + \frac{2}{x} = \frac{3}{x-3}$$
$$\frac{x}{x-3} \cdot x(x-3) + \frac{2}{x} \cdot x(x-3) = \frac{3}{x-3} \cdot x(x-3)$$
$$x^2 + 2(x-3) = 3x$$
$$x^2 - x - 6 = 0$$
$$(x-3)(x+2) = 0$$
$$x = 3, -2$$

Because $x = 3$ makes some of the terms in the original equation undefined, it cannot be a solution of the equation. Thus, we conclude that the only solution of the equation is $x = -2$.

534. a. First, clear the fractions from all terms in the equation by multiplying both sides by the least common denominator. Then, solve the resulting equation using factoring techniques:

$$\frac{3}{x+2} + 1 = \frac{6}{(2-x)(2+x)}$$
$$\frac{3}{x+2} \cdot (2-x)(2+x) + 1 \cdot (2-x)(2+x) =$$
$$\frac{6}{(2-x)(2+x)} \cdot (2-x)(2+x)$$
$$3(2-x) + 4 - x^2 = 6$$
$$10 - 3x - x^2 = 6$$
$$x^2 + 3x - 4 = 0$$
$$(x+4)(x-1) = 0$$
$$x = -4, 1$$

Neither of these values makes any of the expressions in the original equation, or any subsequent step of the solution, undefined, so we conclude that both of them are solutions to the original equation.

535. b. First, clear the fractions from all terms in the equation by multiplying both sides by the least common denominator. Then, solve the resulting equation using factoring techniques:

$$\frac{10}{(2x-1)^2} = 4 + \frac{3}{2x-1}$$

$$\frac{10}{(2x-1)^2} \cdot (2x-1)^2 = 4 \cdot (2x-1)^2 + \frac{3}{2x-1} \cdot (2x-1)^2$$
$$10 = 4(2x-1)^2 + 3(2x-1)$$
$$10 = 16x^2 - 16x + 4 + 6x - 3$$
$$10 = 16x^2 - 10x + 1$$
$$16x^2 - 10x - 9 = 0$$
$$(2x+1)(8x-9) = 0$$
$$x = \frac{-1}{2}, \frac{9}{8}$$

Since neither of these values makes any of the expressions in the original equation, or any subsequent step of the solution, undefined, both of them are solutions to the original equation.

536. b. $\frac{1}{f} = (k-1)\left[\frac{1}{pq} + \frac{1}{q}\right]$

$$\frac{1}{f(k-1)} = \frac{1}{pq} + \frac{1}{q}$$
$$\frac{1}{f(k-1)} \cdot f(k-1)pq = \left[\frac{1}{pq} + \frac{1}{q}\right] \cdot f(k-1)pq$$
$$pq = f(k-1) + f(k-1)p$$
$$q = \frac{f(k-1) + f(k-1)p}{p}$$
$$q = \frac{f(k-1)(1+p)}{p}$$

537. d. First, clear the fractions from all terms in the equation by multiplying both sides by the least common denominator. Then, solve the resulting equation using factoring techniques, as follows:

$$\frac{x-1}{x-5} = \frac{4}{x-5}$$
$$(x-5)\cdot\frac{x-1}{x-5} = (x-5)\cdot\frac{4}{x-5}$$
$$x-1 = 4$$
$$x = 5$$

Because this value of x makes the expressions in the original equation undefined, we conclude that the equation has no solution.

538. d. First, clear the fractions from all terms in the equation by multiplying both sides by the least common denominator. Then, solve the resulting equation using factoring techniques:

$$\frac{22}{2p^2-9p-5} - \frac{3}{2p+1} = \frac{2}{p-5}$$
$$\frac{22}{(2p+1)(p-5)} - \frac{3}{2p+1} = \frac{2}{p-5}$$
$$(2p+1)(p-5)\cdot\frac{22}{(2p+1)(p-5)} -$$
$$(2p+1)(p-5)\cdot\frac{3}{2p+1} = (2p+1)(p-5)\cdot\frac{2}{p-5}$$
$$22 - 3(p-5) = 2(2p+1)$$
$$22 - 3p + 15 = 4p + 2$$
$$-3p + 37 = 4p + 2$$
$$35 = 7p$$
$$p = 5$$

This value of p makes the expressions in the original equation undefined, so the equation has no solution.

539. b. First, clear the fractions from all terms in the equation by multiplying both sides by the least common denominator. Then, solve the resulting equation using factoring techniques:

$$\frac{x+1}{x^3-9x} - \frac{1}{2x^2+x-21} = \frac{1}{2x^2+13x+21}$$
$$\frac{x+1}{x(x-3)(x+3)} - \frac{1}{(2x+7)(x-3)} = \frac{1}{(2x+7)(x+3)}$$
$$x(x-3)(x+3)(2x+7)\cdot$$
$$\left[\frac{x+1}{(x-3)(x+3)} - \frac{1}{(2x+7)(x-3)}\right] =$$
$$x(x-3)(x+3)(2x+7)\cdot\left[\frac{1}{(2x+7)(x-3}\right]$$
$$(x+1)(2x+7) - x(x+3) = x(x-3)$$
$$2x^2 + 9x + 7 - x^2 - 3x = x^2 - 3x$$
$$x^2 + 6x + 7 = x^2 - 3x$$
$$6x + 7 = -3x$$
$$9x = -7$$
$$x = -\frac{7}{9}$$

Since this value does not make any of the expressions in the original equation, or any subsequent step of the solution, undefined, it is indeed a solution of the original equation.

540. c. First, clear the fractions from all terms in the equation by multiplying both sides by the least common denominator. Then, solve the resulting equation using factoring techniques:

$$\frac{x}{x+1} - \frac{3}{x+4} = \frac{3}{x^2+5x+4}$$
$$\frac{x}{x+1} - \frac{3}{x+4} = \frac{3}{(x+1)(x+4)}$$
$$(x+1)(x+4)\cdot\left[\frac{x}{x+1} - \frac{3}{x+4}\right] =$$
$$(x+1)(x+4)\cdot\left[\frac{3}{(x+1)(x+4)}\right]$$
$$x(x+4) - 3(x+1) = 3$$
$$x^2 + 4x - 3x - 3 = 3$$
$$x^2 + x - 6 = 0$$
$$(x+3)(x-2) = 0$$
$$x = -3 \text{ or } x = 2$$

Neither of these values makes any of the expressions in the original equation, or any subsequent step of the solution, undefined. Therefore, we conclude that both are solutions to the original equation.

541. c. First, clear the fractions from all terms in the equation by multiplying both sides by the least common denominator. Then, solve the resulting equation using factoring techniques:

$$1 + \frac{2}{x-3} = \frac{4}{x^2 - 4x + 3}$$

$$1 + \frac{2}{x-3} = \frac{4}{(x-3)(x-1)}$$

$$(x-3)(x-1) \cdot \left[1 + \frac{2}{x-3}\right] =$$

$$(x-3)(x-1) \cdot \left[\frac{4}{(x-3)(x-1)}\right]$$

$$(x-3)(x-1) + 2(x-1) = 4$$

$$x^2 - 4x + 3 + 2x - 2 = 4$$

$$x^2 - 2x - 3 = 0$$

$$(x-3)(x+1) = 0$$

$$x = 3 \text{ or } x = -1$$

Because $x = 3$ makes some of the terms in the original equation undefined, it cannot be a solution of the equation. Thus, we conclude that the only solution of the equation is $x = -1$.

542. d. First, clear the fractions from all terms in the equation by multiplying both sides by the least common denominator. Then, solve the resulting equation using factoring techniques:

$$\frac{3}{x+2} = \frac{x-3}{x-2}$$

$$(x+2)(x-2) \cdot \frac{3}{x+2} = (x+2)(x-2) \cdot \frac{x-3}{x-2}$$

$$3(x-2) = (x+2)(x-3)$$

$$3x - 6 = x^2 - x - 6$$

$$x^2 - 4x = 0$$

$$x(x-4) = 0$$

$$x = 0 \text{ or } x = 4$$

Since neither of these values makes any of the expressions in the original equation, or any subsequent step of the solution, undefined, we conclude that both are solutions to the original equation.

543. c. First, clear the fractions from all terms in the equation by multiplying both sides by the least common denominator. Then, solve the resulting equation using factoring techniques:

$$\frac{t+1}{t-1} = \frac{4}{t^2-1}$$

$$\frac{t+1}{t-1} = \frac{4}{(t-1)(t+1)}$$

$$(t-1)(t+1) \cdot \left[\frac{t+1}{t-1}\right] =$$

$$(t-1)(t+1) \cdot \frac{4}{(t+1)(t-1)}$$

$$(t+1)(t+1) = 4$$

$$t^2 + 2t + 1 = 4$$

$$t^2 + 2t - 3 = 0$$

$$(t+3)(t-1) = 0$$

$$t = -3 \text{ or } t = 1$$

Note that $t = 1$ makes some of the terms in the original equation undefined, so it cannot be a solution of the equation. Thus, we conclude that the only solution of the equation is $t = -3$.

544. a. $v = \dfrac{v_1 + v_2}{1 + \frac{v_1 v_2}{c^2}}$

$$v\left[1 + \frac{v_1 v_2}{c_2}\right] = v_1 + v_2$$

$$v + \frac{v v_1 v_2}{c_2} = v_1 + v_2$$

$$\frac{v v_1 v_2}{c_2} - v_1 = v_2 - v$$

$$v_1\left(\frac{v v_2}{c_2} - 1\right) = v_2 - v$$

$$v_1 = \frac{v_2 - v}{\frac{v v_2}{c^2} - 1} = \frac{v_2 - v}{\frac{v v_2 - c^2}{c^2}} = (v_2 - v) \cdot \frac{c^2}{v v_2 - c^2}$$

$$= \frac{c^2(v_2 - v)}{v v_2 - c^2}$$

Set 35 (Page 86)

545. b. First, determine the *x*-values that make the expression on the left side equal to zero or undefined. Then, assess the sign of the expression on the left side on each subinterval formed using these values. To this end, observe that these values are $x = -3, -2,$ and 1. Now, form a number line, choose a real number in each of the subintervals, and record the sign of the expression above each:

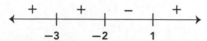

Since the inequality includes "equals," we include those values from the number line that make the numerator equal to zero. The solution set is $[-2, 1]$.

546. c. First, make certain that the numerator and denominator are both completely factored and that all common factors are canceled:

$$\frac{x^2 + 9}{x^2 - 2x - 3} = \frac{x^2 + 9}{(x - 3)(x + 1)}$$

Next, determine the *x*-values that make this expression equal to zero or undefined. Then, assess the sign of the expression on the left side on each subinterval formed using these values. To this end, observe that these values are $x = -1$ and 3. Now, form a number line, choose a real number in each subinterval, and record the sign of the expression above each:

Since the inequality does not include "equals," we do not include those values from the number line that make the numerator equal to zero. Therefore, the solution set is $(-\infty, -1) \cup (3, \infty)$.

547. a. First, make certain that the numerator and denominator are both completely factored and that all common factors are canceled, as follows:

$$\frac{-x^2 - 1}{6x^4 - x^3 - 2x^2} = \frac{-(x^2 + 1)}{x^2(6x^2 - x - 2)} = \frac{-(x^2 + 1)}{x^2(2x + 1)(3x - 2)}$$

Next determine the *x*-values that make this expression equal to zero or undefined. Then, assess the sign of the expression on the left side on each subinterval formed using these values. To this end, observe that these values are $x = -\frac{1}{2}, 0,$ and $\frac{2}{3}$. Now, form a number line, choose a real number in each subinterval, and record the sign of the expression above each:

Since the inequality includes "equals," we include those values from the number line that make the numerator equal to zero. Since none of these values make the numerator equal to zero, we conclude that the solution set is $(-\frac{1}{2}, 0) \cup (0, \frac{2}{3})$.

548. c. First, simplify the complex fraction on the left side of the inequality:

$$\frac{\frac{1}{x} - \frac{1}{x+1}}{x + 2} \geq 0$$

$$\frac{\frac{x+1-x}{x(x+1)}}{x + 2} \geq 0$$

$$\frac{1}{x(x + 1)(x + 2)} \geq 0$$

Next, determine the *x*-values that make this expression equal to zero or undefined.

Then, assess the sign of the expression on the left side on each subinterval formed using these values. Observe that these values are −2, −1, and 0. Now, form a number line, choose a real number in each subinterval, and record the sign of the expression above each.

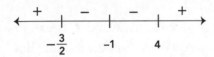

The solution set is $(-2,-1)\cup(0,\infty)$.

549. b. First, factor the numerator and denominator completely and cancel any common factors:

$$\frac{(2z+3)(z+1)}{(z-4)(z+1)} \le 0$$

$$\frac{2z+3}{z-4} \le 0$$

Next, determine the z-values that make the original expression equal to zero or undefined. Then, assess the sign of the expression on the left side on each subinterval formed using these values. Observe that these values are $-\frac{3}{2},-1$ and 4. Now, form a number line, choose a real number in each subinterval, and record the sign of the expression above each.

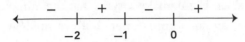

The solution set is $[-\frac{3}{2},-1)\cup(-1,4)$.

550. d. The first step is to make certain that the numerator and denominator are both completely factored and that all common factors are canceled:

$$\frac{25(-x)^4}{x(5x^2)^2} = \frac{25x^4}{x \cdot 25x^4} = \frac{1}{x}$$

Now, determine the x-values that make this expression equal to zero or undefined. Then, assess the sign of the expression on the left side on each subinterval formed using these

values. To this end, observe that the only value for which this is true is $x = 0$. Next, form a number line, choose a real number in each subinterval, and record the sign of the expression above each, as follows:

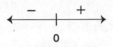

Since the inequality includes "equals," we would include those values from the number line that make the numerator equal to zero. Since none of these values make the numerator equal to zero, we conclude that the solution set is $(-\infty, 0)$.

551. d. First, make certain that the numerator and denominator are both completely factored and that all common terms are canceled:

$$\frac{z^3-16z}{8z-32} = \frac{z(z^2-16)}{8(z-4)} = \frac{z(z-4)(z+4)}{8(z-4)} = \frac{z(z+4)}{8}$$

Next, determine the z-values that make this expression equal to zero or undefined. Then, assess the sign of the expression on the left side on each subinterval formed using these values. To this end, observe that these values are $z = -4, 0, 4$. Now, form a number line, choose a real number in each of the duly formed subintervals, and record the sign of the expression above each:

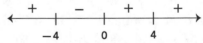

Since the inequality does not include "equals," we do not include those values from the number line that make the numerator equal to zero. The solution set is $(-4, 0)$.

552. b. To begin, make certain that the numerator and denominator are both completely factored and that all common factors are canceled:

$$\frac{y^2 - 64}{8 - y} = \frac{(y-8)(y+8)}{(8-y)} = \frac{(y-8)(y+8)}{-(y-8)} = -(y+8)$$

Now, determine the y-values that make this expression equal to zero or undefined. Then, we assess the sign of the expression on the left side on each subinterval formed using these values. To this end, observe that these values are $y = -8, 8$. Next, form a number line, choose a real number in each subinterval, and record the sign of the expression above each, as follows:

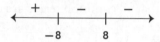

The inequality includes "equals," we include those values from the number line that make the numerator equal to zero. We conclude that the solution set is $[-8,8) \cup (8,\infty)$.

553. a. First, make certain that the numerator and denominator are both completely factored and that all common factors are canceled:

$$\frac{x^2 + 8x}{x^3 - 64x} = \frac{x(x+8)}{x(x^2 - 64)} = \frac{x(x+8)}{x(x+8)(x-8)} = \frac{1}{x-8}$$

Next, determine the x-values that make this expression equal to zero or undefined. Then, assess the sign of the expression on the left side on each subinterval formed using these values. To this end, observe that these values are $x = -8, 8$. Now, form a number line, choose a real number in each subinterval, and record the sign of the expression above each:

Since the inequality does not include "equals," we do not include those values from the number line that make the numerator equal to zero. Therefore, the solution set is $(8,\infty)$.

554. a. To begin, make certain that the numerator and denominator are both completely factored and that all common factors are canceled:

$$\frac{5x^2(x-1) - 3x(x-1) - 2(x-1)}{10x^2(x-1) + 9x(x-1) + 2(x-1)}$$
$$= \frac{(x-1)(5x^2 - 3x - 2)}{(x-1)(10x^2 + 9x + 2)} = \frac{(x-1)(5x+2)(x-1)}{(x-1)(5x+2)(2x+1)}$$
$$= \frac{x-1}{2x+1}$$

Next determine the x-values that make this expression equal to zero or undefined. Then, assess the sign of the expression on the left side on each subinterval formed using these values. To this end, observe that these values are $x = -\frac{1}{2}, -\frac{2}{5}, 1$. Now, form a number line, choose a real number in each subinterval, and record the sign of the expression above each:

The inequality includes "equals," we include those values from the number line that make the numerator equal to zero. We conclude that the solution set is $(-\frac{1}{2}, -\frac{2}{5}) \cup (-\frac{2}{5}, 1)$.

555. c. First, make certain that the numerator and denominator are both completely factored and that all common factors are canceled:

$$\frac{6x^3 - 24x}{24x^2} = \frac{6x(x^2 - 4)}{4 \cdot 6 \cdot x \cdot x} = \frac{x^2 - 4}{4x} = \frac{(x-2)(x+2)}{4x}$$

Now, the strategy is to determine the x-values that make this expression equal to zero or undefined. Then, assess the sign of the expression on the left side on each subinterval formed using these values, which are $x = -2, 0, 2$. Now, form a number line, choose a real number in each subinterval, and record the sign of the expression above each:

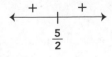

Since the inequality includes "equals," we include those values from the number line that make the numerator equal to zero. The solution set is $[-2,0)\cup[2,\infty)$.

556. c. First, make certain that the numerator and denominator are both completely factored and that all common factors are canceled:

$$\frac{(2x-5)(x+4) - (2x-5)(x+1)}{9(2x-5)} =$$

$$\frac{(2x-5)[(x+4) - (x+1)]}{9(2x-5)} = \frac{3}{9} = \frac{1}{3}$$

Now, determine the x-values that make this expression equal to zero or undefined. Then, assess the sign of the expression on the left side on each subinterval formed using these values. To this end, observe that the only value for which this is true is $x = \frac{5}{2}$. Next, form a number line, choose a real number in each of the duly formed subintervals, and record the sign of the expression above each:

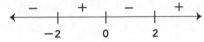

Since the inequality does not include "equals," so we do not include those values from the number line that make the numerator equal to zero. There are no such values, and furthermore, the expression is always positive. Therefore, the solution set is the empty set.

557. c. First, make certain that the numerator and denominator are both completely factored and that all common factors are canceled, as follows:

$$\frac{3-2x}{(x+2)(x-1)} - \frac{2-x}{(x-1)(x+2)} = \frac{3-2x-(2-x)}{(x-1)(x+2)} =$$

$$\frac{1-x}{(x-1)(x+2)} = \frac{-(x-1)}{(x-1)(x+2)} = -\frac{1}{x+2}$$

Next, determine the x-values that make this expression equal to zero or undefined. Then, assess the sign of the expression on the left side on each subinterval formed using these values, which are $x = -2$ and 1. Now, form a number line, choose a real number in each subinterval, and record the sign of the expression above each:

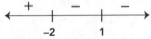

Since the inequality includes "equals," so we include those values from the number line that make the numerator equal to zero. The solution set is $(-\infty,-2)$.

558. b. First make certain that the numerator and denominator are both completely factored and that all common factors are canceled:

$$[\frac{x+5}{x-3} - x] \div \frac{1}{x-3} = [\frac{x+5}{x-3} - \frac{x(x-3)}{x-3}] \div \frac{1}{x-3} =$$

$$\frac{x+5-x^2+3x}{x-3} \cdot (x-3) = -(x^2-4x-5) =$$

$$-(x-5)(x+1)$$

Determine the x-values that make this expression equal to zero or undefined. Then, assess the sign of the expression on the left side on each subinterval formed using these values. To this end, observe that these values are $x = -1, 3, 5$. Now, form a number line, choose a real number in each subinterval, and record the sign of the expression above each:

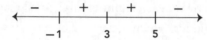

Since the inequality does not include "equals," we would not include those values from the number line that make the numerator equal to zero. As such, we conclude that the solution set is $(-1,3) \cup (3,5)$.

559. d. First, make certain that the numerator and denominator are both completely factored and that all common factors are canceled:

$$\frac{x}{2x+1} - \frac{1}{2x-1} + \frac{2x^2}{4x^2-1} = \frac{x}{2x+1} \cdot \frac{2x-1}{2x-1} -$$

$$\frac{1}{2x-1} \cdot \frac{2x+1}{2x+1} + \frac{2x^2}{4x^2-1}$$

$$= \frac{x(2x-1) - 1(2x+1) + 2x^2}{(2x-1)(2x+1)}$$

$$= \frac{2x^2 - x - 2x - 1 + 2x^2}{(2x-1)(2x+1)} = \frac{4x^2 - 3x - 1}{(2x-1)(2x+1)}$$

$$= \frac{(4x+1)(x-1)}{(2x-1)(2x+1)}$$

Now, the strategy is to determine the x-values that make this expression equal to zero or undefined. Then, assess the sign of the expression on the left side on each

subinterval formed using these values. To this end, observe that these values are $x = -\frac{1}{2}, -\frac{1}{4}, \frac{1}{2}$, and 1. Next, form a number line, choose a real number in each subinterval, and record the sign of the expression above each:

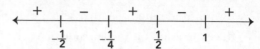

Since the inequality includes "equals," we include those values from the number line that make the numerator equal to zero. The solution set is $(-\frac{1}{2}, -\frac{1}{4}] \cup (\frac{1}{2}, 1]$.

560. d. First, make certain that the numerator and denominator are both completely factored and that all common factors are canceled, as follows:

$$\frac{3y+2}{(y-1)^2} - \frac{7y-3}{(y-1)(y+1)} + \frac{5}{(y+1)} = \frac{3y+2}{(y-1)^2} \cdot$$

$$\frac{y+1}{y+1} - \frac{7y-3}{(y-1)(y+1)} \cdot \frac{y-1}{y-1} + \frac{5}{y+1} \cdot \frac{(y-1)^2}{(y-1)^2} =$$

$$\frac{(3y+2)(y+1) - (7y-3)(y-1) + 5(y-1)^2}{(y-1)^2(y+1)} =$$

$$\frac{3y^2 + 5y + 2 - (7y^2 - 10y + 3) + 5(y^2 - 2y + 1)}{(y-1)^2(y+1)} =$$

$$\frac{3y^2 + 5y + 2 - 7y^2 + 10y - 3 + 5y^2 - 10y + 5}{(y-1)^2(y+1)} =$$

$$\frac{y^2 + 5y + 4}{(y-1)^2(y+1)} = \frac{(y+4)(y+1)}{(y-1)^2(y+1)} = \frac{y+4}{(y-1)^2}$$

Now, the strategy is to determine the y-values that make this expression equal to zero or undefined. Then, assess the sign of the expression on the left side on each subinterval formed using these values. To this end, observe that these values are $y = -4, -1$, and 1. Next, form a number line, choose a real number in each subinterval, and record the sign of the expression above each:

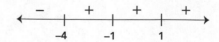

The inequality does not include "equals," so we do not include those values from the number line that make the numerator equal to zero. The solution set is $(-\infty, -4)$.

Section 5—
Radical Expressions
and Quadratic Equations

Set 36 (Page 90)

561. c. -125 since $(-5)^3 = -125$.

562. c. -7 and 7 are both second roots (square roots) since $(-7)^2 = 49$ and $(7)^2 = 49$.

563. a. Note that $625 = 5^4$. So, the principal root of 625 is 5.

564. d. Since $(-2)^5 = -32$, we write $\sqrt[5]{-32} = -2$.

565. a. Since $4^3 = 64$, $b = 64$ satisfies the equation.

566. a. $\sqrt[4]{3^{12}} = \sqrt[4]{(3^3)^4} = 3^3 = 27$

567. c. $\sqrt[5]{5^{15}} = \sqrt[5]{(5^3)^5} = 5^3 = 125$

568. b. Since $\sqrt[4]{2^b} = 2^b$, $b = 3$ satisfies the equation.

569. b. $64^{\frac{1}{6}} = (2^6)^{\frac{1}{6}} = 2$

570. d. We break up the fractional exponent into two separate exponents to obtain $49^{\frac{5}{2}} = (49^{\frac{1}{2}})^5 = 7^5 = 16{,}807$.

571. a. We break up the fractional exponent into two separate exponents to obtain $81^{-\frac{3}{4}} = (81^{\frac{1}{4}})^{-3} = 3^{-3} = \frac{1}{3^3} = \frac{1}{27}$.

572. c. We break up the fractional exponent into two separate exponents to obtain $32^{\frac{3}{5}} = (32^{\frac{1}{5}})^3 = (2)^3 = 8$.

573. c. $\left(\frac{8}{27}\right)^{-\frac{2}{3}} = \left[\left(\frac{2}{3}\right)^3\right]^{-\frac{2}{3}} = \left(\frac{2}{3}\right)^{-2} = \left(\frac{3}{2}\right)^2 = \frac{3^2}{2^2} = \frac{9}{4}$.

574. a. $(-64)^{-\frac{1}{3}} = [(-4)^3]^{-\frac{1}{3}} = (-4)^{-1} = \frac{1}{-4} = -\frac{1}{4}$

575. c. $(4x^{-4})^{-\frac{1}{2}} = \left[(2x^{-2})^2\right]^{-\frac{1}{2}} = (2x^{-2})^{-1} = \frac{1}{2x^{-2}} = \frac{x^2}{2}$

576. b. $4\sqrt{x^{144}} = 4\sqrt{(x^{72})^2} = 4x^{72}$

Set 37 (Page 91)

577. b. $\sqrt[3]{9} \cdot \sqrt[3]{-3} = \sqrt[3]{(9)(-3)} = \sqrt[3]{-27} = \sqrt[3]{(-3)^3} = -3$

578. b. $\frac{\sqrt{x^5}}{\sqrt{x^7}} = \sqrt{\frac{x^5}{x^7}} = \sqrt{\frac{1}{x^2}} = \sqrt{\left(\frac{1}{x}\right)^2} = \frac{1}{x}$

579. a. $a^3\sqrt{a^3} = a^3\sqrt{a^2 a} = a^3 a\sqrt{a} = a^4\sqrt{a}$

580. a. Factor $\sqrt{4g}$ into two radicals. 4 is a perfect square, so factor $\sqrt{4g}$ into $\sqrt{4}\sqrt{g} = 2\sqrt{g}$. Simplify the fraction by dividing the numerator by the and denominator. Cancel the $\sqrt{g}$ terms from the numerator and denominator. That leaves $\frac{4}{2} = 2$.

581. a. The cube root of $27y^3 = 3y$, since $(3y)(3y)(3y) = 27y^3$. Factor the denominator into two radicals: $\sqrt{27y^2} = \sqrt{9y^2} \cdot \sqrt{3}$. The square root of $9y^2 = 3y$, since $(3y)(3y) = 9y^2$. The expression is now equal to $\frac{3y}{3y\sqrt{3}}$. Cancel the $3y$ terms from the numerator and denominator, leaving $\frac{1}{\sqrt{3}}$. Simplify the fraction by multiplying the numerator and denominator by : $\sqrt{3}$: $\left(\frac{1}{\sqrt{3}}\right)\left(\frac{\sqrt{3}}{\sqrt{3}}\right) = \frac{\sqrt{3}}{3}$.

582. c. Factor each term in the numerator: $\sqrt{a^2b} = \sqrt{u^2} \cdot \sqrt{b} = a\sqrt{b}$ and $\sqrt{ab^2} = \sqrt{a} \cdot \sqrt{b^2} = b\sqrt{a}$. Next, multiply the two radicals. Multiply the coefficients of each radical and multiply the radicands of each radical: $(a\sqrt{b})(b\sqrt{a}) = ab\sqrt{ab}$. The expression is now $\frac{ab\sqrt{ab}}{\sqrt{ab}}$. Cancel the $\sqrt{ab}$ terms from the numerator and denominator, leaving ab.

583. c. First, cube the $4g^2$ term. Cube the constant 4 and multiply the exponent of g by 3: $(4g^2)^3 = 64g^6$. Next, multiply $64g^6$ by g^4. Add the exponents of the g terms. $(64g^6)(g^4) = 64g^{10}$ Finally, taking the square root of $64g^{10}$ yields $8g^5$, since $(8g^5)(8g^5) = 64g^{10}$.

584. e. First, find the square root of $9pr$. $\sqrt{9pr} = \sqrt{9} \cdot \sqrt{pr} = 3\sqrt{pr}$. The denominator $(pr)^{\frac{3}{2}}$ has a negative exponent, so it can be rewritten in the numerator with a positive exponent. The expression $\sqrt{pr}$ can be written as $(pr)^{\frac{1}{2}}$ since a value raised to the exponent $\frac{1}{2}$ is another way of representing the square root of the value. The expression is now $3(pr)^{\frac{3}{2}}(pr)^{\frac{1}{2}}$. To multiply the pr terms, add the exponents. $\frac{1}{2} + \frac{3}{2} = \frac{4}{2} = 2$, so $= 3(pr)^{\frac{3}{2}}(pr)^{\frac{1}{2}} = 3(pr)^2 = 3p^2r^2$.

585. e. Substitute 20 for n: $\frac{\sqrt{20+5}}{\sqrt{20}}(\frac{20}{2}\sqrt{5}) = \frac{\sqrt{25}}{\sqrt{20}}(10\sqrt{5}) = \frac{5}{2\sqrt{5}}(10\sqrt{5})$. Cancel the $\sqrt{5}$ terms and multiply the fraction by 10: $\frac{5}{2\sqrt{5}}(10\sqrt{5}) = \frac{5(10)}{2} = \frac{50}{2} = 25$

586. c. $\sqrt{\frac{125}{9}} = \frac{\sqrt{125}}{\sqrt{9}} = \frac{\sqrt{5^2 \cdot 5}}{\sqrt{3^2}} = \frac{5\sqrt{5}}{3}$

587. d. $\frac{\sqrt[4]{243}}{\sqrt[4]{3}} = \sqrt[4]{\frac{243}{3}} = \sqrt[4]{81} = \sqrt[4]{3^4} = 3$

588. d. $\sqrt{x^2 + 4x + 4} = \sqrt{(x+2)^2} = x + 2$

589. d. $\sqrt[4]{32x^8} = \sqrt[4]{2^4 \cdot 2 \cdot (x^2)^4} = \sqrt[4]{2^4} \sqrt[4]{2} \sqrt[4]{(x^2)^4} = 2x^2\sqrt[4]{2}$

590. b. $\sqrt[4]{x^{21}} = \sqrt[4]{(x^5)^4 x} = \sqrt[4]{(x^5)^4} \cdot \sqrt[4]{x} = x^5 \sqrt[4]{x}$

591. b. $\sqrt[3]{54x^5} = \sqrt[3]{2 \cdot 3^3 \cdot x^3 \cdot x^2} = 3x\sqrt[3]{2x^2}$

592. a. $\sqrt{x^3 + 40x^2 + 400x} = \sqrt{x(x^2 + 40x + 400)} = \sqrt{x(x+20)^2} = (x+20)\sqrt{x}$

Set 38 (Page 93)

593. b. $\sqrt{-25} = \sqrt{25 \cdot (-1)} = \sqrt{5^2 \cdot i^2} = \sqrt{5^2}\sqrt{i^2} = 5i$

594. a. $\sqrt{-32} = \sqrt{32(-1)} = \sqrt{32}\sqrt{-1} = (4\sqrt{2})(i) = 4i\sqrt{2}$

595. a. $-\sqrt{48} + 2\sqrt{27} - \sqrt{75} = -\sqrt{4^2 \cdot 3} + 2\sqrt{3^2 \cdot 3} - \sqrt{5^2 \cdot 3} = -4\sqrt{3} + 6\sqrt{3} - 5\sqrt{3} = (-4 + 6 - 5)\sqrt{3} = -3\sqrt{3}$

596. d. $3\sqrt{3} + 4\sqrt{5} - 8\sqrt{3} = (3 - 8)\sqrt{3} + 4\sqrt{5} = -5\sqrt{3} + 4\sqrt{5}$

597. d. First, simplify each radical expression. Then, because the variable/radical parts are alike, we can add the coefficients: $xy\sqrt{8xy^2} + 3y^2\sqrt{18x^3} = xy(2y)\sqrt{2x} + 3y^2(3x)\sqrt{2x} = 2xy^2\sqrt{2x} + 9xy^2\sqrt{2x} = 11xy^2\sqrt{2x}$

598. c. We first simplify each fraction. Then, we find the LCD and add.
$\sqrt{\frac{18}{25}} + \sqrt{\frac{32}{9}} = \frac{\sqrt{18}}{\sqrt{25}} + \frac{\sqrt{32}}{\sqrt{9}} = \frac{3\sqrt{2}}{5} + \frac{4\sqrt{2}}{3} = \frac{3\sqrt{2}}{5} \cdot \frac{3}{3} + \frac{4\sqrt{2}}{3} \cdot \frac{5}{5} = \frac{9\sqrt{2} + 20\sqrt{2}}{15} = \frac{29\sqrt{2}}{15}$

599. a. $(5 - \sqrt{3})(7 + \sqrt{3}) = 5(7) + 5(\sqrt{3}) - 7(\sqrt{3}) - \sqrt{3^2} = 35 + (5 - 7)\sqrt{3} - 3 = 32 - 2\sqrt{3}$

600. b. $(4 + \sqrt{6})(6 - \sqrt{15}) = 24 - 4\sqrt{15} + 6\sqrt{6} - \sqrt{90} = 24 - 4\sqrt{15} + 6\sqrt{6} - 3\sqrt{10}$

601. a. $\frac{-10 + \sqrt{-25}}{5} = \frac{-10 + \sqrt{25}\sqrt{-1}}{5} = \frac{-10 + 5i}{5} = \frac{5(-2 + i)}{5} = -2 + i$

602. c. $(4 + 2i)(4 - 2i) = 16 - (2i)^2 = 16 - 2^2i^2 = 16 - (4)(-1) = 16 + 4 = 20$

603. d. $(4 + 2i)^2 = 16 + (4)(2i) + (2i)(4) + (2i)^2 = 16 + 16i + 2^2i^2 = 16 + 16i - 4 = 12 + 16i$

604. b. $\sqrt{21}\left(\frac{\sqrt{3}}{\sqrt{7}} + \frac{\sqrt{7}}{\sqrt{3}}\right) = 21\left(\frac{\sqrt{3}\sqrt{3} + \sqrt{7}\sqrt{7}}{\sqrt{7}\sqrt{3}}\right) = \sqrt{21}\left(\frac{3 + 7}{\sqrt{21}}\right) = 10$

605. d. $(2 + \sqrt{3x})^2 = 4 + (2)(\sqrt{3x}) + (\sqrt{3x})(2) + (\sqrt{3x})^2 = 4 + 4\sqrt{3x} + 3x$

606. a. $(\sqrt{3} + \sqrt{7})(2\sqrt{3} - 5\sqrt{7}) = (\sqrt{3})(2\sqrt{3}) + (\sqrt{7})(2\sqrt{3}) - (\sqrt{3})(5\sqrt{7}) - (\sqrt{7})(5\sqrt{7}) = 2(\sqrt{3})^2 = 2\sqrt{7} \cdot \sqrt{3} - 5\sqrt{3 \cdot 7} - 5(\sqrt{7})^2 = 2 \cdot 3 + 2\sqrt{21} - 5\sqrt{21} - 5 \cdot 7 = 6 - 3\sqrt{21} - 35 = -29 - 3\sqrt{21}$

607. d. $\frac{1}{3 - 5\sqrt{2}} = \frac{1}{3 - 5\sqrt{2}} \cdot \frac{3 + 5\sqrt{2}}{3 + 5\sqrt{2}} = \frac{3 + 5\sqrt{2}}{3^2 - (5\sqrt{2})^2} = \frac{3 + 5\sqrt{2}}{9 - 5^2(\sqrt{2})^2} = \frac{3 + 5\sqrt{2}}{9 - 25(2)} = -\frac{3 + 5\sqrt{2}}{41}$

608. d. $\dfrac{\sqrt{2x}}{2-3\sqrt{x}} = \dfrac{\sqrt{2x}}{2-3\sqrt{x}} \cdot \dfrac{2+3\sqrt{x}}{2+3\sqrt{x}}$

$= \dfrac{2(\sqrt{2x})+(3\sqrt{x})(\sqrt{2x})}{2^2-(3\sqrt{x})^2} = \dfrac{2\sqrt{2x}+3x\sqrt{2}}{4-3^2(\sqrt{x})^2}$

$= \dfrac{2\sqrt{2x}+3x\sqrt{2}}{4-9x}$

Set 39 (Page 94)

609. a. Square both sides of the equation and then solve for x:

$\sqrt{7+3x} = 4$

$(\sqrt{7+3x})^2 = (4)^2$

$7+3x = 16$

$3x = 9$

$x = 3$

Substituting this value into the original equation yields the true statement $4 = 4$, so we know that it is indeed a solution.

610. a. Square both sides of the equation and then solve for x:

$\sqrt{4x+33} = 2x-1$

$(\sqrt{4x+33})^2 = (2x-1)^2$

$4x+33 = 4x^2-4x+1$

$0 = 4x^2-8x-32$

$0 = 4(x^2-2x-8)$

$0 = 4(x-4)(x+2)$

$x = 4, -2$

Substituting $x = 4$ into the original equation yields the true statement $7 = 7$, but substituting $x = -2$ into the original equation results in the false statement $5 = -5$. So, only $x = 4$ is a solution to the original equation.

611. e. $a^{\frac{4}{3}} = (a^{\frac{2}{3}})^2 = 6^2 = 36.$

612. d. $q^{-3} = -\frac{1}{2}$

$(q^{-3})^{-\frac{1}{3}} = (-\frac{1}{2})^{-\frac{1}{3}}$

$q = (-2)^{\frac{1}{3}} = \sqrt[3]{-2}$

613. d. To eliminate the radical term, raise both sides to the third power and solve for x:

$\sqrt[3]{5x-8} = 3$

$5x-8 = 3^3 = 27$

$5x = 35$

$x = 7$

Substituting this value into the original equation yields the true statement $3 = 3$, so it is indeed a solution.

614. b. To eliminate the radical term, raise both sides to the third power and solve for x:

$\sqrt[3]{7-3x} = -2$

$7-3x = (-2)^3 = -8$

$-3x = -15$

$x = 5$

Substituting this value into the original equation yields the true statement $-2 = -2$, so it is indeed a solution.

615. a. Take the square root of both sides and solve for x:

$(x-3)^2 = -28$

$(x-3)^2 = \pm\sqrt{-28}$

$x-3 = \pm 2i\sqrt{7}$

$x = 3 \pm 2i\sqrt{7}$

616. c. Square both sides of the equation and then solve for x:

$\sqrt{10-3x} = x-2$

$(\sqrt{10-3x})^2 = (x-2)^2$

$10-3x = x^2-4x+4$

$0 = x^2-x-6$

$0 = (x-3)(x+2)$

$x = 3, -2$

Substituting $x = 3$ into the original equation yields the true statement $1 = 1$, but substituting $x = -2$ into the original equation results in the false statement $4 = -4$. Only $x = 3$ is a solution to the original equation.

617. d. Square both sides of the equation and then solve for x:

$$\sqrt{3x + 4} + x = 8$$
$$\sqrt{3x + 4} = 8 - x$$
$$(\sqrt{3x + 4})^2 = (8 - x)^2$$
$$3x + 4 = 64 - 16x + x^2$$
$$0 = x^2 - 19x + 60$$
$$0 = (x - 4)(x - 15)$$
$$x = 4, 15$$

Substituting $x = 4$ into the original equation yields the true statement $8 = 8$, but substituting $x = 15$ into the original equation results in the false statement $22 = 8$. Therefore, only $x = 4$ is a solution to the original equation.

618. b. Isolate the squared expression on one side and then, take the square root of both sides and solve for x:

$$(x - 1)^2 + 16 = 0$$
$$(x - 1)^2 = -16$$
$$\sqrt{(x - 1)^2} = \pm\sqrt{-16}$$
$$x - 1 = \pm 4i$$
$$x = 1 \pm 4i$$

619. b. Take the cube root of both sides.

$$x^3 = -27$$
$$\sqrt[3]{x^3} = \sqrt[3]{-27}$$
$$x = \sqrt[3]{(-3)^3}$$
$$x = -3$$

620. c. Take the square root of both sides.

$$x^2 = 225$$
$$\sqrt{x^2} = \pm\sqrt{225}$$
$$x = \pm 15$$

621. a. Take the cube root of both sides.

$$x^3 = -125$$
$$\sqrt[3]{x^3} = \sqrt[3]{-125}$$
$$x = -5$$

622. c. Take the square root of both sides.

$$(x + 4)^2 = 81$$
$$\sqrt{(x + 4)^2} = \pm\sqrt{81}$$
$$x + 4 = \pm 9$$
$$x = -4 \pm 9$$
$$x = 5, -13$$

623. d. Isolate x^2 and then take square root of both sides.

$$x^2 + 1 = 0$$
$$x^2 = -1$$
$$\sqrt{x^2} = \pm\sqrt{-1}$$
$$x = \pm i$$

624. b. Isolate x^2 and then take square root of both sides.

$$x^2 + 81 = 0$$
$$x^2 = -81$$
$$\sqrt{x^2} = \pm\sqrt{-81}$$
$$x = \pm 9i$$

Set 40 (Page 95)

625. d. Apply the quadratic formula with $a = 1$, $b = 0$, and $c = -7$ to obtain:

$$x = \frac{-b \pm \sqrt{b^2 - 4ac}}{2a} = \frac{-(0) \pm \sqrt{(0)^2 - 4(1)(-7)}}{2(1)} =$$
$$\pm\frac{\sqrt{28}}{2} = \pm\frac{2\sqrt{7}}{2} = \pm\sqrt{7}$$

626. a. Apply the quadratic formula with $a = 2$, $b = 0$, and $c = -1$ to obtain:

$$x = \frac{-b \pm \sqrt{b^2 - 4ac}}{2a} = \frac{-(0) \pm \sqrt{(0)^2 - 4(2)(-1)}}{2(2)} =$$
$$\pm\frac{\sqrt{8}}{4} = \pm\frac{2\sqrt{2}}{4} = \pm\frac{\sqrt{2}}{2}$$

627. a. Apply the quadratic formula with $a = 4$, $b = 3$, and $c = 0$ to obtain:

$$x = \frac{-b \pm \sqrt{b^2 - 4ac}}{2a} = \frac{-(3) \pm \sqrt{(3)^2 - 4(4)(0)}}{2(4)}$$
$$= \frac{-3 \pm \sqrt{9}}{8} = \frac{-3 \pm 3}{8} = 0, -\frac{3}{4}$$

628. b. Apply the quadratic formula with $a = -5$, $b = 20$, and $c = 0$ to obtain:

$$x = \frac{-b \pm \sqrt{b^2 - 4ac}}{2a} = \frac{-(20) \pm \sqrt{(20)^2 - 4(-5)(0)}}{2(-5)} =$$
$$\frac{-20 \pm \sqrt{20^2}}{-10} = \frac{-20 \pm 20}{-10} = 0, 4$$

629. c. Apply the quadratic formula with $a = 1$, $b = 4$, and $c = 4$ to obtain:

$$x = \frac{-b \pm \sqrt{b^2 - 4ac}}{2a} = \frac{-(4) \pm \sqrt{(4)^2 - 4(1)(4)}}{2(1)}$$

$$= \frac{-4 \pm \sqrt{0}}{2} = -2 \text{ (repeated solution)}$$

630. c. Apply the quadratic formula with $a = 1$, $b = -5$, and $c = -6$ to obtain:

$$x = \frac{-b \pm \sqrt{b^2 - 4ac}}{2a} = \frac{-(-5) \pm \sqrt{(-5)^2 - 4(1)(-6)}}{2(1)} =$$

$$\frac{5 \pm \sqrt{49}}{2} = \frac{5 \pm 7}{2} = -1, 6$$

631. b. Apply the quadratic formula with $a = 3$, $b = 5$, and $c = 2$ to obtain:

$$x = \frac{-b \pm \sqrt{b^2 - 4ac}}{2a} = \frac{-(5) \pm \sqrt{(5)^2 - 4(3)(2)}}{2(3)} =$$

$$\frac{-5 \pm \sqrt{1}}{6} = \frac{-5 \pm 1}{6} = -1, -\frac{2}{3}$$

632. a. Apply the quadratic formula with $a = 5$, $b = 0$, and $c = -24$ to obtain:

$$x = \frac{-b \pm \sqrt{b^2 - 4ac}}{2a} = \frac{-(0) \pm \sqrt{(0)^2 - 4(5)(-24)}}{2(5)} =$$

$$\pm \frac{\sqrt{480}}{10} = \pm \frac{4\sqrt{30}}{10} = \pm \frac{2\sqrt{30}}{5}$$

633. a. First, put the equation into standard form by moving all terms to the left side of the equation to obtain the equivalent equation $2x^2 + 5x + 4 = 0$. Now, apply the quadratic formula with $a = 2$, $b = 5$, and $c = 4$ to obtain:

$$x = \frac{-b \pm \sqrt{b^2 - 4ac}}{2a} = \frac{-(5) \pm \sqrt{(5)^2 - 4(2)(4)}}{2(2)} =$$

$$\frac{-5 \pm \sqrt{-7}}{4} = \frac{-5 \pm i\sqrt{7}}{4}$$

634. a. Apply the quadratic formula with $a = 1$, $b = -2\sqrt{2}$, and $c = 3$ to obtain:

$$x = \frac{-b \pm \sqrt{b^2 - 4ac}}{2a} =$$

$$\frac{-(-2\sqrt{2}) \pm \sqrt{(-2\sqrt{2})^2 - 4(1)(3)}}{2(1)} = \frac{2\sqrt{2} \pm \sqrt{8 - 12}}{2} =$$

$$\frac{2\sqrt{2} \pm \sqrt{-4}}{2} = \frac{2\sqrt{2} \pm 2i}{2} = \sqrt{2} \pm i$$

635. b. First, put the equation into standard form by moving all terms to the left side of the equation to obtain the equivalent equation $x^2 + 2x = 0$. Now, apply the quadratic formula with $a = 1$, $b = 2$, and $c = 0$ to obtain:

$$x = \frac{-b \pm \sqrt{b^2 - 4ac}}{2a} = \frac{-(2) \pm \sqrt{(2)^2 - 4(1)(0)}}{2(1)} =$$

$$\frac{-2 \pm \sqrt{4}}{2} = \frac{-2 \pm 2}{2} = -2, 0$$

636. c. First, put the equation into standard form by expanding the expression on the left side, and then moving all terms to the left side of the equation:

$$(3x - 8)^2 = 45$$
$$9x^2 - 48x + 64 = 45$$
$$9x^2 - 48x + 19 = 0$$

Now, apply the quadratic formula with $a = 9$, $b = -48$, and $c = 19$ to obtain:

$$x = \frac{-b \pm \sqrt{b^2 - 4ac}}{2a} = \frac{-(-48) \pm \sqrt{(-48)^2 - 4(9)(19)}}{2(9)} =$$

$$\frac{-48 \pm \sqrt{1620}}{18} = \frac{-48 \pm 18\sqrt{5}}{18} = \frac{-8 \pm 3\sqrt{5}}{3}$$

637. d. We first multiply both sides of the equation by 100, then divide both sides by 20 in order to make the coefficients integers; this will help with the simplification process. Doing so yields the equivalent equation $x^2 - 11x + 10 = 0$. Now, apply the quadratic formula with $a = 1$, $b = -11$, and $c = 10$ to obtain:

$$x = \frac{-b \pm \sqrt{b^2 - 4ac}}{2a} = \frac{-(-11) \pm \sqrt{(-11)^2 - 4(1)(10)}}{2(1)}$$

$$= \frac{11 \pm \sqrt{81}}{2} = \frac{11 \pm 9}{2} = 1, 10$$

638. d. Apply the quadratic formula with $a = 1$, $b = -3$, and $c = -3$ to obtain:

$$x = \frac{-b \pm \sqrt{b^2 - 4ac}}{2a} = \frac{-(-3) \pm \sqrt{(-3)^2 - 4(1)(-3)}}{2(1)} =$$

$$\frac{3 \pm \sqrt{21}}{2}$$

639. b. The simplification process will be easier if we first eliminate the fractions by multiplying both sides of the equation $\frac{1}{6}x^2 - \frac{5}{3}x + 1 = 0$ by 6. Doing so yields the equivalent equation $x^2 - 10x + 6 = 0$. Now, apply the quadratic formula with $a = 1$, $b = -10$, and $c = 6$ to obtain:

$$x = \frac{-b \pm \sqrt{b^2 - 4ac}}{2a} = \frac{-(-10) \pm \sqrt{(-10)^2 - 4(1)(6)}}{2(1)} =$$

$$\frac{10 \pm \sqrt{76}}{2} = \frac{10 \pm 2\sqrt{19}}{2} = 5 \pm \sqrt{19}$$

640. b. First, put the equation into standard form by expanding the expression on the left side, and then moving all terms to the left side of the equation:

$$(x - 3)(2x + 1) = x(x - 4)$$
$$2x^2 - 6x + x - 3 = x^2 - 4x$$
$$x^2 - x - 3 = 0$$

Now, apply the quadratic formula with $a = 1$, $b = -1$, and $c = -3$ to obtain:

$$x = \frac{-b \pm \sqrt{b^2 - 4ac}}{2a} = \frac{-(-1) \pm \sqrt{(-1)^2 - 4(1)(-3)}}{2(1)} =$$

$$\frac{1 \pm \sqrt{13}}{2}$$

Set 41 (Page 97)

641. a. Isolate the squared expression on one side, take the square root of both sides, and solve for x, as follows:

$$4x^2 = 3$$
$$x^2 = \frac{3}{4}$$
$$x = \pm\sqrt{\frac{3}{4}} = \pm\frac{\sqrt{3}}{\sqrt{4}} = \pm\frac{\sqrt{3}}{2}$$

642. c. Isolate the squared expression on one side, take the square root of both sides, and solve for x:

$$-3x^2 = -9$$
$$x^2 = 3$$
$$x = \pm\sqrt{3}$$

643. b. Take the square root of both sides, and solve for x:

$$(4x + 5)^2 = -49$$
$$4x + 5 = \pm\sqrt{-49} = \pm 7i$$
$$4x = -5 \pm 7i$$
$$x = \frac{-5 \pm 7i}{4}$$

644. c. Take the square root of both sides, and solve for x:

$$(3x - 8)^2 = 45$$
$$3x - 8 = \pm\sqrt{45}$$
$$3x = 8 \pm\sqrt{45}$$
$$x = \frac{8 \pm \sqrt{45}}{3} = \frac{8 \pm 3\sqrt{5}}{3}$$

645. c. Isolate the squared expression on one side, take the square root of both sides, and solve for x:

$$(-2x + 1)^2 - 50 = 0$$
$$(-2x + 1)^2 = 50$$
$$-2x + 1 = \pm\sqrt{50}$$
$$-2x = -1 \pm\sqrt{50}$$
$$x = \frac{-1 \pm \sqrt{50}}{-2} = \frac{1 \pm \sqrt{50}}{2} = \frac{1 \pm 5\sqrt{2}}{2}$$

646. b. Isolate the squared expression on one side, take the square root of both sides, and solve for x:

$$-(1 - 4x)^2 - 121 = 0$$
$$(1 - 4x)^2 = -121$$
$$1 - 4x = \pm\sqrt{-121}$$
$$-4x = -1 \pm\sqrt{-121}$$
$$x = \frac{-1 \pm \sqrt{-121}}{-4} = \frac{1 \pm \sqrt{-121}}{4} = \frac{1 \pm 11i}{4}$$

647. a. To solve the given equation graphically, let $y_1 = 5x^2 - 24$, $y_2 = 0$. Graph these on the same set of axes and identify the points of intersection:

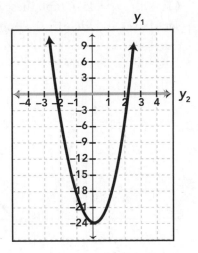

The x-coordinates of the points of intersection are the solutions of the original equation. We conclude that the solutions are approximately ± 2.191.

648. d. To solve the equation graphically, let $y_1 = 2x^2$, $y_2 = -5x - 4$. Graph these on the same set of axes and identify the points of intersection:

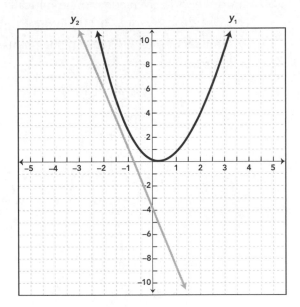

The x-coordinates of the points of intersection are the solutions of the original equation. Since the curves do not intersect, the solutions are imaginary.

649. a. To solve the given equation graphically, let $y_1 = 4x^2$, $y_2 = 20x - 24$. Graph these on the same set of axes and identify the points of intersection:

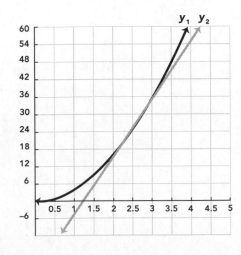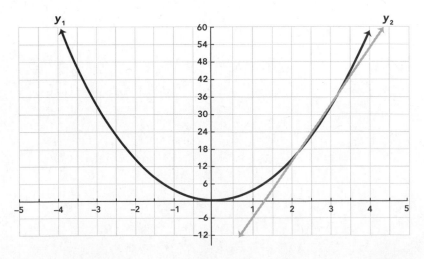

The x-coordinates of the points of intersection are the solutions of the original equation. The solutions are $x = 2, 3$.

650. c. To solve the given equation graphically, let $y_1 = 12x - 15x^2$, $y_2 = 0$. Graph these on the same set of axes and identify the points of intersection:

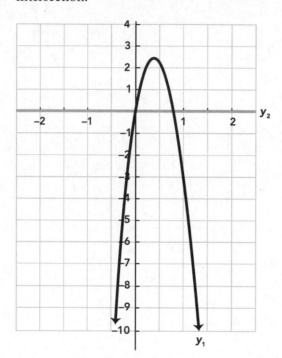

The x-coordinates of the points of intersection are the solutions of the original equation, so the solutions are $x = 0, 0.8$.

651. c. To solve the equation graphically, let $y_1 = (3x - 8)^2$, $y_2 = 45$. Graph these on the same set of axes and identify the points of intersection:

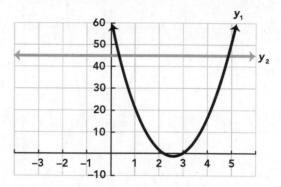

The x-coordinates of the points of intersection are the solutions of the original equation. We conclude that the solutions are approximately $x = 3.875, 4.903$.

652. b. To solve the given equation graphically, let $y_1 = 0.20x^2 - 2.20x + 2$, $y_2 = 0$. Graph these on the same set of axes and identify the points of intersection:

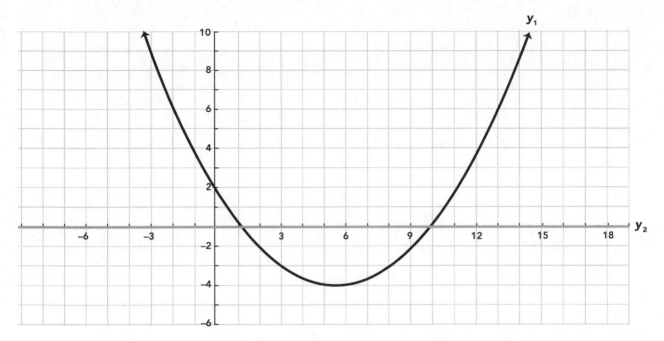

The x-coordinates of the points of intersection are the solutions of the original equation. We conclude that the solutions are $x = 1, 10$.

653. a. To solve the equation graphically, let $y_1 = x^2 - 3x - 3$, $y_2 = 0$. Graph these on the same set of axes and identify the points of intersection:

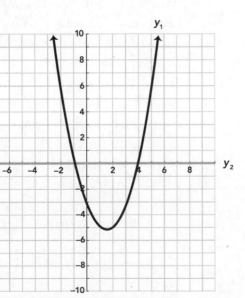

The x-coordinates of the points of intersection are the solutions of the original equation: approximately $x = -0.791, 3.791$.

654. b. To solve the given equation graphically, let $y_1 = x^2$ and $y_2 = -2x$. Graph these on the same set of axes and identify the points of intersection:

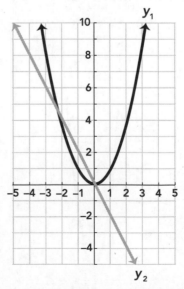

The x-coordinates of the points of intersection are the solutions of the original equation: $x = -2, 0$.

655. b. To solve the equation graphically, let $y_1 = \frac{1}{6}x^2 - \frac{5}{3}x + 1$, $y_2 = 0$. Graph these on the same set of axes and identify the points of intersection:

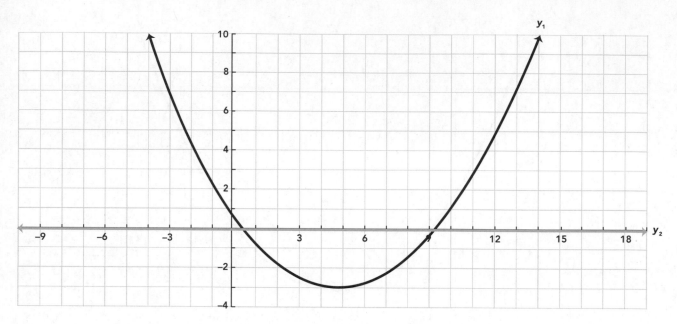

The x-coordinates of the points of intersection are the solutions of the original equation. We conclude that the solutions are approximately $x = 0.641, 9.359$.

656. c. To solve the equation graphically, let $y_1 = (2x + 1)^2 - 2(2x + 1) - 3$, $y_2 = 0$. Graph these on the same set of axes and identify the points of intersection:

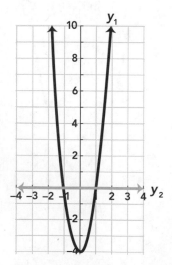

The x-coordinates of the points of intersection are the solutions of the original equation. The solutions are $x = -1, 1$.

Set 42 (Page 99)

657. b. Observe that $b^4 - 7b^2 + 12 = 0$ can be written as $(b^2)^2 - 7(b)^2 + 12 = 0$. Let $u = b^2$. Then, rewriting the previous equation yields the equation $u^2 - 7u + 12 = 0$, which is quadratic. Factoring the left side results in the equivalent equation $(u - 4)(u - 3) = 0$. Solving this equation for u yields the solutions $u = 4$ or $u = 3$. In order to solve the original equation, we must go back to the substitution and write u in terms of the original variable b:

$u = 4$ is the same as $b^2 = 4$, which gives us $b = \pm 2$

$u = 3$ is the same as $b^2 = 3$, which gives us $b = \pm\sqrt{3}$

The solutions of the original equation are $b = \pm 2, \pm\sqrt{3}$.

658. a. Let $u = b^2$. Observe that $(3b^2 - 1)(1 - 2b^2) = 0$ can be written as $(3u - 1)(1 - 2u) = 0$, which is quadratic. Solving this equation for u yields the solutions $u = \frac{1}{3}$ or $u = \frac{1}{2}$. In order to solve the original equation, we must go back to the substitution and write u in terms of the original variable b: $u = \frac{1}{3}$ is the same as $b^2 = \frac{1}{3}$, which gives us $b = \pm\sqrt{\frac{1}{3}} = \pm\frac{\sqrt{3}}{3}$ $u = \frac{1}{2}$ is the same as $b^2 = \frac{1}{2}$, which gives us $b = \pm\sqrt{\frac{1}{2}} = \pm\frac{\sqrt{2}}{2}$

The solutions of the original equation are $b = \pm\frac{\sqrt{2}}{2}, \pm\frac{\sqrt{3}}{3}$.

659. d. Note that $4b^4 + 20b^2 + 25 = 0$ can be written as $4(b^2)^2 + 20(b)^2 + 25 = 0$. Let $u = b^2$. Rewriting the original equation yields the equation, $4u^2 + 20u + 25 = 0$, which is quadratic. Factoring the left side results in the equivalent equation $(2u + 5)^2 = 0$. Solving this equation for u yields the solution $u = -\frac{5}{2}$. Solving the original equation requires that we go back to the substitution and write u in terms of the original variable b: $u = -\frac{5}{2}$ is the same as $b^2 = -\frac{5}{2}$, which gives us $b = \pm\sqrt{-\frac{5}{2}} = \pm i\sqrt{\frac{5}{2}} = \pm i(\frac{\sqrt{10}}{2})$

Therefore, the solutions of the original equation are $b = \pm i(\frac{\sqrt{10}}{2})$.

660. b. Observe that $16b^4 - 1 = 0$ can be written as $16(b^2)^2 - 1 = 0$. Let $u = b^2$. Rewriting the original equation yields the equation $16u^2 - 1 = 0$, which is quadratic. Factoring the left side results in the equivalent equation $(4u - 1)(4u + 1) = 0$. Solving this equation for u yields the solution $u = \pm\frac{1}{4}$. In order to solve the original equation, we must go back to the substitution and write u in terms of the original variable b: $u = -\frac{1}{4}$ is the same as $b^2 = -\frac{1}{4}$, which gives us $b = \pm\sqrt{-\frac{1}{4}} = \pm i\sqrt{\frac{1}{4}} \pm i(\frac{1}{2})$ $u = \frac{1}{4}$ is the same as $b^2 = \frac{1}{4}$, which gives us $b = \pm\sqrt{\frac{1}{4}} = \pm\frac{1}{2}$

The solutions of the original equation are $b = \pm i(\frac{1}{2}), \pm\frac{1}{2}$.

661. c. Note that, $x + 21 = 10x^{\frac{1}{2}}$ or equivalently, $x - 10x^{\frac{1}{2}} + 21 = 0$ can be written as $(x^{\frac{1}{2}})^2 - 10(x^{\frac{1}{2}}) + 21 = 0$. Let $u = x^{\frac{1}{2}}$. Then, rewriting the original equation yields the equation $u^2 - 10u + 21 = 0$, which is quadratic. Factoring the left side results in the equivalent equation $(u - 3)(u - 7) = 0$. Solving this equation for u yields the solution $u = 3$ or $u = 7$. In order to solve the original equation, we go back to the substitution and write u in terms of the original variable x: $u = 3$ is the same as $x^{\frac{1}{2}} = 3$, which gives $x = 3^2 = 9$ $u = 7$ is the same as $x^{\frac{1}{2}} = 7$, which gives $x = 7^2 = 49$

The solutions of the original equation are $x = 9, 49$.

662. a. Observe that $16 - 56\sqrt{x} + 49x = 0$ can be written as $16 - 56\sqrt{x} + 49(\sqrt{x})^2 = 0$. Let $u = \sqrt{x}$. Rewriting the original equation yields the equation $16 - 56u + 49u^2 = 0$, which is quadratic. Factoring the left side results in the equivalent equation $(4 - 7u)^2 = 0$. Solving this equation for u yields the solution $u = \frac{4}{7}$. In order to solve the original equation, we go back to the substitution and write u in terms of the original variable x: $u = \frac{4}{7}$ is the same as $\sqrt{x} = \frac{4}{7}$, which gives us $x = (\frac{4}{7})^2 = \frac{16}{49}$

Therefore, solution of the original equation is $x = \frac{16}{49}$.

663. a. Note that $x - \sqrt{x} = 6$, or equivalently $x - \sqrt{x} - 6 = 0$, can be written as $(\sqrt{x})^2 - (\sqrt{x}) - 6 = 0$. Let $u = \sqrt{x}$. Then, rewriting the *above* equation yields the equation $u^2 - u - 6 = 0$, which is quadratic. Factoring yields the equivalent equation $(u - 3)(u + 2) = 0$. Solving this equation for u yields the solutions $u = -2$ or $u = 3$. In order to solve the original equation, we must go back to the substitution and write u in terms of the original variable x:

$u = -2$ is the same as $\sqrt{x} = -2$, which does not have a real solution

$u = 3$ is the same as $\sqrt{x} = 3$, so that $x = 9$

Therefore, the solution of the original equation is $x = 9$.

664. c. We must first write the equation in the correct form:

$$2x^{\frac{1}{6}} - x^{\frac{1}{3}} - 1 = 0$$
$$x^{\frac{1}{3}} - 2x^{\frac{1}{6}} + 1 = 0$$
$$(x^{\frac{1}{6}})^2 - 2(x^{\frac{1}{6}}) + 1 = 0$$

Next, let $u = x^{\frac{1}{6}}$. Then, we must solve the equation $u^2 - 2u + 1 = 0$. Observe that factoring this equation yields $(u - 1)^2 = 0$. Consequently, $u = 1$. Next, we must go back to the actual substitution and solve the new equations obtained by substituting in this value of u. Specifically, we must solve $x^{\frac{1}{6}} = 1$. This is easily solved by raising both sides to the power 6. The result is that $x = 1$.

665. b. We must first rewrite the equation in a nicer form. Observe that

$$3 + x^{\frac{-1}{4}} - x^{\frac{-1}{2}} = 0$$
$$-x^{\frac{-1}{2}} + x^{\frac{-1}{4}} + 3 = 0$$
$$-(x^{\frac{-1}{4}})^2 + (x^{\frac{-1}{4}}) + 3 = 0$$

Let $u = x^{\frac{-1}{4}}$. Then, solve the quadratic equation $-u^2 + u + 3 = 0$. Using the quadratic formula yields

$$u = \frac{-1 \pm \sqrt{1 - 4(-1)(3)}}{2(-1)} = \frac{-1 \pm \sqrt{13}}{-2} = \frac{1 \pm \sqrt{13}}{2}.$$

Now, we must go back to the actual substitution and solve the following equations involving the original variable x:

$$x^{\frac{-1}{4}} = \frac{1 + \sqrt{13}}{2} \qquad x^{\frac{-1}{4}} = \frac{1 - \sqrt{13}}{2}$$

$$(x^{\frac{-1}{4}})^{-4} = (\frac{1 + \sqrt{13}}{2})^{-4} \quad (x^{\frac{-1}{4}})^{-4} = (\frac{1 - \sqrt{13}}{2})^{-4}$$

$$x = (\frac{2}{1 + \sqrt{13}})^4 = \frac{16}{(1 + \sqrt{13})^4}$$

$$x = (\frac{2}{1 - \sqrt{13}})^4 = \frac{16}{(1 - \sqrt{13})^4}$$

So, the two solutions to the original equation are $x = \frac{16}{(1 + \sqrt{13})^4}, \frac{16}{(1 - \sqrt{13})^4}$.

666. d. Let $u = x^3 + 5$. Then, the equation $(x^3 + 5)^2 - 5(x^3 + 5) + 6 = 0$ can be written equivalently as $u^2 - 5u + 6 = 0$. This factors as $(u - 3)(u - 2) = 0$, so we conclude that $u = 3, 2$. Next, we must solve the following equations obtained by going back to the actual substitution:

$$x^3 + 5 = 3 \qquad\qquad x^3 + 5 = 2$$
$$x^3 = -2 \qquad\qquad x^3 = -3$$
$$(x^3)^{\frac{1}{3}} = (-2)^{\frac{1}{3}} \qquad (x^3)^{\frac{1}{3}} = (-3)^{\frac{1}{3}}$$
$$x = \sqrt[3]{-2} \qquad\qquad x = \sqrt[3]{-3}$$

So, the two solutions to the original equation are $x = \sqrt[3]{-2}, \sqrt[3]{-3}$.

667. c. Observe that $4x^6 + 1 = 5x^3$, or equivalently $4x^6 - 5x^3 + 1 = 0$, can be written as $4(x^3)^2 - 5(x^3) + 1 = 0$. Let $u = x^3$. Rewriting the original equation yields the equation $4u^2 - 5u + 1 = 0$, which is quadratic. Factoring yields the equivalent equation $(4u - 1)(u - 1) = 0$. Solving this equation for u yields the solutions $u = \frac{1}{4}$ or $u = 1$. Solving the original equation requires that we go back to the substitution and write u in terms of the original variable x:

$u = \frac{1}{4}$ is the same as $x^3 = \frac{1}{4}$, so that $x = \sqrt[3]{\frac{1}{4}}$

$u = 1$ is the same as $x^3 = 1$, so that $x = 1$

The solutions of the original equation are $x = \sqrt[3]{\frac{1}{4}}, 1$.

668. b. Let $u = x^2 + x$. Observe that $(x^2 + x)^2 + 12 = 8(x^2 + x)$ or equivalently $(x^2 + x)^2 - 8(x^2 + x) + 12 = 0$, can be written as $u^2 - 8u + 12 = 0$, which is quadratic. Factoring yields the equivalent equation $(u - 6)(u - 2) = 0$. Solving this equation for u yields the solutions $u = 2$ or $u = 6$. To solve the original equation, we go back to the substitution and write u in terms of the original variable x. Doing so yields two more quadratic equations, this time in x, that must be solved: First, $u = 2$ is the same as $x^2 + x = 2$, or equivalently $x^2 + x - 2 = 0$. Factoring yields the equation $(x + 2)(x - 1) = 0$, so that $x = -2$ or 1. Similarly, $u = 6$ is the same as $x^2 + x = 6$, or equivalently $x^2 + x - 6 = 0$. Factoring yields the equation $(x + 3)(x - 2) = 0$, so that $x = -3$ or 2. Therefore, the solutions of the original equation are $x = -3, -2, 1,$ or 2.

669. a. Let $u = 1 + \sqrt{w}$. Observe that $2\left(1 + \sqrt{w}\right)^2 = 13\left(1 + \sqrt{w}\right) - 6$, or equivalently $2\left(1 + \sqrt{w}\right)^2 - 13\left(1 + \sqrt{w}\right) + 6 = 0$, can be written as $2u^2 - 13u + 6 = 0$, which is quadratic. Factoring yields the equivalent equation $(2u - 1)(u - 6) = 0$. Solving this equation for u yields the solutions $u = \frac{1}{2}$ or $u = 6$. Solving the original equation requires that we go back to the substitution and write u in terms of the original variable w. Doing so yields two more radical equations, this time in w, that must be solved. First, $u = \frac{1}{2}$ is the same as $1 + \sqrt{w} = \frac{1}{2}$. Isolating the radical term yields $\sqrt{w} = -\frac{1}{2}$, which has no real solution. Similarly, $u = 6$ is the same as $1 + \sqrt{w} = 6$. Isolating the radical term yields $\sqrt{w} = 5$, so that $w = 25$. The solution of the original equation is $w = 25$.

670. a. Let $u = r - \frac{3}{r}$. Observe that $(r - \frac{3}{r})^2 - (r - \frac{3}{r}) - 6 = 0$ can be written as $u^2 - u - 6 = 0$, which is quadratic. Factoring yields the equivalent equation $(u - 3)(u + 2) = 0$. Solving this equation for u yields the solutions $u = -2$ or $u = 3$. In order to solve the original equation, we must go back to the substitution and write u in terms of the original variable r. Doing so yields two more equations involving rational expressions, this time in r, that must be solved. First, $u = -2$ is the same as $r - \frac{3}{r} = -2$. Multiply both sides by r and solve for r:

$r - \frac{3}{r} = -2$

$r^2 - 3 = -2r$

$r^2 + 2r - 3 = 0$

$(r + 3)(r - 1) = 0$

$r = -3, 1$

Similarly, $u = 3$ is the same as $r - \frac{3}{r} = 3$. Multiply both sides by r and solve for r:

$r - \frac{3}{r} = 3$

$r^2 - 3 = 3r$

$r^2 - 3r - 3 = 0$

Using the quadratic formula then yields

$r = \frac{-(-3) \pm \sqrt{(-3)^2 - 4(1)(-3)}}{2(1)} = \frac{3 \pm \sqrt{21}}{2}$.

The solutions of the original equation are $r = -3, 1, \frac{3 \pm \sqrt{21}}{2}$.

671. b. Observe that $6\sqrt{x} - 13\sqrt[4]{x} + 6 = 0$ can be written as $6(\sqrt[4]{x})^2 - 13(\sqrt[4]{x}) + 6 = 0$. Let $u = \sqrt[4]{x}$. Rewriting the original equation yields the equation $6u^2 - 13u + 6 = 0$, which is quadratic. Factoring yields the equivalent equation $(2u - 3)(3u - 2) = 0$. Solving this equation for u yields the solutions $u = \frac{2}{3}$ or $u = \frac{3}{2}$. Solving the original equation requires that we go back to the substitution and write u in terms of the original variable x:

$u = \frac{2}{3}$ is the same as $\sqrt[4]{x} = \frac{2}{3}$, so that $x = (\frac{2}{3})^4 = \frac{16}{81}$.

$u = \frac{3}{2}$ is the same as $\sqrt[4]{x} = \frac{3}{2}$, so that $x = (\frac{3}{2})^4 = \frac{81}{16}$.

Therefore, the solutions of the original equation are $x = \frac{16}{81}, \frac{81}{16}$.

672. c. Let $u = a^{\frac{1}{3}}$. Observe that $2a^{\frac{2}{3}} - 11a^{\frac{1}{3}} + 12 = 0$ can be written as $2(a^{\frac{1}{3}})^2 - 11(a^{\frac{1}{3}}) + 12 = 0$.

Rewriting the original equation yields the equation $2u^2 - 11u + 12 = 0$, which is quadratic. Factoring yields the equivalent equation $(2u - 3)(u - 4) = 0$. Solving this equation for u yields the solutions $u = \frac{3}{2}$ or $u = 4$. In order to solve the original equation, we go back to the substitution and write u in terms of the original variable a:

$u = \frac{3}{2}$ is the same as $a^{\frac{1}{3}} = \frac{3}{2}$, so $a = (\frac{3}{2})^3 = \frac{27}{8}$

$u = 4$ is the same as $a^{\frac{1}{3}} = 4$, so $a = (4)^3 = 64$

The solutions of the original equation are $a = 64, \frac{27}{8}$.

Section 6—
Elementary Functions

Set 43 (Page 102)

673. **b.** Draw a horizontal line across the coordinate plane where $y = 3$. This line touches the graph of $f(x)$ in exactly one place. Therefore, there is one value for which $f(x) = 3$.

674. **d.** The x-axis is the graph of the line $y = 0$. The graph of $f(x)$ touches the x-axis in 5 places. Therefore, there are 5 values for which $f(x) = 0$.

675. **b.** Draw a horizontal line across the coordinate plane where $y = 10$. The arrowheads on the ends of the curve imply that the graph extends upward, without bound, as x tends toward both positive and negative infinity. The line $y = 10$ touches the graph of $f(x)$ in 2 places. Therefore, there are 2 values for which $f(x) = 10$.

676. **e.** The domain of a real-valued function is the set of all values that, when substituted for the variable, produce a meaningful output, while the range of a function is the set of all possible outputs. All real numbers can be substituted for x in the function $f(x) = x^2 - 4$, so the domain of the function is the set of all real numbers. Since the x term is squared, the smallest value that this term can equal is 0 (when $x = 0$). Therefore, the smallest value that $f(x)$ can attain occurs when $x = 0$. Observe that $f(0) = 0^2 - 4 = -4$. The range of $f(x)$ is the set of all real numbers greater than or equal to -4.

677. **a.** The domain of the function is the set of all real numbers, so any real number can be substituted for x. The range of a function is the set of all possible outputs of the function. Since the x term is squared, then made negative, the largest value that this term can equal is 0 (when $x = 0$). Every other x value will result in a negative value for $f(x)$. As such, the range of $f(x)$ is the set of all real numbers less than or equal to 0.

678. **c.** You must identify all possible y-values that are attained within the graph of f. The graph of f is comprised of three distinct components, each of which contributes an interval of values to the range of f. The set of y-values corresponding to the bottommost segment is $(-2,-1]$; note that -2 is excluded due to the open hole at $(5,-2)$ on the graph, and there is no other x-value in $[-5,5]$ whose functional value is -2. Next, the portion of the range corresponding to the middle segment is $[0,2)$; note that 2 is excluded from the range for the same reason -2 is excluded. Finally, the horizontal segment contributes the singleton $\{3\}$ to the range; even though there is a hole in the graph at $(0,3)$, there are infinitely many other x-values in $[-5,5]$ whose functional value is 3, thereby requiring that it be included in the range. Thus, the range is $(-2, -1]\cup[0,2)\cup\{3\}$.

679. **c.** The graph of g is steadily decreasing from left to right, beginning at the point $(-5,4)$ and ending at $(5,-4)$, with the only gap occurring in the form of a hole at $(0,1)$. Since there is no x-value in $[-5,5]$ whose functional value is 1, this value must be excluded from the range. All other values in the interval $[-4,4]$ do belong to the range. Thus, the range is $[-4,1)\cup(1, 4]$.

680. d. Using the graphs yields $f(0) = 0$, $f(2) = -1$, and $g(5) = -4$. Substituting these values into the given expression yields
$2 \cdot f(0) + [f(2) \cdot g(5)]^2 = 2(0) + [(-1)(-4)]^2$
$= 0 + 4^2 = 16$

681. b. The zeros of a polynomial are precisely its x-intercepts, which are –3,1, and 3.

682. c. The lowest point on the graph of $y = p(x)$ occurs at $(2, -1)$, so the smallest possible y-value attained is –1. Further, every real number greater than –1 is also attained at some x-value. Hence, the range is $[-1, \infty)$.

683. a. The domain of any polynomial function is $\mathbb{R}$ because any real number can be substituted in for x in $p(x)$ and yield another real number.

684. d. We must identify the x-values of the portion of the graph of $y = p(x)$ that lies between the horizontal lines $y = -1$ and $y = 0$ (i.e., the x-axis). Once this is done, we exclude the x-values of the points where the graph of $y = p(x)$ intersects the line $y = -1$ (because of the strict inequality), and we include those x-values of the points where the graph of $y = p(x)$ intersects the line $y = 0$. This yields the set $[1, 2) \cup (2, 3] \cup \{-3\}$.

685. b. The graph of f has a vertical asymptote at $x = 1$ and a horizontal asymptote at $y = -2$. Since the graph follows the vertical asymptote up to positive infinity as x approaches $x = 1$ from the left and down to negative infinity as x approaches $x = 1$ from the right, and it does not cross the horizontal asymptote, we conclude that the graph attains all y-values except –2. Hence, the range is $(-\infty, -2) \cup (-2, \infty)$.

686. b. $\frac{9f(x)}{g(x)} = \frac{9[-(2x-(-1-x^2))]}{3(1+x)} = \frac{-9(x^2+2x+1)}{3(x+1)}$
$= \frac{-9(x+1)^2}{3(x+1)} = -3(x+1) = -g(x)$

687. d. Since the function $2g(x)h(x) = \frac{6(x+1)}{x^2+1}$ is a rational function, its domain is the set of all those x-values for which the denominator is not equal to zero. There is no real number x that satisfies the equation $x^2 + 1 = 0$. Therefore, the domain is the set of all real numbers.

688. b. $3f(x) - 2xg(x) - \frac{1}{h(x)}$
$= 3[-(2x-(-1-x^2))] - 2x[3(x+1)] - \frac{1}{\frac{1}{1+x^2}}$
$= -3(x^2+2x+1) - 6x(x+1) - (x^2+1)$
$= -3x^2 - 6x - 3 - 6x^2 - 6x - x^2 - 1$
$= -10x^2 - 12x - 4$
$= -2(5x^2 + 6x + 2)$

Set 44 (Page 105)

689. a. The graph of the equation in diagram A is not a function. A function requires that each unique input yields no more than one output. The graph in diagram A fails the vertical line test for all x-values where $-2 < x < 2$. For each of these x-values (inputs), there are two y-values (outputs).

690. d. The range of a function is the set of all possible outputs of the function. In each of the five equations, the set of possible y-values that can be generated for the equation is the range of the equation. We must identify the coordinate planes that show a graph that extends below the x-axis. These equations have negative y-values, which means that the range contains negative values. The graphs of the equations in diagrams A, B, and D extend below the x-axis. However, the graph in diagram A is not a function. It fails the vertical line test for all x-values where $-2 < x < 2$. The equations graphed in diagrams B and D are functions whose ranges contain negative values.

691. e. The equation of the graph in diagram B is $y = |x| - 3$. Any real number can be substituted into this equation. There are no x-values that will generate an undefined or imaginary y-value. The equation of the graph in diagram E is $y = (x - 3)^2 + 1$. With this equation as well, any real number can be substituted for x—there are no x-values that will generate an undefined or imaginary y-value. The equation of the graph in diagram D is $y = \frac{1}{x}$. If $x = 0$, this function will be undefined. Therefore, the domain of this function is all real numbers excluding 0. Only the functions in diagrams B and E have a domain of all real numbers with no exclusions.

692. b. The equation of the graph in diagram C is $y = \sqrt{x}$. Since the square root of a negative number is imaginary, the domain of this equation is the set of all real numbers greater than or equal to 0. The square roots of real numbers greater than or equal to 0 are also real numbers that are greater than or equal to 0. Therefore, the range of $y = \sqrt{x}$ is the set of all real numbers greater than or equal to 0; the domain and range of the equation are the same. The equation of the graph in diagram D is $y = \frac{1}{x}$. If $x = 0$, this function will be undefined. Therefore, the domain of this function is all real numbers excluding 0. Dividing 1 by a real number (excluding 0) will yield real numbers, excluding 0. Therefore, the range of $y = \frac{1}{x}$ is all real numbers excluding 0, and so the domain and range of the equation are the same. The equation of the graph in diagram B is $y = |x| - 3$. Any real number can be substituted into this equation. There are no x-values that will generate an undefined or imaginary y-value. However, it is impossible to generate a y-value that is less than -3. Any x-value will generate a y-value that is

greater than -3. Therefore, the range of $y = |x| - 3$ is the set of all real numbers greater than or equal to -3. So, the domain and range of $y = |x| - 3$ are not the same. The equation of the graph in diagram E is $y = (x - 3)^2 + 1$. With this equation as well, any real number can be substituted for x— there are no x-values that will generate an undefined or imaginary y-value. However, it is impossible to generate a y-value that is less than 1. Any x-value will generate a y-value that is greater than 1. Therefore, the range of the equation $y = (x - 3)^2 + 1$ is the set of all real numbers greater than or equal to 1. So, the domain and range of $y = (x - 3)^2 + 1$ are not the same.

693. a. Substitute the expression $-\frac{2}{x}$ for every occurrence of x in the definition of the function $f(x)$, and then simplify:

$$f\left(-\tfrac{2}{x}\right) = -\frac{1}{\left(-\frac{2}{x}\right)^3}$$

$$= -\frac{1}{\left(-\frac{8}{x^3}\right)}$$

$$= -1 \cdot \left(-\tfrac{x^3}{8}\right)$$

$$= \tfrac{x^3}{8}$$

694. a. Substitute the expression $2y - 1$ for every occurrence of x in the definition of the function $f(x)$, and then simplify:
$$f(2y - 1) =$$
$$(2y - 1)^2 + 3(2y - 1) - 2 =$$
$$4y^2 - 4y + 1 + 6y - 3 - 2 =$$
$$4y^2 + 2y - 4$$

695. c. Simplifying $f(x + h)$ requires we substitute the expression $x + h$ for every occurrence of x in the definition of the function $f(x)$, and then simplify:
$$f(x + h) =$$
$$-[(x + h) - 1]^2 + 3 =$$
$$-[(x + h)^2 - 2(x + h) + 1] + 3 =$$
$$-[x^2 + 2hx + h^2 - 2x - 2h + 1] + 3 =$$
$$-x^2 - 2hx - h^2 + 2x + 2h - 1 + 3 =$$
$$-x^2 - 2hx - h^2 + 2x + 2h + 2$$

Next, in anticipation of simplifying $f(x + h) - f(x)$, we expand the expression for $f(x) = -(x-1)^2 + 3$ in order to facilitate combining like terms:

$f(x) =$
$-(x-1)^2 + 3 =$
$-(x^2 - 2x + 1) + 3 =$
$-x^2 + 2x - 1 + 3 =$
$-x^2 + 2x + 2$

Finally, simplify the original expression $f(x + h) - f(x)$:

$f(x + h) - f(x) =$
$(-x^2 - 2hx - h^2 + 2x + 2h + 2) - (-x^2 + 2x + 2) =$
$-x^2 - 2hx - h^2 + 2x + 2h + 2 + x^2 - 2x - 2 =$
$-2hx - h^2 + 2h =$
$-h(2x + h - 2)$

696. c. By definition, $(g \circ h)(4) = g(h(4))$. Observe that $h(4) = 4 - 2\sqrt{4} = 4 - 2(2) = 0$, so $g(h(4)) = g(0) = 2(0)^2 - 0 - 1 = -1$. Thus, we conclude that $(g \circ h)(4) = -1$.

697. d. By definition, $(f \circ f \circ f)(2x) = f(f(f(2x)))$. Working from the inside outward, we first note that $f(2x) = -(2x)^2 = -4x^2$. Then, $f(f(2x)) = f(-4x^2) = -(-4x^2)^2 = -16x^4$. Finally, $f(f(f(2x))) = f(-16x^4) = -(-16x^4)^2 = -256x^8$. Thus, we conclude that $(f \circ f \circ f)(2x) = -256x^8$.

698. b. Begin with the innermost function: find $f(-2)$ by substituting -2 for x in the function $f(x)$:

$f(-2) = 3(-2) + 2 = -6 + 2 = -4$

Then, substitute the result for x in $g(x)$.

$g(-4) = 2(-4) - 3 = -8 - 3 = -11$
Thus, $g(f(-2)) = -11$

699. e. Begin with the innermost function: Find $f(3)$ by substituting 3 for x in the function $f(x)$:

$f(3) = 2(3) + 1 = 6 + 1 = 7$

Next, substitute that result for x in $g(x)$.

$g(7) = 7 - 2 = 5$
Finally, substitute this result for x in $f(x)$:
$f(5) = 2(5) + 1 = 10 + 1 = 11$
Thus, $f(g(f(3))) = 11$.

700. c. Begin with the innermost function. You are given the value of $f(x)$: $f(x) = 6x + 4$. Substitute this expression for x in the equation $g(x)$, and then simplify:

$g(6x + 4) =$
$(6x + 4)^2 - 1 =$
$36x^2 + 24x + 24x + 16 - 1 =$
$36x^2 + 48x + 15$
Therefore, $g(f(x)) = 36x^2 + 48x + 15$.

701. b. Since $g(0) = 2$ and $f(2) = -1$, we have $(f \circ g)(0) = f(g(0)) = f(2) = -1$.

702. b. Since $f(5) = 0$ and $f(0) = 0$, we work from the inside outward to obtain $f(f(f(f(5)))) = f(f(f(0))) = f(f(0)) = f(0) = 0$

703. c. Replace x by $x + 2$ in $f(x)$ to obtain:

$f(x + 2) =$
$\sqrt{(x + 2)^2 - 4(x + 2)} =$
$\sqrt{x^2 + 4x + 4 - 4x - 8} =$
$\sqrt{x^2 - 4}$

704. b. The domain of $g \circ f$ consists of only those values of x for which the quantity $f(x)$ is defined (that is, x belongs to the domain of f) and for which $f(x)$ belongs to the domain of g. For the present scenario, the domain of f consists of only those x-values for which $-3x \geq 0$, which is equivalent to $x \leq 0$. Since the domain of $g(x) = \sqrt{2x^2 + 18}$ is the set of all real numbers, it follows that all x-values in the interval $(-\infty,0]$ are permissible inputs in the composition function $(g \circ f)(x)$, and that, in fact, these are the only permissible inputs. Therefore, the domain of $g \circ f$ is $(-\infty,0]$.

Set 45 (Page 107)

705. a. The radicand of an even-indexed radical term (e.g., a square root) must be nonnegative if it occurs in the numerator of a fraction and strictly positive if it is in the denominator of a fraction. For the present function, this restriction takes the form of the inequality $-x \geq 0$, which upon multiplication on both sides by -1, is equivalent to $x \leq 0$. Hence, the domain of the function $f(x) = \sqrt{-x}$ is $(-\infty,0]$.

706. d. There is no restriction on the radicand of an odd-indexed radical term (e.g., a cube root) if it is in the numerator of a fraction, whereas the radicand of such a radical term must be nonzero if it occurs in the denominator of a fraction. For the present function, this restriction takes the form of the statement $-1 - x \neq 0$, which is equivalent to $x \neq -1$. Hence, the domain of the function $g(x) = \frac{1}{\sqrt[3]{-1-x}}$ is $(-\infty,-1)\cup(-1,\infty)$.

707. b. The graph of the equation $y = 2$ is a horizontal line that crosses the y-axis at $(0,2)$. Horizontal lines have a slope of 0. This line is a function, since it passes the vertical line test: A vertical line can be drawn through the graph of $y = 2$ at any point and will cross the graphed function in only one place. The domain of the function is the set of all real numbers, but all x-values yield the same y-value: 2. Therefore, the range of $y = 2$ is $\{2\}$.

708. b. The graph of $f(x) = |x|$ has its lowest point at the origin, which is both an x-intercept and a y-intercept. Since $f(x) > 0$ for any nonzero real number x, it cannot have another x-intercept. Moreover, a function can have only one y-intercept, since if it had more than one, it would not pass the vertical line test.

709. b. The intersection of the graph of $f(x) = x^3$ and the graph of the horizontal line $y = a$ can be found by solving the equation $x^3 = a$. Taking the cube root of both sides yields a single solution of $x = \sqrt[3]{a}$, which is meaningful for any real number a.

710. c. The graph of $f(x) = \frac{1}{x}$, in fact, decreasing on its entire domain, not just $(0,\infty)$. Its graph is given here:

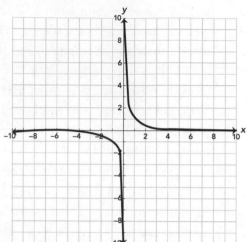

711. c. The square root of a negative value is imaginary, so the value of $4x - 1$ must be greater than or equal to 0. Symbolically, we have:

$4x - 1 \geq 0$

$4x \geq 1$

$x \geq \frac{1}{4}$

Hence, the domain of $f(x)$ is the set of all real numbers greater than or equal to $\frac{1}{4}$. The smallest value of $f(x)$ occurs at $x = \frac{1}{4}$, and its value is $\sqrt{4(\frac{1}{4}) - 1} = \sqrt{0} = 0$.

So, the range of the function is the set of all real numbers greater than or equal to 0.

712. c. The minimum value for each of these functions occurs at its vertex, which is at $(0,0)$. Also, for any positive real number a, the graphs of both f and g intersect the horizontal line $y = a$ (twice). Therefore, the range of both functions is $[0,\infty)$.

713. c. The radicand of an odd-indexed radical term (e.g., a fifth root) must be nonzero if it occurs in the denominator of a fraction, which is presently the case. As such, the restriction takes the form of the statement $2 - x \neq 0$, which is equivalent to $x \neq 2$. Thus, the domain is $(-\infty,2) \cup (2,\infty)$.

714. c. The x-intercepts of f are those values of x satisfying the equation $1 - |2x - 1| = 0$, which is equivalent to $|2x - 1| = 1$. Using the fact that $|a| = b$ if and only if $a = \pm b$, we solve the two equations $2x - 1 = \pm 1$ separately:

$2x - 1 = -1 \qquad\qquad 2x - 1 = 1$

$2x = 0 \qquad\qquad\quad\; 2x = 2$

$x = 0 \qquad\qquad\qquad x = 1$

Thus, there are precisely two x-intercepts of the given function.

715. d. The x-values of the points of intersection of the graphs of $f(x) = x^2$ and $g(x) = x^4$ must satisfy the equation $x^4 = x^2$. This equation is solved as follows:

$x^4 = x^2$

$x^4 - x^2 = 0$

$x^2(x^2 - 1) = 0$

$x^2(x - 1)(x + 1) = 0$

$x = -1, 0, 1$

The points of intersection are $(-1,1)$, $(0,0)$, and $(1,1)$. So, there are more than two points of intersection.

716. d. The x-values of the points of intersection of the graphs of $f(x) = 2x$ and $g(x) = 4x^3$ must satisfy the equation $4x^3 = 2x$. This equation is solved as follows:

$4x^3 = 2x$

$4x^3 - 2x = 0$

$4x(x^2 - \frac{1}{2}) = 0$

$4x(x - \sqrt{\frac{1}{2}})(x + \sqrt{\frac{1}{2}}) = 0$

$x = 0, \pm\sqrt{\frac{1}{2}}$

So, the points of intersection are $(0,0)$, $(\sqrt{\frac{1}{2}}, 2\sqrt{\frac{1}{2}})$, and $(-\sqrt{\frac{1}{2}}, -2\sqrt{\frac{1}{2}})$.

So, there are more than two points of intersection.

717. b. The x-values of the points of intersection of the graphs of $f(x) = \frac{3}{4}x^2$ and $g(x) = \frac{5}{16}x^2$ must satisfy the equation $\frac{3}{4}x^2 = \frac{5}{16}x^2$. This equation is solved as follows:

$\frac{3}{4}x^2 = \frac{5}{16}x^2$

$\frac{3}{4}x^2 - \frac{5}{16}x^2 = 0$

$(\frac{3}{4} - \frac{5}{16})x^2 = 0$

$x^2 = 0$

$x = 0$

Hence, there is only one point of intersection, namely $(0,0)$.

718. b. The y-intercept for a function $y = f(x)$ is the point $(0, f(0))$. Observe that $f(0)$

$= \frac{-2 - |2 - 3(0)|}{4 - 2(0)^2 |-0|} = \frac{-2 - 2}{4 - 0} = \frac{-4}{4} = -1$.

So, the y-intercept is $(0, -1)$.

719. a. If there is no x-value that satisfies the equation $f(x) = 0$ then the graph of $y = f(x)$ does not cross the horizontal line $y = 0$, which is the x-axis. So, there are no x-intercepts.

720. b. If there is no x-value that satisfies the equation $f(x) = 3$, then there is no point on the graph of $y = f(x)$ when $y = 3$. Therefore, 3 is not in the range of f.

Set 46 (Page 109)

721. c. The domain of a rational function is the set of all real numbers that do not make the denominator equal to zero. For this function, the values of x that must be excluded from the domain are the solutions of the equation $x^3 - 4x = 0$. Factoring the left side yields the equivalent equation $x^3 - 4x = x(x^2 - 4) = x(x - 2)(x + 2) = 0$ The solutions of this equation are $x = -2, 0$, and 2. Hence, the domain is $(-\infty, -2) \cup (-2, 0) \cup (0, 2) \cup (2, \infty)$.

722. d. First, simplify the expression for $f(x)$ as follows:

$\frac{(x - 3)(x^2 - 16)}{(x^2 + 9)(x - 4)} = \frac{(x - 3)(x - 4)(x + 4)}{(x^2 + 9)(x - 4)} =$

$\frac{(x - 3)(x + 4)}{(x^2 + 9)} = \frac{x^2 + x - 12}{x^2 + 9}$

While there is a hole in the graph of f at $x = 4$, there is no x-value that makes the denominator of the simplified expression equal to zero. Hence, there is no vertical asymptote. But, since the degrees of the numerator and denominator are equal, there is a horizontal asymptote given by $y = 1$ (since the quotient of the coefficients of the terms of highest degree in the numerator and denominator is $1 \div 1 = 1$).

723. a. We must identify the intervals in the domain of $p(x)$ on which the graph of $y = p(x)$ rises from left to right. This happens on the intervals $(-3, 0) \cup (2, \infty)$.

724. a. The graph of f passes the horizontal line test on this interval, so, it has an inverse.

725. c. Determining the inverse function for f requires that we solve for x in the expression $y = \frac{x - 1}{5x + 2}$:

$y = \frac{x - 1}{5x + 2}$

$y(5x + 2) = x - 1$

$5xy + 2y = x - 1$

$5xy - x = -2y - 1$

$x(5y - 1) = -2y - 1$

$x = \frac{-2y-1}{5y-1}$

Now, we conclude that the function $f^{-1}(y) = \frac{-2y-1}{5y-1}, y \uparrow \frac{1}{5}$ is the inverse function of f.

726. d. Remember that the domain of f is equal to the range of f^{-1}. As such, since 0 does not belong to the range of f^{-1}, it does not belong to the domain of f, so $f(0)$ is undefined. Also, the fact that $(1,4)$ is on the graph of $y = f(x)$ means that $f(1) = 4$; this is equivalent to saying that $(4,1)$ is on the graph of f^{-1}, or that $f^{-1}(4) = 1$. So, all three statements are true.

727. a. Determining the inverse function for f requires that we solve for x in the expression $y = x^3 + 2$:

$y = x^3 + 2$

$y - 2 = x^3$

$x = \sqrt[3]{y - 2}$

Hence, the inverse is $f^{-1}(y) = \sqrt[3]{y - 2}$.

728. b. First, note that $x^2 + 1$ does not factor, so $x = 1$ is a vertical asymptote for the graph of f. Since the degree of the numerator of the fraction is exactly one more than that of the denominator, we can conclude that the graph has no horizontal asymptote, but does have an oblique asymptote. Hence, II is a characteristic of the graph of f.

Next, while there is a y-intercept, $(0,3)$, there is no x-intercept. To see this, we must consider the equation $f(x) = 0$ which is equivalent to $2 - \frac{x^2+1}{x-1} = \frac{2(x-1)-(x^2+1)}{x-1}$

$= \frac{-(x^2 - 2x + 3)}{x-1} = 0.$

The x-values that satisfy this equation are those that make the numerator equal to zero, but do not make the denominator equal to zero. Since the numerator does not factor, we know that $(x - 1)$ is not a factor of it, so we need only solve the equation. $x^2 - 2x + 3 = 0$. Using the quadratic formula yields

$x = \frac{-(-2) \pm \sqrt{(-2)^2 - 4(1)(3)}}{2(1)} = \frac{2 \pm \sqrt{-8}}{2} =$

$1 \pm i\sqrt{2}$

Since the solutions are imaginary, we conclude that there are no x-intercepts. Hence, III is not a characteristic of the graph of f.

729. b. The expression for f can be simplified as follows:

$\frac{(2-x)^2(x+3)}{} = \frac{(-(x-2))^2(x+3)}{x(x-2)^2} =$

$\frac{(x-2)^2(x+3)}{x(x-2)^2} = \frac{x+3}{x}$

Since $x = 2$ makes both the numerator and denominator of the unsimplified expression equal to zero, there is a hole in the graph of f at this value. So, statement I holds. Next, since the degrees of the numerator and denominator are equal, there is a horizontal asymptote given by $y = 1$ (since the quotient of the coefficients of the terms of highest degree in the numerator and denominator is $1 \div 1 = 1$). Because $x = 0$ makes the denominator equal to zero, but does not make the numerator equal to zero, it is a vertical asymptote. So, statement II holds. Finally, since $x = 0$ is a vertical asymptote, the graph of f cannot intersect this line, and there is no y-intercept. So, statement III does not hold.

730. c. The y-values of $f(x) = \frac{1}{x}$ get smaller as x-values move from left to right on $(-\infty, 0)$. The graph is as follows:

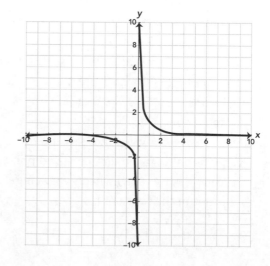

731. d. The y-values on all three graphs increase as the x-values move from left to right in the interval $(0,\infty)$. Their graphs are as follows:

$f(x) = x^3$

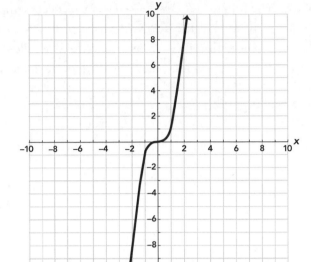

$f(x) = 2x + 5$

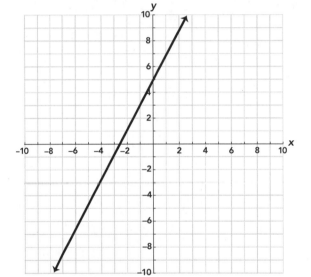

$f(x) = |x|$

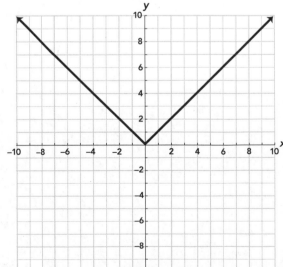

732. c. The function $f(x) = \frac{x^2+1}{x^2+3}$ has no vertical asymptote since no value of x makes the denominator equal to zero, and has the horizontal asymptote $y = 1$.

733. d. The function $f(x) = x^3$ is an example that illustrates the truth of statements **a** and **b**. And the function $g(x) = -1 - x^2$ illustrates the truth of statement **c**. Their graphs are as follows:

$f(x) = x^3$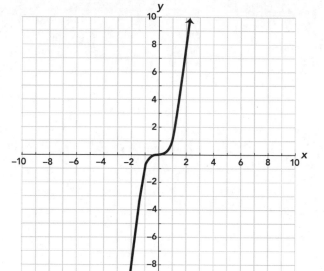

$g(x) = -1 - x^2$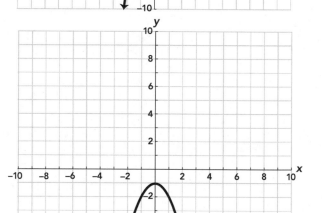

734. d. Statement **a** is true because, by definition, a positive slope of m means that the graph of the line rises vertically m units for every positive unit increase in x. Next, the graph of the function $f(x) = \frac{1}{x^2}$ intersects both Quadrants I and II, illustrating the truth of statement **b**. (See the following graph.) Statement **c** is true because the vertex is the point at which the maximum or minimum value of a quadratic function occurs. These graphs resemble the letter U or an upside down U, so that the graph is indeed decreasing on one side of the vertex and increasing on the other side of the vertex.

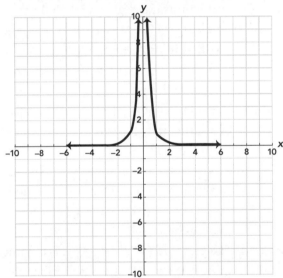

735. b. The x-values of the points of intersection of the graphs of $f(x) = -4x$ and $g(x) = 2\sqrt{x}$ must satisfy the equation $-4x = 2\sqrt{x}$, which is solved as follows:
$$-4x = 2\sqrt{x}$$
$$(-4x)^2 = (2\sqrt{x})^2$$
$$16x^2 = 4x$$
$$16x^2 - 4x = 0$$
$$4x(4x - 1) = 0$$
$$x = 0, \tfrac{1}{4}$$

736. a. The x-values of the points of intersection of the graphs of $f(x) = \sqrt{x}$ and $g(x) = 3\sqrt{x}$ must satisfy the equation $\sqrt{x} = 3\sqrt{x}$, which is solved as follows:
$$\sqrt{x} = 3\sqrt{x}$$
$$\sqrt{x} - 3\sqrt{x} = 0$$
$$-2\sqrt{x} = 0$$
$$x = 0$$

Set 47 (Page 112)

737. d. We apply the general principle that the graph of $y = g(x + h) + k$ is obtained by shifting the graph of $y = g(x)$ right (resp. left) h units if $h < 0$ (resp. $h > 0$), and up (resp.) down k units if $k > 0$ (resp. $k < 0$). Here, observe that $f(x) = (x + 2)^3 - 3 = g(x + 2) - 3$, so the correct choice is **d**.

738. d. In order for the turning point of the parabola to be in Quadrant II, its x-coordinate must be negative and its y-coordinate must be positive. Note that the turning point of $y = -(x + 2)^2 + 1$ is $(-2, 1)$, so that the correct choice is **d**.

739. a. The turning point of $y = (x - 2)^2 - 2$ is $(2, -2)$, while the turning point of $y = x^2$ is $(0,0)$. Therefore, we would shift the graph of $y = x^2$ to the right 2 units and down 2 units.

740. c. The graph of $y = g(x + h) + k$ is obtained by shifting the graph of $y = g(x)$ right (resp. left) h units if $h < 0$ (resp. $h > 0$), and up (resp. down) k units if $k > 0$ (resp. $k < 0$). Here, observe that $f(x) = (x - 4)^3 + 1 = g(x - 4) + 1$, so the correct choice is **c**.

741. d. The graph of $y = g(x + h) + k$ is obtained by shifting the graph of $y = g(x)$ right (resp. left) h units if $h < 0$ (resp. $h > 0$), and up (resp. down) k units if $k > 0$ (resp. $k < 0$). Here, observe that $f(x) = (x - 2)^2 - 4 = g(x - 2) - 4$, so the correct choice is **d**.

742. b. The graph of $y = g(x + h) + k$ is obtained by shifting the graph of $y = g(x)$ right (resp. left) h units if $h < 0$ (resp. $h > 0$), and up (resp. down) k units if $k > 0$ (resp. $k < 0$). Here, observe that $f(x) = (x - 2)^3 - 1 = g(x - 2) - 1$, so the correct choice is **b.**

743. d. The graph of $y = g(x + h) + k$ is obtained by shifting the graph of $y = g(x)$ right (resp. left) h units if $h < 0$ (resp. $h > 0$), and up (resp. down) k units if $k > 0$ (resp. $k < 0$). Here, observe that $f(x) = \sqrt{x - 5} - 3 = g(x - 5) - 3$, so the correct choice is **d.**

744. d. The graph of $y = g(x + h) + k$ is obtained by shifting the graph of $y = g(x)$ right (resp. left) h units if $h < 0$ (resp. $h > 0$), and up (resp. down) k units if $k > 0$ (resp. $k < 0$). Here, observe that $f(x) = 2\sqrt{x + 3} = g(x + 3)$, so the correct choice is **d.**

745. a. The graph of $y = g(x + h) + k$ is obtained by shifting the graph of $y = g(x)$ right (resp. left) h units if $h < 0$ (resp. $h > 0$), and up (resp. down) k units if $k > 0$ (resp. $k < 0$). Here, observe that, $f(x) = |x + 6| + 4 = g(x + 6) + 4$, so that the correct choice is **a.**

746. b. The graph of $y = -g(x + h) + k$ is obtained by shifting the graph of $y = g(x)$ right (resp. left) h units if $h < 0$ (resp. $h > 0$), then reflecting the graph over the x-axis, and finally shifting the graph up (resp. down) k units if $k > 0$ (resp. $k < 0$). Here, observe that $f(x) = -|x - 1| + 5 = -g(x - 1) + 5$, so the correct choice is **b.**

747. a. The graph of $y = -g(x + h) + k$ is obtained by shifting the graph of $y = g(x)$ right (resp. left) h units if $h < 0$ (resp. $h > 0$), then reflecting the graph over the x-axis, and finally shifting the graph up (resp. down) k units if $k > 0$ (resp. $k < 0$). Here, observe that $f(x) = -(x + 3)^3 + 5 = -g(x + 3) + 5$, so the correct choice is **a.**

748. d. The graph of $y = -g(x + h) + k$ is obtained by shifting the graph of $y = g(x)$ right (resp. left) h units if $h < 0$ (resp. $h > 0$), then reflecting the graph over the x-axis, and finally shifting the graph up (resp. down) k units if $k > 0$ (resp. $k < 0$). Hence, the correct choice is **d.**

749. a. The graph of $y = g(x + h) + k$ is obtained by shifting the graph of $y = g(x)$ right (resp. left) h units if $h < 0$ (resp. $h > 0$), and up (resp. down) k units if $k > 0$ (resp. $k < 0$). Hence, the correct choice is **a.**

750. c. The graph of $y = -g(x + h) + k$ is obtained by shifting the graph of $y = g(x)$ right (resp. left) h units if $h < 0$ (resp. $h > 0$), then reflecting the graph over the x-axis, and finally shifting the graph up (resp. down) k units if $k > 0$ (resp. $k < 0$). Hence, the correct choice is **c.**

751. b. The graph of $y = -g(x + h) + k$ is obtained by shifting the graph of $y = g(x)$ right (resp. left) h units if $h < 0$ (resp. $h > 0$), then reflecting the graph over the x-axis, and finally shifting the graph up (resp. down) k units if $k > 0$ (resp. $k < 0$). Hence, the correct choice is **b.**

752. d. The graph of $y = g(x + h) + k$ is obtained by shifting the graph of $y = g(x)$ right (resp. left) h units if $h < 0$ (resp. $h > 0$), then reflecting the graph over the x-axis, and finally shifting the graph up (resp. down) k units if $k > 0$ (resp. $k < 0$). Hence, the correct choice is **d.**

Set 48 (Page 115)

753. b. $e^{3x-2y} = e^{3x}e^{-2y} = (e^x)^3(e^y)^{-2} = 2^3 3^{-2} = \frac{8}{9}$

754. b. $2^x \cdot 2^{x+1} = 2^{x+x+1} = 2^{2x+1}$

755. c. $(4^{x-1})^2 \cdot 16 = 4^{2(x-1)} \cdot 16 = 4^{2x-2} \cdot 4^2 = 4^{2x-2+2} = 4^{2x}$

756. c. $\left(\frac{5^{4x}}{5^{2x-6}}\right)^{\frac{1}{2}} = (5^{4x-(2x-6)})^{\frac{1}{2}} = (5^{2x+6})^{\frac{1}{2}} = 5^{\frac{1}{2}(2x+6)} = 5^{x+3}$

757. b. $(e^x + e^{-x})^2 = (e^x)^2 + 2(e^x)(e^{-x}) + (e^{-x})^2 = e^{2x} + 2 + e^{-2x} = e^{2x} + e^{-2x} + 2$

758. d. $\frac{(5^{3x-1})^3 \cdot 5^{x-1}}{5^{2x}} = \frac{5^{9x-3}5^{x-1}}{5^{2x}} = \frac{5^{10x-4}}{5^{2x}} = 5^{8x-4}$
$= (5^4)^{2x-1} = 625^{2x-1}$

759. b. $e^x(e^x - 1) - e^{-x}(e^x - 1) = (e^x)^2 - e^x - (e^{-x})(e^x) - (e^{-x})(-1) = e^{2x} - e^x - 1 + e^{-x}$

760. b. $\frac{e^x(e^x - e^{-x}) + e^{-x}(e^x + e^{-x})}{e^{-2x}}$

$= \frac{(e^x)(e^x) - (e^x)(e^{-x}) + (e^{-x})(e^x) + (e^{-x})(e^{-x})}{e^{-2x}}$

$= \frac{e^{2x} - 1 + 1 + e^{-2x}}{e^{-2x}}$

$= \frac{e^{2x} + e^{-2x}}{e^{-2x}}$

$= e^{4x} + 1$

761. b. If $b > 1$, the graph of $y = b^x$ gets very close to the x-axis as the x-values move to the left. A typical graph is as follows:

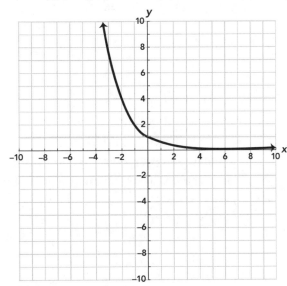

762. d. If $0 < b < 1$, the graph of $y = b^x$ gets very close to the x-axis as the x-values move to the right, and the y-values grow very rapidly as the x-values move to the left. A typical graph is as follows:

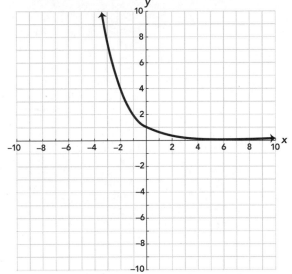

763. c. If $b > 0$, then $b^x > 0$, so the equation $b^x = 0$ has no solution.

764. c. Note that $\left(\frac{1}{2}\right)^{-x} = 2^x$. The graph of $y = 2^x$ is always above the x-axis, so statement **c** is true.

765. b. Observe that $1 - 3^x \le 0$ is equivalent to $3^x \ge 1$. The graph of $y = 3^x$ is always increasing and is equal to 1 when $x = 0$, so for all $x > 0$, $3^x > 1$. Hence, the solution set is $[0, \infty)$.

766. c. Observe that $\left(-\frac{2}{3}\right)^{2x} = \left(\frac{4}{9}\right)^x$, which is strictly positive, for any real number x. Therefore, the solution set for the inequality $\left(-\frac{2}{3}\right)^{2x} \le 0$ is the empty set.

767. c. The graph of $f(x) = -\left(\frac{3}{4}\right)^x$ is always below the x-axis and increases as x moves from left to right, as shown by its graph:

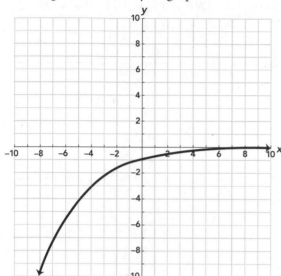

768. d. The graph of $f(x) = -\left(\frac{15}{7}\right)^{3x}$ is always below the x-axis and is decreasing as x moves from left to right, as shown by its graph:

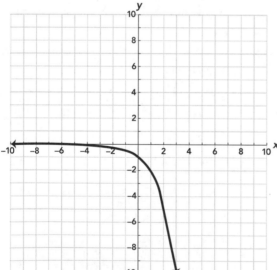

Set 49 (Page 116)

769. b. The graph of $f(x) = 1 - 2e^x$ is obtained by reflecting the graph of $g(x) = e^x$ over the x-axis, scaling it by a factor of 2, and then translating it up one unit. In doing so, the original horizontal asymptote $y = 0$ for g becomes $y = 1$, and the graph of f always stays below this asymptote. Hence, the range is $(-\infty, 1)$. See the following graph.

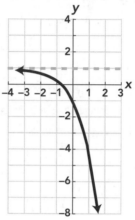

770. c. Rewrite the expression on the right side of the equation as a power of 2, as $4^{3x} = (2^2)^{3x} = 2^{6x}$, and substitute into the original equation to obtain the equivalent equation $2^{7x^2-1} = 2^{6x}$. Now, equate the exponents and solve for x:
$$7x^2 - 1 = 6x$$
$$7x^2 - 6x - 1 = 0$$
$$(7x + 1)(x - 1) = 0$$
The solutions are $x = -\frac{1}{7}$ and $x = 1$.

771. d. Since $\frac{1}{25} = 5^{-2}$, the original equation is equivalent to $5^{\sqrt{x+1}} = 5^{-2}$. The x-values that satisfy this equation must satisfy the one obtained by equating the exponents of the expressions on both sides of the equation, which is $\sqrt{x+1} = -2$. But the left side of this equation is nonnegative for any x-value that does not make the radicand negative. Hence, the equation has no solution.

772. d. The graph of $f(x) = -e^{2-x} - 3$ can be obtained by reflecting the graph of $g(x) = e^x$ about the y-axis, then shifting it to the left 2 units, then reflecting it over the x-axis, and finally shifting it down 3 units. The graph is as follows:

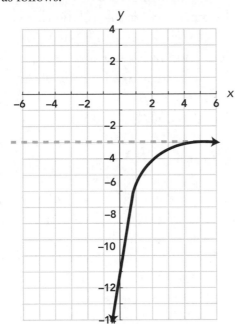

It is evident that all three characteristics provided in choices, **a, b,** and **c** hold from this graph.

773. d. Since 8 can be expressed as a power of 2, we can write the following:
$$2^{x-5} = 8$$
$$2^{x-5} = 2^3$$
Equating exponents yields $x - 5 = 3$, so $x = 8$ is the solution.

774. a. First, rewrite 9 as a power of 3. Then, simplify the right side, as follows:
$$3^{2x} = 9 \cdot 3^{x-1}$$
$$3^{2x} = 3^2 \cdot 3^{x-1}$$
$$3^{2x} = 3^{2+x-1}$$
$$3^{2x} = 3^{x+1}$$
Equating exponents yields $2x = x + 1$. Solving this equation yields $x = 1$.

775. c. Multiply both sides by the denominator on the right side. Then, observe that 1 can be written as 4^0.
$$4^{2x-3} = \frac{1}{4^x}$$
$$4^{2x-3} \cdot 4^x = \frac{1}{4^x} \cdot 4^x$$
$$4^{2x-3+x} = 1$$
$$4^{3x-3} = 4^0$$
$$3x - 3 = 0$$
$$x = 1$$

776. d. The power to which 125 should be raised in order to yield 25 is not obvious. However, note that 125 and 25 are both powers of 5. We use this fact in order to first rewrite the equation in a more convenient form, and then solve for x:
$$125^x = 25$$
$$(5^3)^x = 5^2$$
$$5^{3x} = 5^2$$
Equating exponents yields $3x = 2$. Thus, we conclude that the solution is $x = \frac{2}{3}$.

777. b. First, simplify the left side. Then, equate the exponents and solve the resulting equation.
$$(e^x)^{x-3} = e^{10}$$
$$e^{x^2-3x} = e^{10}$$
$$x^2 - 3x = 10$$
$$x^2 - 3x - 10 = 0$$
$$(x - 5)(x + 2) = 0$$
$$x = 5, x = -2$$

778. d. Rewrite the left side of the equation as a power of 4, then equate the exponents and solve for x:
$$16^{3x-1} = 4^{2x+3}$$
$$(4^2)^{3x-1} = 4^{2x+3}$$
$$4^{2(3x-1)} = 4^{2x+3}$$
$$2(3x - 1) = 2x + 3$$
$$6x - 2 = 2x + 3$$
$$4x = 5$$
$$x = \frac{5}{4}$$

779. c. Rewrite both sides of the equation as a power of 2, then equate the exponents and solve for x:
$$4^{x+1} = (\tfrac{1}{2})^{2x}$$
$$(2^2)^{x+1} = (2^{-1})^{2x}$$
$$2^{2(x+1)} = 2^{-2x}$$
$$2(x+1) = -2x$$
$$2x + 2 = -2x$$
$$4x = -2$$
$$x = -\tfrac{1}{2}$$

780. a. Factor the left side of the equation, set each factor equal to zero, and solve for x, as follows:
$$x \cdot 3^x + 5 \cdot 3^x = 0$$
$$3^x(x + 5) = 0$$
$$3^x = 0 \text{ or } x + 5 = 0$$
Since $3^x = 0$ has no solutions, we conclude that the only solution is $x = -5$.

781. a. Rewrite both sides of the equation as a power of 10, then equate the exponents and solve for x using the quadratic formula:
$$(10^{x+1})^{2x} = 100$$
$$10^{2x(x+1)} = 10^2$$
$$2x(x + 1) = 2$$
$$2x^2 + 2x = 2$$
$$2x^2 + 2x - 2 = 0$$
$$x^2 + x - 1 = 0$$
$$x = \frac{-1 \pm \sqrt{1 - 4(1)(-1)}}{2(1)} = \frac{-1 \pm \sqrt{5}}{2}$$

782. c. Rewrite both sides of the equation as a power of 2, then equate the exponents and solve for x using the quadratic formula:
$$2^{\sqrt{x}} \cdot 2 = 8$$
$$2^{\sqrt{x}+1} = 2^3$$
$$\sqrt{x} + 1 = 3$$
$$\sqrt{x} = 2$$
$$x = 4$$

783. c. Factor the left side of the equation, set each factor equal to zero, and solve for x:
$$2x^2 \cdot e^x - 7x \cdot e^x + 6 \cdot e^x = 0$$
$$e^x(2x^2 - 7x + 6) = 0$$
$$e^x(2x - 3)(x - 2) = 0$$
$$e^x = 0 \text{ or } 2x - 3 = 0 \text{ or } x - 2 = 0$$
Since $e^x = 0$ has no solutions, we conclude that the solutions are $\tfrac{3}{2}$ and 2.

784. b. Factor the left side of the equation, set each factor equal to zero, and solve for x:
$$e^{2x} + 5e^x - 6 = 0$$
$$(e^x)^2 + 5(e^x) - 6 = 0$$
$$(e^x + 6)(e^x - 1) = 0$$
$$e^x + 6 = 0 \text{ or } e^x - 1 = 0$$
$$e^x = -6 = 0 \text{ or } e^x = 1$$
Since e^x is always positive, the first equation has no solution. However, the second equation is satisfied when $x = 0$.

Set 50 (Page 118)

785. d. Finding x such that $\log_3 27 = x$ is equivalent to finding x such that $3^x = 27$. Since $27 = 3^3$, the solution of this equation is 3.

786. b. Finding x such that $\log_3\left(\tfrac{1}{9}\right) = x$ is equivalent to finding x such that $3^x = \tfrac{1}{9}$. Since $\tfrac{1}{9} = 3^{-2}$, we conclude that the solution of this equation is -2.

787. c. Finding x such that $\log_{\frac{1}{2}} 8 = x$ is equivalent to finding x such that $\left(\tfrac{1}{2}\right)^x = 8$. Since $8 = 2^3$ and $\left(\tfrac{1}{2}\right)^x = (2^{-1})^x = 2^{-x}$, this equation is equivalent to $2^3 = 2^{-x}$, the solution of which is -3.

788. c. Finding x such that $\log_7 \sqrt{7} = x$ is equivalent to finding x such that $7^x = \sqrt{7}$. Since $\sqrt{7} = 7^{\frac{1}{2}}$, we conclude that the solution of this equation is $\tfrac{1}{2}$.

ANSWERS AND EXPLANATIONS

789. a. Finding x such that $\log_5 1 = x$ is equivalent to finding x such that $5^x = 1$. We conclude that the solution of this equation is 0.

790. d. Finding x such that $\log_{16} 64 = x$ is equivalent to finding x such that $16^x = 64$. We rewrite the expressions on both sides of the equation using the same base of 2. Indeed, observe that $16^x = (2^4)^x = 2^{4x}$ and $64 = 2^6$. Hence, the value of x we seek is the solution of the equation $4x = 6$, which is $x = \frac{6}{4} = \frac{3}{2}$.

791. c. The equation $\log_6 x = 2$ is equivalent to $x = 6^2 = 36$. So, the solution is $x = 36$.

792. b. $\log_a x = \log_a(5\sqrt{a}) = \log_a 5 + \log_a(\sqrt{a}) = \log_a 5 + \log_a(a^{\frac{1}{2}}) = \log_a 5 + \frac{1}{2}\log_a a = \log_a 5 + \frac{1}{2}$

793. b. $\log_3(3^4 \cdot 9^3) = \log_3(3^4) + \log_3(9^3) = 4\log_3 3 + 3\log_3 9 = 4(1) + 3(2) = 10$

794. d. The equation $5^{3x-1} = 7$ is equivalent to $\log_5 7 = 3x - 1$. This equation is solved as follows:
$$\log_5 7 = 3x - 1$$
$$3x = 1 + \log_5 7$$
$$x = \frac{1}{3}(1 + \log_5 7)$$

795. b. The given expression can be written as one involving the terms $\log_a x$ and $\log_a y$:
$$\log_a\left(\frac{x}{y^3}\right) = \log_a x - \log_a(y^3) = \log_a x - 3\log_a y$$
Substituting $\log_a x = 2$ and $\log_a y = -3$ into this expression yields
$$\log_a\left(\frac{x}{y^3}\right) = \log_a x - 3\log_a(y) = 2 - 3(-3) = 2 + 9 = 11$$

796. a. The equation $3^{\log_3 2} = x$ is equivalent to $\log_3 x = \log_3 2$. So, the solution is $x = 2$.

797. c. Since $f(x) = \log_a x$ and $g(x) = a^x$ are inverses, it follows by definition that $f(g(x)) = \log_a(a^x) = x$.

798. a. First, write the expression on the left side as the ln of a single expression.
$$3\ln\left(\frac{1}{x}\right) = \ln 8$$
$$\ln\left(\frac{1}{x}\right)^3 = \ln 8$$
$$\left(\frac{1}{x}\right)^3 = 8$$
$$x^{-3} = 8$$
$$(x^{-3})^{\frac{-1}{3}} = 8^{\frac{-1}{3}}$$
$$x = \frac{1}{2}$$

799. b. $e^{\frac{-1}{2}\ln 3} = e \ln(3)^{\frac{-1}{2}} = 3^{\frac{-1}{2}} = \frac{1}{\sqrt{3}} = \frac{\sqrt{3}}{3}$

800. c. The given expression can be written as one involving the terms $\ln x$ and $\ln y$:
$$\ln\left(\frac{e^2 y}{\sqrt{x}}\right) =$$
$$\ln(e^2 y) - \ln(\sqrt{x}) =$$
$$\ln(e^2) + \ln y - \ln(x^{\frac{1}{2}}) =$$
$$2\ln(e) + \ln y - \frac{1}{2}\ln x =$$
$$2 + \ln y - \frac{1}{2}\ln x$$
Substituting $\ln x = 3$ and $\ln y = 2$ into this expression yields
$$\ln\left(\frac{e^2 y}{\sqrt{x}}\right) = 2 + \ln y - \frac{1}{2}\ln x = 2 + 2 - \frac{1}{2}(3) = 4 - \frac{3}{2} = \frac{5}{2}$$

Set 51 (Page 119)

801. a. Using the logarithm rules yields
$$3\ln(xy^2) - 4\ln(x^2 y) + \ln(xy) =$$
$$\ln(xy^2)^3 - \ln(x^2 y)^4 + \ln(xy) =$$
$$\ln\left[\frac{(xy^2)^3}{(x^2 y)^4}\right] + \ln(xy) =$$
$$\ln\left[\frac{(xy^2)^3(xy)}{(x^2 y)^4}\right] =$$
$$\ln\left[\frac{x^3 y^6 xy}{x^8 y^4}\right] =$$
$$\ln\frac{y^3}{x^4}$$

802. a. $\log_8 2 + \log_8 4 = \log_8(2 \cdot 4) = \log_8 8 = 1$

803. c. $4\log_9 3 = \log_9(3^4) = \log_9 81 = 2$

804. a. $\ln(18x^3) - \ln(6x) = \ln(\frac{18x^3}{6x}) = \ln(3x^2)$

805. b. $\log_7 \frac{2}{49} - \log_7 \frac{2}{7} = \log_7(\frac{2}{49} \div \frac{2}{7}) = \log_7(\frac{2}{49} \cdot \frac{7}{2})$
$= \log_7 \frac{1}{7} = -1$

806. d. $3 \log_4 \frac{2}{3} + \log_4 27 = \log_4 (\frac{2}{3})^3 + \log_4 27 =$
$\log_4(\frac{8}{27}) + \log_4 27 = \log_4 (\frac{8}{27} \cdot 27) =$
$\log_4 8 = \frac{3}{2}$

807. c. $\log(2x^3) = \log 2 + \log x^3 = \log 2 + 3 \log x$

808. d. $\log_2 (\frac{8yz^4}{x^2}) = \log_2 (8yz^4) - \log_2 (x^2) =$
$\log_2 8 + \log_2 y + \log_2 (z^4) - \log_2 (x^2) =$
$3 + \log_2 y + 4\log_2 z - 2\log_2 x$

809. d. $\frac{3}{2} \log_2 4 - \frac{2}{3}\log_2 8 + \log_2 2 =$
$\frac{3}{2} \log_2 (2^2) - \frac{2}{3}\log_2 (2^3) + \log_2 2 =$
$\frac{3}{2}(2) - \frac{2}{3}(3) + 1 =$
$3 - 2 + 1 =$
2

810. d. $3 \log_b (x + 3)^{-1} - 2\log_b x + \log_b (x + 3)^3 =$
$-3 \log_b (x + 3) - 2\log_b x + 3\log_b (x + 3) =$
$- 2\log_b x =$
$\log_b (x^{-2}) =$
$\log_b (\frac{1}{x^2})$

811. c. $\ln \left[(2\sqrt{x + 1})(x^2 + 3)^4 \right] =$
$\ln (2\sqrt{x + 1}) + \ln(x^2 + 3)^4 =$
$\ln (2\sqrt{x + 1}) + 4\ln(x^2 + 3) =$
$\ln 2 + \ln (\sqrt{x + 1}) + 4\ln (x^2 + 3) =$
$\ln 2 + \ln (x + 1)^{\frac{1}{2}} + 4\ln (x^2 + 3) =$
$\ln 2 + \frac{1}{2}\ln (x + 1) + 4\ln (x^2 + 3)$

812. c. $\log_3 \frac{x^2\sqrt{2x - 1}}{(2x + 1)^{\frac{3}{2}}}$

$= \log_3 (x^2\sqrt{2x - 1}) - \log_3 (2x + 1)^{\frac{3}{2}}$

$= \log_3 (x^2) + \log_3 (\sqrt{2x - 1}) - \log_3 (2x + 1)^{\frac{3}{2}}$

$= \log_3 (x^2) + \log_3 (2x - 1)^{\frac{1}{2}} - \log_3 (2x + 1)^{\frac{3}{2}}$

$= 2\log_3 (x) + \frac{1}{2} \log_3 (2x - 1) - \frac{3}{2} \log_3 (2x + 1)$

813. a. As the *x*-values decrease toward zero, the *y*-values on the graph of $f(x) = \ln x$ plunge downward very sharply, as can be seen in the following graph:

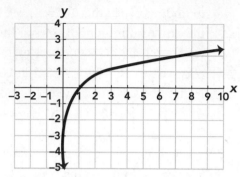

814. a. The inputs for logarithmic functions must be positive. Since the input in this case is $-x$, then the inequality $-x > 0$ must be satisfied. Clearly $-x > 0$ is equivalent to $x < 0$. Thus the domain of *k* is $(-\infty,0)$.

815. d. The inputs of a logarithm must be positive. Since $x^2 + 1 > 0$ for any real number *x*, it follows that the domain of 6(*x*) is the set of all real numbers.

816. c. The *x*-intercept is the point of the form $(x,0)$. Since $\log_2 1 = 0$, we conclude that the *x*-intercept is $(1,0)$.

Set 52 (Page 121)

817. d. The range of the function $g(x) = \ln x$ is $\mathbb{R}$. Since using $2x - 1$ as the input for g for all $x > \frac{1}{2}$ covers the same set of inputs as g, the range of g is also $\mathbb{R}$. The graph is provided here.

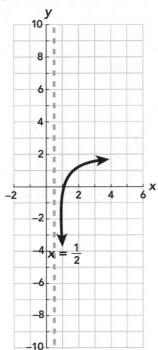

818. c. The x-intercepts of the function $f(x) = \ln (x^2 - 4x + 4)$ are those x-values that satisfy the equation $x^2 - 4x + 4 = 1$, solved as follows:
$$x^2 - 4x + 4 = 1$$
$$x^2 - 4x + 3 = 0$$
$$(x - 3)(x - 1) = 0$$
The solutions of this equation are $x = 3$ and $x = 1$. So, the x-intercepts are $(1,0)$ and $(3,0)$.

819. c. The domain of the function $f(x) = \ln(x^2 - 4x + 4)$ is the set of x-values for which $x^2 - 4x + 4 > 0$. This inequality is equivalent to $(x - 2)^2 > 0$, which is satisfied by all real numbers x except 2. So, the domain is $(-\infty,2) \cup (2,\infty)$.

820. b. The vertical asymptote for $f(x) = \ln(x + 1) + 1$ occurs at those x-values that make the input of the ln portion equal to zero, namely $x = -1$.

821. b. Using exponent and logarithm rules yields
$$f(g(x)) = e^{2(\ln \sqrt{x})} = e^{2\ln x^{\frac{1}{2}}} = e^{2(\frac{1}{2})\ln x}$$
$$= e^{\ln x} = x, \text{ for all } x > 0, g(f(x)) = \ln \sqrt{e^{2x}}$$
$$= \ln \sqrt{(e^x)^2} = \ln(e^x) = x, \text{ for all real}$$
numbers x.

Hence, we conclude that these choices for f and g are inverses.

822. a. Apply the logarithm properties and then solve for x:
$$\ln(x - 2) - \ln(3 - x) = 1$$
$$\ln\left(\frac{x-2}{3-x}\right) = 1$$
$$\frac{x-2}{3-x} = e$$
$$x - 2 = e(3 - x)$$
$$x - 2 = 3e - ex$$
$$ex + x = 3e + 2$$
$$(e + 1)x = 3e + 2$$
$$x = \frac{3e+2}{e+1}$$

Substituting this value for x in the original equation reveals that it is indeed a solution.

823. b.
$$5 \leq 4e^{2-3x} + 1 < 9$$
$$4 \leq 4e^{2-3x} < 8$$
$$1 \leq e^{2-3x} < 2$$
$$\ln 1 \leq 2 - 3x < \ln 2$$
$$-2 + 0 \geq -3x < -2 + \ln 2$$
$$\frac{2}{3} \geq x > \frac{-2 + \ln 2}{-3}$$
$$\frac{2 - \ln 2}{3} < x \geq \frac{2}{3}$$

So, the solution set is $\left(\frac{2 - \ln 2}{3}, \frac{2}{3}\right]$.

ANSWERS AND EXPLANATIONS

824. c. First, note that the expression $\ln(1-x^2)$ is only defined when $1-x^2 > 0$, which is equivalent to $-1 < x < 1$. Since the only values of y for which $\ln y \le 0$ are $0 < y \le 1$, we must determine which x-values satisfy the more restrictive inequality $0 < 1 - x^2 \le 1$. However, that is true for every x in the interval $(-1,1)$. Hence, the solution set of the inequality $\ln(1-x^2) \le 0$ is $(-1,1)$.

825. d. First, rewrite the left side as a single logarithm. Then, convert the resulting equation to an equivalent exponential equation and solve:
$\log x + \log(x+3) = 1$
$\log[x(x+3)] = 1$
$x(x+3) = 10^1$
$x^2 + 3x - 10 = 0$
$(x+5)(x-2) = 0$
$x = -5, 2$
However, $x = -5$ cannot be a solution since -5 is not in the domain of $\log x$. Thus, $x = 2$ is the only solution.

826. b. We combine the two logarithms, and then convert to an equivalent exponential equation, which is solved as follows:
$\log_2(2x) + \log_2(x+1) = 2$
$\log_2[2x(x+1)] = 2$
$2x(x+1) = 2^2$
$2x^2 + 2x = 4$
$2x^2 + 2x - 4 = 0$
$2(x-1)(x+2) = 0$
$x = 1, -2$
We must exclude $x = -2$ because it is not in the domain of either logarithm in the original equation. So, $x = 1$ is the only solution.

827. a. We combine the two logarithms, and then convert to an equivalent exponential equation, which is solved as follows:
$\log(x-2) = 2 + \log(x+3)$
$\log(x+2) - \log(x+3) = 2$
$\log\left(\frac{x+2}{x-3}\right) = 2$
$\frac{x+2}{x-3} = 10^2$
$x+2 = 100(x-3)$
$x+2 = 100x - 300$
$302 = 99x$
$x = \frac{302}{99}$

828. d. Beause b to a power is equal to one if and only if the power is zero, we must determine the value of x such that $3\log_b x = 0$, or equivalently $\log_b x = 0$. This equation can be written in the exponential form $x = b^0 = 1$. Thus, the solution is 1.

829. a. $y = e^{-a(b+x)}$
$\ln y = -a(b+x)$
$\ln y = -ab - ax$
$ab + \ln y = -ax$
$-\frac{1}{a}(ab + \ln y) = x$

830. a. $3\ln 4y + \ln A = \ln B$
$3\ln 4y = \ln B - \ln A$
$\ln 4y = \frac{1}{3}(\ln B - \ln A)$
$4y = e^{\frac{1}{3}(\ln B - \ln A)}$
$y = \frac{1}{4}e^{\frac{1}{3}(\ln B - \ln A)}$

831. c. $1 + \ln(x \cdot y) = \ln z$
$1 + \ln x + \ln y = \ln z$
$\ln x = \ln z - \ln y - 1$
$x = e^{\ln z - \ln y - 1}$

832. a. $P = P_0 e^{-kt}$
$\frac{P}{P_0} = e^{-kt}$
$\ln\left(\frac{P}{P_0}\right) = -kt$
$-\frac{1}{k}\ln\left(\frac{P}{P_0}\right) = t$

ANSWERS AND EXPLANATIONS

Section 7—Matrix Algebra

Set 53 (Page 124)

833. b. The matrix $[1 \quad 2 \quad -1 \quad 0]$ has one row and four columns, so its dimensions are 1×4.

834. b. The matrix $\begin{bmatrix} 0 & -2 \\ 0 & 1 \\ 0 & -2 \\ 0 & 0 \end{bmatrix}$ has four rows and two columns, so its dimensions are 4×2.

835. a. A matrix with dimensions 3×2 has three rows and two columns. The matrix in choice **a** is the correct choice.

836. c. $-3 \begin{bmatrix} -1 & -1 & 0 \\ 0 & -3 & 1 \end{bmatrix} = \begin{bmatrix} 3 & 3 & 0 \\ 0 & 9 & -3 \end{bmatrix}$

837. d. These matrices cannot be added because they do not have the same dimensions.

838. c. $2 \begin{bmatrix} -3 & -1 \\ 0 & 1 \end{bmatrix} - 3 \begin{bmatrix} 2 & -1 \\ 2 & 2 \end{bmatrix} = \begin{bmatrix} -6 & -2 \\ 0 & 2 \end{bmatrix} - \begin{bmatrix} 6 & -3 \\ 6 & 6 \end{bmatrix} = \begin{bmatrix} -12 & 1 \\ -6 & -4 \end{bmatrix}$

839. c. $\frac{2}{5} \begin{bmatrix} -1 & 0 & 1 \\ 1 & 0 & 1 \\ 0 & 1 & -1 \end{bmatrix} = \begin{bmatrix} -\frac{2}{5} & 0 & \frac{2}{5} \\ \frac{2}{5} & 0 & \frac{2}{5} \\ 0 & \frac{2}{5} & -\frac{2}{5} \end{bmatrix}$

840. a. Two matrices are equal if and only if each pair of corresponding entries is the same. Note that

$\frac{1}{2} \begin{bmatrix} 6 & -4 \\ 0 & 4 \end{bmatrix} = \begin{bmatrix} 3 & -2 \\ 0 & 2 \end{bmatrix}$. Therefore, the equation $\begin{bmatrix} x & -2 \\ 0 & 2 \end{bmatrix} = \frac{1}{2} \begin{bmatrix} 6 & -4 \\ 0 & 4 \end{bmatrix}$ is equivalent to

$\begin{bmatrix} x & -2 \\ 0 & 2 \end{bmatrix} = \begin{bmatrix} 3 & -2 \\ 0 & 2 \end{bmatrix}$, which occurs only when $x = 3$.

841. d. Two matrices are equal if and only if each pair of corresponding entries are the same. Therefore, both $x^2 = 4$ and $3x = 6$ must hold simultaneously. The solutions of the first equation are $x = -2$ and 2, but only $x = 2$ also satisfies the second equation. Thus, we conclude that the only x-value that makes the equality true is $x = 2$.

842. d. Two matrices are equal if and only if each pair of corresponding entries are the same. Observe that

$$3\begin{bmatrix} 0 & 2 \\ 1 & 1 \\ 1 & x \end{bmatrix} = \begin{bmatrix} 0 & 6 \\ 3 & 3 \\ 3 & 3x \end{bmatrix} \text{ and } -1\begin{bmatrix} 0 & -6 \\ -3 & -3 \\ -3 & 3x \end{bmatrix} = \begin{bmatrix} 0 & 6 \\ 3 & 3 \\ 3 & -6y \end{bmatrix}. \text{ The two matrices are equal if and}$$

only if $3x = -6y$.

This implies that for any given real value of x, if we choose $y = -\frac{1}{2}x$, the ordered pair (x,y) makes the equality true. Since there are infinitely many possible values of x, we conclude that infinitely many ordered pairs make this equality true.

843. b. First, simplify the left side of the equation:

$$-4\begin{bmatrix} 3x-2 & 0 & -2 \\ -2 & 2y & -1 \end{bmatrix} - 2\begin{bmatrix} 4x+2 & -5 & 1 \\ 2 & 4-3y & -1 \end{bmatrix} = \begin{bmatrix} 2x & 10 & 6 \\ 4 & y & 6 \end{bmatrix}$$

$$\begin{bmatrix} -12x+8 & 0 & 8 \\ 8 & -8y & 4 \end{bmatrix} - \begin{bmatrix} 8x+4 & -10 & 2 \\ 4 & 8-6y & -2 \end{bmatrix} = \begin{bmatrix} 2x & 10 & 6 \\ 4 & y & 6 \end{bmatrix}$$

$$\begin{bmatrix} -20x+4 & 10 & 6 \\ 4 & -2y-8 & 6 \end{bmatrix} = \begin{bmatrix} 2x & 10 & 6 \\ 4 & y & 6 \end{bmatrix}$$

Equating corresponding entries reveals that the following equations must hold simultaneously: $-20x + 4 = 2x$ and $-2y - 8 = y$. Solving these equations, we find that $x = \frac{2}{11}$ and $y = -\frac{8}{3}$. Hence, the ordered pair that makes the equality true is $(\frac{2}{11}, -\frac{8}{3})$.

844. c. Simplifying the left side of the equation yields the equivalent equation $\begin{bmatrix} -x & -y \\ -z & 0 \end{bmatrix} = \begin{bmatrix} 3x & 4y \\ -2z & 0 \end{bmatrix}$.

Equating corresponding entries reveals that the following equations must hold simultaneously: $-x = 3x$, $-y = 4y$, and $-z = -2z$. Solving these equations, we find that $x = y = z = 0$. Hence, the ordered triple that makes the equality true is $(0,0,0)$.

845. a. The sum of two matrices is defined only when both matrices have the same number of rows and columns (i.e., the same dimensions); the resulting matrix is one with the same number of rows and columns. Statement **a** is true.

846. c. $3\begin{bmatrix} -1 & 0 & 0 \end{bmatrix} - 2\begin{bmatrix} 0 & 1 & 0 \end{bmatrix} + \begin{bmatrix} 0 & 0 & -1 \end{bmatrix}$

$= \begin{bmatrix} -3 & 0 & 0 \end{bmatrix} + \begin{bmatrix} 0 & -2 & 0 \end{bmatrix} + \begin{bmatrix} 0 & 0 & -1 \end{bmatrix}$

$= \begin{bmatrix} -3 & -2 & -1 \end{bmatrix}$

$= \begin{bmatrix} 3 & 2 & 1 \end{bmatrix}$

847. d. Statement **a** is false because you cannot add a real number to a matrix. As for statement **b**, simplifying the left side of the equation in choice **b** yields the equivalent equation.

$$\begin{bmatrix} -3X & -3 & 0 & 0 \\ 0 & -3X & -3 & 0 \\ 0 & 0 & -3X & -3 \\ 0 & 0 & 0 & -3X \end{bmatrix} = \begin{bmatrix} 15 & 1 & 0 & 0 \\ 0 & 15 & 15 & 0 \\ 0 & 0 & 0 & 1 \\ 0 & 0 & 0 & 15 \end{bmatrix}$$

While $X = -5$ makes the corresponding entries along the diagonals of the two matrices the same, the entries to their immediate right are not equal. There is no X-value that makes these two matrices equal. And finally, statement c is false since you cannot subtract two matrices that have different dimensions.

848. a. Equating corresponding entries reveals that the following three equations must hold simultaneously: $x - 2 = -x^2$, $2\sqrt{y} = y^2$, and $4z^2 = 8z$. First, solve each equation:

$$x - 2 = -x^2 \qquad\qquad 2\sqrt{y} = y^2 \qquad\qquad 4z^2 = 8z$$
$$x^2 + x - 2 = 0 \qquad\qquad 4y = y^4 \qquad\qquad 4z^2 - 8z = 0$$
$$(x + 2)(x - 1) = 0 \qquad\qquad y^4 - 4y = 0 \qquad\qquad 4z(z - 2) = 0$$
$$x = -2, 1 \qquad\qquad y(y^3 - 4) = 0 \qquad\qquad z = 0, 2$$
$$y = 0, \sqrt[3]{4}$$

We must form all combinations of x, y, and z values to form the ordered triples that make the equality true. There are eight such ordered pairs:

$(-2, 0, 0), (-2, 0, 2), (-2, \sqrt[3]{4}, 0), (-2, \sqrt[3]{4}, 2)$
$(1, 0, 0), (1, 0, 2), (1, \sqrt[3]{4}, 0), (1, \sqrt[3]{4}, 2)$

Set 54 (Page 127)

849. b.
$$CF = \begin{bmatrix} 0 & 1 \\ 1 & -4 \end{bmatrix}\begin{bmatrix} 0 \\ 0 \end{bmatrix} = \begin{bmatrix} 0 \\ 0 \end{bmatrix}$$

850. d. The matrix $2G$ has dimensions 3×4 and the matrix $-3E$ has dimensions 1×3. Since the inner dimensions are not equal in the terms of the product, $(2G) \cdot (-3E)$, the product is not defined.

851. c.
$$AB = \begin{bmatrix} -1 & 2 \\ 0 & 2 \\ -1 & -1 \end{bmatrix}\begin{bmatrix} 1 & -2 & -1 \\ 3 & 5 & 0 \end{bmatrix} = \begin{bmatrix} 5 & 12 & 1 \\ 6 & 10 & 0 \\ -4 & -3 & 1 \end{bmatrix}$$

852. c.
$$4AB = 4\begin{bmatrix} 1 & -2 & -1 \\ 3 & 5 & 0 \end{bmatrix}\begin{bmatrix} -1 & 2 \\ 0 & 2 \\ -1 & -1 \end{bmatrix} = 4\begin{bmatrix} 0 & -1 \\ -3 & 16 \end{bmatrix} = \begin{bmatrix} 0 & -4 \\ -12 & 64 \end{bmatrix}$$

853. a.

$$(-2D)(3D) = -6D \cdot D = -6 \begin{bmatrix} 3 & 2 & 1 \\ 0 & 1 & 2 \\ -1 & -1 & 0 \end{bmatrix} \begin{bmatrix} 3 & 2 & 1 \\ 0 & 1 & 2 \\ -1 & -1 & 0 \end{bmatrix} = -6 \begin{bmatrix} 8 & 7 & 7 \\ -2 & -1 & 2 \\ -3 & -3 & -3 \end{bmatrix}$$

$$= \begin{bmatrix} -48 & -42 & -42 \\ 12 & 6 & -12 \\ 18 & 18 & 18 \end{bmatrix}$$

854. d. Note that the dimensions of the matrix F are 2×1. The product FF is not defined because the inner dimensions do not match.

855. b.

$$IE + D = \begin{bmatrix} 2 \\ 2 \\ 1 \end{bmatrix} \begin{bmatrix} -4 & -2 & 0 \end{bmatrix} + \begin{bmatrix} 3 & 2 & 1 \\ 0 & 1 & 2 \\ -1 & -1 & 0 \end{bmatrix} = \begin{bmatrix} -8 & -4 & 0 \\ -3 & -4 & 0 \\ -4 & -2 & 0 \end{bmatrix} + \begin{bmatrix} 3 & 2 & 1 \\ 0 & 1 & 2 \\ -1 & -1 & 0 \end{bmatrix}$$

$$= \begin{bmatrix} -5 & -2 & 1 \\ -8 & -3 & 2 \\ -5 & -3 & 0 \end{bmatrix}$$

856. a.

$$(BG)H = \left(\begin{bmatrix} 1 & -2 & -1 \\ 3 & 5 & 0 \end{bmatrix} \cdot \begin{bmatrix} -2 & -1 & 0 & 1 \\ -1 & -2 & -1 & 0 \\ 1 & -1 & -2 & -1 \end{bmatrix} \right) \cdot \begin{bmatrix} 3 & 1 & -1 \\ 1 & -2 & 1 \\ 0 & 0 & -2 \\ -2 & 1 & 0 \end{bmatrix}$$

$$= \begin{bmatrix} -1 & 4 & 4 & 2 \\ -11 & -13 & -5 & 3 \end{bmatrix} \cdot \begin{bmatrix} 3 & 1 & -1 \\ 1 & -2 & 1 \\ 0 & 0 & -2 \\ -2 & 1 & 0 \end{bmatrix} = \begin{bmatrix} -3 & -7 & -3 \\ -52 & 18 & 8 \end{bmatrix}$$

857. c.

$$(EG)(HI) = \left(\begin{bmatrix} -4 & -2 & 0 \end{bmatrix} \cdot \begin{bmatrix} -2 & -1 & 0 & 1 \\ -1 & -2 & -1 & 0 \\ 1 & -1 & -2 & -1 \end{bmatrix} \right) \cdot \left(\begin{bmatrix} 3 & 1 & -1 \\ 1 & -2 & 1 \\ 0 & 0 & -2 \\ -2 & 1 & 0 \end{bmatrix} \cdot \begin{bmatrix} 2 \\ 2 \\ 1 \end{bmatrix} \right)$$

$$= \begin{bmatrix} 10 & 8 & 2 & -4 \end{bmatrix} \cdot \begin{bmatrix} 7 \\ -1 \\ -2 \\ -2 \end{bmatrix} = 66$$

(Note that a real number is indeed a matrix with dimensions 1×1.)

858. b. $(ED)(AC) = \left(\begin{bmatrix} -4 & -2 & 0 \end{bmatrix} \cdot \begin{bmatrix} 3 & 2 & 1 \\ 0 & 1 & 2 \\ -1 & -1 & 0 \end{bmatrix} \right) \left(\begin{bmatrix} -1 & 2 \\ 0 & 2 \\ -1 & -1 \end{bmatrix} \cdot \begin{bmatrix} 0 & 1 \\ 1 & -4 \end{bmatrix} \right)$

$= \begin{bmatrix} -12 & -10 & -8 \end{bmatrix} \cdot \begin{bmatrix} 2 & -9 \\ 2 & -8 \\ -1 & 3 \end{bmatrix} = \begin{bmatrix} -36 & 164 \end{bmatrix}$

859. d. The sum $G + A$ is not defined because G and A have different dimensions. So, this entire computation is not well-defined.

860. d. The product EF is not defined because the inner dimensions of F and E are not the same. Therefore, this entire computation is not well-defined.

861. a. $(2C)(2C)(2C)F = 8(C \cdot C \cdot C \cdot F) = 8 \left(\begin{bmatrix} 0 & 1 \\ 1 & -4 \end{bmatrix} \cdot \begin{bmatrix} 0 & 1 \\ 1 & -4 \end{bmatrix} \cdot \begin{bmatrix} 0 & 1 \\ 1 & -4 \end{bmatrix} \cdot \begin{bmatrix} 0 \\ 0 \end{bmatrix} \right)$

$= \left(\begin{bmatrix} 0 & 1 \\ 1 & -4 \end{bmatrix} \cdot \begin{bmatrix} 0 & 1 \\ 1 & -4 \end{bmatrix} \cdot \begin{bmatrix} 0 \\ 0 \end{bmatrix} \right) = 8 \begin{bmatrix} 0 \\ 0 \end{bmatrix} = \begin{bmatrix} 0 \\ 0 \end{bmatrix}$

862. c. First, we insert the actual matrices into the expression:

$(EAF)(CF) = \left(\begin{bmatrix} -4 & -2 & 0 \end{bmatrix} \cdot \begin{bmatrix} -1 & 2 \\ 0 & 2 \\ -1 & -1 \end{bmatrix} \cdot \begin{bmatrix} 0 \\ 0 \end{bmatrix} \right) \left(\begin{bmatrix} 0 & 1 \\ 1 & -4 \end{bmatrix} \cdot \begin{bmatrix} 0 \\ 0 \end{bmatrix} \right)$

Note that both products enclosed by parentheses are well-defined when computed from left to right since the inner dimensions are the same for each product to be computed. Moreover, the product enclosed in the first set of parentheses results in a real number:

$\begin{bmatrix} -4 & -2 & 0 \end{bmatrix} \cdot \begin{bmatrix} -1 & 2 \\ 0 & 2 \\ -1 & -1 \end{bmatrix} \cdot \begin{bmatrix} 0 \\ 0 \end{bmatrix} = \begin{bmatrix} 4 & -12 \end{bmatrix} \cdot \begin{bmatrix} 0 \\ 0 \end{bmatrix} = 0$

The original product becomes $0 \left(\begin{bmatrix} 0 & 1 \\ 1 & -4 \end{bmatrix} \cdot \begin{bmatrix} 0 \\ 0 \end{bmatrix} \right)$, which is simplified as follows:

$0 \left(\begin{bmatrix} 0 & 1 \\ 1 & -4 \end{bmatrix} \cdot \begin{bmatrix} 0 \\ 0 \end{bmatrix} \right) = 0 \begin{bmatrix} 0 \\ 0 \end{bmatrix} = \begin{bmatrix} 0 \\ 0 \end{bmatrix}$

863. **b.** $3D - 2AB + GH$

$$= 3\begin{bmatrix} 3 & 2 & 1 \\ 0 & 1 & 2 \\ -1 & -1 & 0 \end{bmatrix} - 2\begin{bmatrix} -1 & 2 \\ 0 & 2 \\ -1 & -1 \end{bmatrix} \cdot \begin{bmatrix} 1 & -2 & -1 \\ 3 & 5 & 0 \end{bmatrix} + \begin{bmatrix} -2 & -1 & 0 & 1 \\ -1 & -2 & -1 & 0 \\ 1 & -1 & -2 & -1 \end{bmatrix} \cdot \begin{bmatrix} 3 & 1 & -1 \\ 1 & -2 & 1 \\ 0 & 0 & -2 \\ -2 & 1 & 0 \end{bmatrix}$$

$$= \begin{bmatrix} -1 & 2 \\ 0 & 2 \\ -1 & -1 \end{bmatrix} - 2\begin{bmatrix} 5 & 12 & 1 \\ 6 & 10 & 0 \\ -4 & -3 & 1 \end{bmatrix} + \begin{bmatrix} -9 & 1 & 1 \\ -5 & 3 & 1 \\ 4 & 2 & 2 \end{bmatrix}$$

$$= \begin{bmatrix} 9 & 6 & 3 \\ 0 & 3 & 6 \\ -3 & -3 & 0 \end{bmatrix} + \begin{bmatrix} -10 & -24 & -2 \\ -12 & -20 & 0 \\ 8 & 6 & -2 \end{bmatrix} + \begin{bmatrix} -9 & 1 & 1 \\ -5 & 3 & 1 \\ 4 & 2 & 2 \end{bmatrix} = \begin{bmatrix} -10 & -17 & 2 \\ -17 & -14 & 7 \\ 9 & 5 & 0 \end{bmatrix}$$

864. **d.** Note that $(2E)(-2F) = -4EF$. Since the inner dimensions of F and E do not match, this product is not defined. As such, the entire computation is not well-defined.

Set 55 (Page 130)

865. **d.** $\det\begin{bmatrix} -3 & 7 \\ 1 & 5 \end{bmatrix} = (-3)(5) - (1)(7) = -22$

866. **c.** $\det\begin{bmatrix} a & 0 \\ 0 & b \end{bmatrix} = (a)(b) - (0)(0) = ab$

867. **a.** $\det\begin{bmatrix} 1 & 2 \\ 2 & 3 \end{bmatrix} = (1)(3) - (2)(2) = -1$

868. **d.** $\det\begin{bmatrix} 2 & 3 \\ 1 & 1 \end{bmatrix} = (2)(1) - (1)(3) = -1$

869. **d.** $\det\begin{bmatrix} -1 & 2 \\ 2 & -4 \end{bmatrix} = (-1)(-4) - (2)(2) = 0$

870. **b.** $\det\begin{bmatrix} 6 & 3 \\ 2 & 1 \end{bmatrix} = (6)(1) - (2)(3) = 0$

871. **c.** $\det\begin{bmatrix} -3 & 4 \\ 4 & 2 \end{bmatrix} = (-3)(2) - (4)(4) = -22$

872. **a.** $\det\begin{bmatrix} 1 & -4 \\ 0 & 25 \end{bmatrix} = (1)(25) - (0)(-4) = 25$

873. **c.** $\det\begin{bmatrix} 3 & -1 \\ 1 & -2 \end{bmatrix} = (3)(-2) - (1)(-1) = -5$

874. **c.** $\det\begin{bmatrix} -2 & 0 \\ -12 & 3 \end{bmatrix} = (-2)(3) - (-12)(0) = -6$

875. **b.** $\det\begin{bmatrix} 0 & 1 \\ -2 & -1 \end{bmatrix} = (0)(-1) - (-2)(1) = 2$

876. **b.** $\det\begin{bmatrix} -1 & 0 \\ 2 & -1 \end{bmatrix} = (-1)(-1) - (2)(0) = 1$

877. **a.** $\det\begin{bmatrix} 3 & 2 \\ 3 & 2 \end{bmatrix} = (3)(2) - (3)(2) = 0$

878. **a.** $\det\begin{bmatrix} 3 & -2 \\ 9 & -6 \end{bmatrix} = (3)(-6) - (9)(-2) = 0$

879. **c.** $\det\begin{bmatrix} -1 & -1 \\ -1 & 0 \end{bmatrix} = (-1)(0) - (-1)(-1) = -1$

880. **d.** $\det\begin{bmatrix} 0 & 2 \\ 4 & 0 \end{bmatrix} = (0)(0) - (4)(2) = -8$

Set 56 (Page 131)

881. d. First, extract the coefficients from the variable terms on the left sides of the equations to form a 2×2 coefficient matrix. Multiplying this by $\begin{bmatrix} x \\ y \end{bmatrix}$ and identifying the right side as a 2×1 constant matrix, we can rewrite the system $\begin{cases} -3 + 7y = 2 \\ x + 5y = 8 \end{cases}$ as the following matrix equation:

$$\begin{bmatrix} -3 & 7 \\ 1 & 5 \end{bmatrix} \begin{bmatrix} x \\ y \end{bmatrix} = \begin{bmatrix} 2 \\ 8 \end{bmatrix}$$

882. a. Extract the coefficients from the variable terms on the left sides of the equations to form a 2×2 coefficient matrix. Multiplying this by $\begin{bmatrix} x \\ y \end{bmatrix}$ and identifying the right side as a 2×1 constant matrix, we can rewrite the system $\begin{cases} x = a \\ y = b \end{cases}$ as the following matrix equation:

$$\begin{bmatrix} 1 & 0 \\ 0 & 1 \end{bmatrix} \begin{bmatrix} x \\ y \end{bmatrix} = \begin{bmatrix} a \\ b \end{bmatrix}$$

883. b. Begin by extracting the coefficients from the variable terms on the left sides of the equations to form a 2×2 coefficient matrix. Multiplying this by $\begin{bmatrix} x \\ y \end{bmatrix}$ and identifying the right side as a 2×1 constant matrix, we can rewrite the system $\begin{cases} x + 2y = 4 \\ 2x + 3y = 2 \end{cases}$ system as the following matrix equation:

$$\begin{bmatrix} 1 & 2 \\ 2 & 3 \end{bmatrix} \begin{bmatrix} x \\ y \end{bmatrix} = \begin{bmatrix} 4 \\ 2 \end{bmatrix}$$

884. c. Extract the coefficients from the variable terms on the left sides of the equations to form a 2×2 coefficient matrix. Multiplying this by $\begin{bmatrix} x \\ y \end{bmatrix}$ and identifying the right side as a 2×1 constant matrix, we can rewrite the system $\begin{cases} 2x + 3y = 1 \\ x + y = -2 \end{cases}$ as the following matrix equation:

$$\begin{bmatrix} 2 & 3 \\ 1 & 1 \end{bmatrix} \begin{bmatrix} x \\ y \end{bmatrix} = \begin{bmatrix} 1 \\ -2 \end{bmatrix}$$

885. c. First, extract the coefficients from the variable terms on the left sides of the equations to form a 2×2 coefficient matrix. Multiplying this by $\begin{bmatrix} x \\ y \end{bmatrix}$ and identifying the right side as a 2×1 constant matrix, we can rewrite the system $\begin{cases} -x + 2y = 3 \\ 2x - 4 = -6 \end{cases}$ as the following matrix equation:

$$\begin{bmatrix} -1 & 2 \\ 2 & -4 \end{bmatrix} \begin{bmatrix} x \\ y \end{bmatrix} = \begin{bmatrix} 3 \\ -6 \end{bmatrix}$$

886. d. Extract the coefficients from the variable terms on the left sides of the equations to form a 2×2 coefficient matrix. Multiplying this by $\begin{bmatrix} x \\ y \end{bmatrix}$ and identifying the right side as a 2×1 constant matrix, we can rewrite the system $\begin{cases} 6x + 3y = 8 \\ 2x + y = 3 \end{cases}$ as the following matrix equation:

$$\begin{bmatrix} 6 & 3 \\ 2 & 1 \end{bmatrix} \begin{bmatrix} x \\ y \end{bmatrix} = \begin{bmatrix} 8 \\ 3 \end{bmatrix}$$

887. c. First, rewrite the system by moving all variable terms to the left sides of the equations and all constant terms to the right sides to obtain the following equivalent system:

$$\begin{cases} -3x + 4y = 1 \\ 4x + 2y = -3 \end{cases}$$

Now, extract the coefficients from the variable terms on the left sides of the equations to form a 2×2 coefficient matrix. Multiplying this by $\begin{bmatrix} x \\ y \end{bmatrix}$ and identifying the right side as a 2×1 constant matrix, we can rewrite this system as the following matrix equation:

$$\begin{bmatrix} -3 & 4 \\ 4 & 2 \end{bmatrix} \begin{bmatrix} x \\ y \end{bmatrix} = \begin{bmatrix} 1 \\ -3 \end{bmatrix}$$

888. a. First, rewrite the system by moving all variable terms to the left sides of the equations and all constant terms to the right sides to obtain the following equivalent system:

$$\begin{cases} x - 4y = -2 \\ 5y = \frac{1}{5} \end{cases}$$

Now, extract the coefficients from the variable terms on the left sides of the equations to form a 2×2 coefficient matrix. Multiplying this by $\begin{bmatrix} x \\ y \end{bmatrix}$ and identifying the right side as a 2×1 constant matrix, we can rewrite this system as the following matrix equation:

$$\begin{bmatrix} 1 & -4 \\ 0 & 5 \end{bmatrix} \begin{bmatrix} x \\ y \end{bmatrix} = \begin{bmatrix} -2 \\ \frac{1}{5} \end{bmatrix}$$

889. b. First, rewrite the system by moving all variable terms to the left sides of the equations and all constant terms to the right sides to obtain the following equivalent system:

$$\begin{cases} -3x + y = 5 \\ x - 2y = -9 \end{cases}$$

Now, extract the coefficients from the variable terms on the left sides of the equations to form a 2×2 coefficient matrix. Multiplying this by $\begin{bmatrix} x \\ y \end{bmatrix}$ and identifying the right side as a 2×1 constant matrix, we can rewrite this system as the following matrix equation:

$$\begin{bmatrix} -3 & 1 \\ 1 & -2 \end{bmatrix} \begin{bmatrix} x \\ y \end{bmatrix} = \begin{bmatrix} 5 \\ -9 \end{bmatrix}$$

890. b. First, rewrite the system by moving all variable terms to the left sides of the equations and all constant terms to the right sides to obtain the following equivalent system:

$$\begin{cases} 2x = -2 \\ 12x - 3y = 4 \end{cases}$$

Now, extract the coefficients from the variable terms on the left sides of the equations to form a 2×2 coefficient matrix. Multiplying this by $\begin{bmatrix} x \\ y \end{bmatrix}$ and identifying the right side as a 2×1 constant matrix, we can rewrite this system as the following matrix equation:

$$\begin{bmatrix} 2 & 0 \\ 12 & -3 \end{bmatrix} \begin{bmatrix} x \\ y \end{bmatrix} = \begin{bmatrix} -2 \\ 4 \end{bmatrix}$$

891. a. First, rewrite the system by moving all variable terms to the left sides of the equations and all constant terms to the right sides to obtain the following equivalent system:

$$\begin{cases} y = -4 \\ -2x - y = 0 \end{cases}$$

Now, extract the coefficients from the variable terms on the left sides of the equations to form a 2×2 coefficient matrix. Multiplying this by $\begin{bmatrix} x \\ y \end{bmatrix}$ and identifying the right side as a 2×1 constant matrix, we can rewrite this system as the following matrix equation:

$$\begin{bmatrix} 0 & 1 \\ -2 & -1 \end{bmatrix} \begin{bmatrix} x \\ y \end{bmatrix} = \begin{bmatrix} -4 \\ 0 \end{bmatrix}$$

892. c. Compute the product on the left side to write it as a single matrix, and then equate corresponding entries in the matrices on the left and right sides of the equation to obtain the desired system:

$$\begin{bmatrix} -1 & 0 \\ 2 & -1 \end{bmatrix} \begin{bmatrix} x \\ y \end{bmatrix} = \begin{bmatrix} -2 \\ 1 \end{bmatrix}$$

$$\begin{bmatrix} -x \\ 2x - y \end{bmatrix} = \begin{bmatrix} -2 \\ 1 \end{bmatrix}$$

$$\begin{cases} -x = -2 \\ 2x - y = 1 \end{cases}$$

893. d. First, compute the product on the left side to write it as a single matrix, and then equate corresponding entries in the matrices on the left and right sides of the equation to obtain the desired system:

$$\begin{bmatrix} 3 & 2 \\ 3 & 2 \end{bmatrix} \begin{bmatrix} x \\ y \end{bmatrix} = \begin{bmatrix} -2 \\ 1 \end{bmatrix}$$

$$\begin{bmatrix} 3x + 2y \\ 3x + 2y \end{bmatrix} = \begin{bmatrix} -2 \\ 1 \end{bmatrix}$$

$$\begin{cases} 3x + 2y = -2 \\ 3x + 2y = 1 \end{cases}$$

894. c. Compute the product on the left side to write it as a single matrix, and then equate corresponding entries in the matrices on the left and right sides of the equation to obtain the desired system:

$$\begin{bmatrix} 3 & -2 \\ 9 & -6 \end{bmatrix} \begin{bmatrix} x \\ y \end{bmatrix} = \begin{bmatrix} 4 \\ 12 \end{bmatrix}$$

$$\begin{bmatrix} 3x - 2y \\ 9x + 6y \end{bmatrix} = \begin{bmatrix} 4 \\ 12 \end{bmatrix}$$

$$\begin{cases} 3x - 2y = 4 \\ 9x + 6y = 12 \end{cases}$$

895. b. First, compute the product on the left side to write it as a single matrix, and then equate corresponding entries in the matrices on the left and right sides of the equation to obtain the desired system:

$$\begin{bmatrix} -1 & -1 \\ -1 & 0 \end{bmatrix}\begin{bmatrix} x \\ y \end{bmatrix} = \begin{bmatrix} -1 \\ 1 \end{bmatrix}$$

$$\begin{bmatrix} -x - y \\ -x \end{bmatrix} = \begin{bmatrix} -1 \\ 1 \end{bmatrix}$$

$$\begin{cases} -x - y = -1 \\ -x = 1 \end{cases}$$

896. b. Compute the product on the left side to write it as a single matrix, and then equate corresponding entries in the matrices on the left and right sides of the equation to obtain the desired system:

$$\begin{bmatrix} 0 & 2 \\ 4 & 0 \end{bmatrix}\begin{bmatrix} x \\ y \end{bmatrix} = \begin{bmatrix} 14 \\ -20 \end{bmatrix}$$

$$\begin{bmatrix} 2y \\ 4x \end{bmatrix} = \begin{bmatrix} 14 \\ -20 \end{bmatrix}$$

$$\begin{cases} 2y = 14 \\ 4x = -20 \end{cases}$$

Set 57 (Page 135)

897. b. It is known that if $A = \begin{bmatrix} a & b \\ c & d \end{bmatrix}$ is such that det $A = ad - bc \neq 0$, then $A^{-1} = \frac{1}{\det A}\begin{bmatrix} d & -b \\ -c & a \end{bmatrix}$. Applying this formula with $A = \begin{bmatrix} -3 & 7 \\ 1 & 5 \end{bmatrix}$ yields the following:

$$A^{-1} = \frac{1}{-22}\begin{bmatrix} 5 & -7 \\ -1 & -3 \end{bmatrix} = \begin{bmatrix} -\frac{5}{22} & \frac{7}{22} \\ \frac{1}{22} & \frac{3}{22} \end{bmatrix}$$

898. a. It is known that if $A = \begin{bmatrix} a & b \\ c & d \end{bmatrix}$ is such that det $A = ad - bc \neq 0$, then $A^{-1} = \frac{1}{\det A}\begin{bmatrix} d & -b \\ -c & a \end{bmatrix}$. Applying this formula with $A = \begin{bmatrix} a & 0 \\ 0 & b \end{bmatrix}$ yields the following:

$$A^{-1} = \frac{1}{ab}\begin{bmatrix} b & 0 \\ 0 & a \end{bmatrix} = \begin{bmatrix} \frac{1}{a} & 0 \\ 0 & \frac{1}{b} \end{bmatrix}$$

899. a. It is known that if $A = \begin{bmatrix} a & b \\ c & d \end{bmatrix}$ is such that $\det A = ad - bc \neq 0$, then $A^{-1} = \frac{1}{\det A}\begin{bmatrix} d & -b \\ -c & a \end{bmatrix}$. Applying this formula with $A = \begin{bmatrix} 1 & 2 \\ 2 & 3 \end{bmatrix}$ yields the following:

$$A^{-1} = \frac{1}{-1}\begin{bmatrix} 3 & -2 \\ -2 & 1 \end{bmatrix} = \begin{bmatrix} -3 & 2 \\ 2 & -1 \end{bmatrix}$$

900. a. It is known that if $A = \begin{bmatrix} a & b \\ c & d \end{bmatrix}$ is such that $\det A = ad - bc \neq 0$, then $A^{-1} = \frac{1}{\det A}\begin{bmatrix} d & -b \\ -c & a \end{bmatrix}$. Applying this formula with $A = \begin{bmatrix} 2 & 3 \\ 1 & 1 \end{bmatrix}$ yields the following:

$$A^{-1} = \frac{1}{-1}\begin{bmatrix} 1 & -3 \\ -1 & 2 \end{bmatrix} = \begin{bmatrix} -1 & 3 \\ 1 & -2 \end{bmatrix}$$

901. a. It is known that if $A = \begin{bmatrix} a & b \\ c & d \end{bmatrix}$ is such that $\det A = ad - bc \neq 0$, then $A^{-1} = \frac{1}{\det A}\begin{bmatrix} d & -b \\ -c & a \end{bmatrix}$. Note that the determinant of $A = \begin{bmatrix} -1 & 2 \\ 2 & -4 \end{bmatrix}$ is zero, so the matrix does not have an inverse.

902. d. It is known that if $A = \begin{bmatrix} a & b \\ c & d \end{bmatrix}$ is such that $\det A = ad - bc \neq 0$, then $A^{-1} = \frac{1}{\det A}\begin{bmatrix} d & -b \\ -c & a \end{bmatrix}$. Note that the determinant of $A = \begin{bmatrix} 6 & 3 \\ 2 & 1 \end{bmatrix}$ is zero, so the matrix does not have an inverse.

903. a. It is known that if $A = \begin{bmatrix} a & b \\ c & d \end{bmatrix}$ is such that $\det A = ad - bc \neq 0$, then $A^{-1} = \frac{1}{\det A}\begin{bmatrix} d & -b \\ -c & a \end{bmatrix}$. Applying this formula with $A = \begin{bmatrix} -3 & 4 \\ 4 & 2 \end{bmatrix}$ yields the following:

$$A^{-1} = \frac{1}{-22}\begin{bmatrix} 2 & -4 \\ -4 & -3 \end{bmatrix} = \begin{bmatrix} -\frac{1}{11} & \frac{2}{11} \\ \frac{2}{11} & \frac{3}{22} \end{bmatrix}$$

904. a. It is known that if $A = \begin{bmatrix} a & b \\ c & d \end{bmatrix}$ is such that $\det A = ad - bc \neq 0$, then $A^{-1} = \frac{1}{\det A}\begin{bmatrix} d & -b \\ -c & a \end{bmatrix}$. Applying this formula with $A = \begin{bmatrix} 1 & -4 \\ 0 & 25 \end{bmatrix}$ yields the following:

$$A^{-1} = \frac{1}{25}\begin{bmatrix} 25 & 4 \\ 0 & 1 \end{bmatrix} = \begin{bmatrix} 1 & \frac{4}{25} \\ 0 & \frac{1}{25} \end{bmatrix}$$

ANSWERS AND EXPLANATIONS

905. c. It is known that if $A = \begin{bmatrix} a & b \\ c & d \end{bmatrix}$ is such that det $A = ad - bc \neq 0$, then

$A^{-1} = \frac{1}{\det A}\begin{bmatrix} d & -b \\ -c & a \end{bmatrix}$. Applying

this formula with $A = \begin{bmatrix} -3 & 1 \\ 1 & -2 \end{bmatrix}$ yields the following:

$A^{-1} = \frac{1}{5}\begin{bmatrix} -2 & -1 \\ -1 & -3 \end{bmatrix} = \begin{bmatrix} -\frac{2}{5} & \frac{1}{5} \\ -\frac{1}{5} & -\frac{3}{5} \end{bmatrix}$

906. b. It is known that if $A = \begin{bmatrix} a & b \\ c & d \end{bmatrix}$ is such that det $A = ad - bc \neq 0$, then

$A^{-1} = \frac{1}{\det A}\begin{bmatrix} d & -b \\ -c & a \end{bmatrix}$. Applying

this formula with $A = \begin{bmatrix} 2 & 0 \\ 12 & -3 \end{bmatrix}$ yields the following:

$A^{-1} = \frac{1}{-6}\begin{bmatrix} -3 & 0 \\ -12 & 2 \end{bmatrix} = \begin{bmatrix} \frac{1}{2} & 0 \\ 2 & -\frac{1}{3} \end{bmatrix}$

907. a. It is known that if $A = \begin{bmatrix} a & b \\ c & d \end{bmatrix}$ is such that det $A = ad - bc \neq 0$, then

$A^{-1} = \frac{1}{\det A}\begin{bmatrix} d & -b \\ -c & a \end{bmatrix}$. Applying

this formula with $A = \begin{bmatrix} 0 & 1 \\ -2 & -1 \end{bmatrix}$ yields the following:

$A^{-1} = \frac{1}{2}\begin{bmatrix} -1 & -1 \\ 2 & 0 \end{bmatrix} = \begin{bmatrix} -\frac{1}{2} & -\frac{1}{2} \\ 1 & 0 \end{bmatrix}$

908. a. It is known that if $A = \begin{bmatrix} a & b \\ c & d \end{bmatrix}$ is such that det $A = ad - bc \neq 0$, then

$A^{-1} = \frac{1}{\det A}\begin{bmatrix} d & -b \\ -c & a \end{bmatrix}$. Applying

this formula with $A = \begin{bmatrix} -1 & 0 \\ 2 & -1 \end{bmatrix}$ yields the following:

$A^{-1} = \frac{1}{1}\begin{bmatrix} -1 & 0 \\ -2 & -1 \end{bmatrix} = \begin{bmatrix} -1 & 0 \\ -2 & -1 \end{bmatrix}$

909. d. It is known that if $A = \begin{bmatrix} a & b \\ c & d \end{bmatrix}$ is such that det $A = ad - bc \neq 0$, then

$A^{-1} = \frac{1}{\det A}\begin{bmatrix} d & -b \\ -c & a \end{bmatrix}$. Note that

the determinant of $A = \begin{bmatrix} 3 & 2 \\ 3 & 2 \end{bmatrix}$ is

zero, so the matrix does not have an inverse.

910. d. It is known that if $A = \begin{bmatrix} a & b \\ c & d \end{bmatrix}$ is such

that det $A = ad - bc \neq 0$, then

$A^{-1} = \frac{1}{\det A} \begin{bmatrix} d & -b \\ -c & a \end{bmatrix}$. Note that

the determinant of $A = \begin{bmatrix} 3 & -2 \\ 9 & -6 \end{bmatrix}$ is

zero, so the matrix does not have an inverse.

911. a. It is known that if $A = \begin{bmatrix} a & b \\ c & d \end{bmatrix}$ is such

that det $A = ad - bc \neq 0$, then

$A^{-1} = \frac{1}{\det A} \begin{bmatrix} d & -b \\ -c & a \end{bmatrix}$. Applying

this formula with $A = \begin{bmatrix} -1 & -1 \\ -1 & 0 \end{bmatrix}$ yields

the following:

$A^{-1} = \frac{1}{-1} \begin{bmatrix} 0 & -1 \\ -1 & 1 \end{bmatrix} = \begin{bmatrix} 0 & -1 \\ -1 & 1 \end{bmatrix}$

912. b. It is known that if $A = \begin{bmatrix} a & b \\ c & d \end{bmatrix}$ is such

that det $A = ad - bc \neq 0$, then

$A^{-1} = \frac{1}{\det A} \begin{bmatrix} d & -b \\ -c & a \end{bmatrix}$. Applying

this formula with $A = \begin{bmatrix} 0 & 2 \\ 4 & 0 \end{bmatrix}$ yields

the following:

$A^{-1} = \frac{1}{-8} \begin{bmatrix} 0 & -2 \\ -4 & 0 \end{bmatrix} = \begin{bmatrix} 0 & \frac{1}{4} \\ \frac{1}{2} & 0 \end{bmatrix}$

Set 58 (Page 138)

913. a. The solution to the matrix equation

$\begin{bmatrix} a & b \\ c & d \end{bmatrix} \begin{bmatrix} x \\ y \end{bmatrix} = \begin{bmatrix} e \\ f \end{bmatrix}$, where $a, b, c, d,$

$e,$ and f are real numbers, is given by

$\begin{bmatrix} x \\ y \end{bmatrix} = \begin{bmatrix} a & b \\ c & d \end{bmatrix}^{-1} \begin{bmatrix} e \\ f \end{bmatrix}$, provided that

the inverse matrix on the right side exists.
From Problem 881, the given system can be
written as the equivalent matrix equation

$\begin{bmatrix} -3 & 7 \\ 1 & 5 \end{bmatrix} \begin{bmatrix} x \\ y \end{bmatrix} = \begin{bmatrix} 2 \\ 8 \end{bmatrix}$. The solution

is therefore given by

$\begin{bmatrix} x \\ y \end{bmatrix} = \begin{bmatrix} -3 & 7 \\ 1 & 5 \end{bmatrix}^{-1} \begin{bmatrix} 2 \\ 8 \end{bmatrix}$. Using the

calculation for the inverse from Problem
897 yields the following solution:

$\begin{bmatrix} x \\ y \end{bmatrix} = \begin{bmatrix} -\frac{5}{22} & \frac{7}{22} \\ \frac{1}{22} & \frac{3}{22} \end{bmatrix} \begin{bmatrix} 2 \\ 8 \end{bmatrix} = \begin{bmatrix} \frac{46}{22} \\ \frac{26}{22} \end{bmatrix}$

$= \begin{bmatrix} \frac{23}{11} \\ \frac{13}{11} \end{bmatrix}$

So, the solution of the system is $x = \frac{23}{11}$,
$y = \frac{13}{11}$.

914. c. The solution of this system is clearly $x = a$,
$y = b$, without needing to go through the
formal procedure of solving the matrix
equation.

915. b. The solution to the matrix equation

$$\begin{bmatrix} a & b \\ c & d \end{bmatrix}\begin{bmatrix} x \\ y \end{bmatrix} = \begin{bmatrix} e \\ f \end{bmatrix}, \text{ where } a, b, c, d,$$

e, and f are real numbers, is given by

$$\begin{bmatrix} x \\ y \end{bmatrix} = \begin{bmatrix} a & b \\ c & d \end{bmatrix}^{-1}\begin{bmatrix} e \\ f \end{bmatrix}, \text{ provided that}$$

the inverse matrix on the right side exists. From Problem 883, the given system can be written as the equivalent matrix equation

$$\begin{bmatrix} 1 & 2 \\ 2 & 3 \end{bmatrix}\begin{bmatrix} x \\ y \end{bmatrix} = \begin{bmatrix} 4 \\ 2 \end{bmatrix}. \text{ The solution}$$

is therefore given by

$$\begin{bmatrix} x \\ y \end{bmatrix} = \begin{bmatrix} 1 & 2 \\ 2 & 3 \end{bmatrix}^{-1}\begin{bmatrix} 4 \\ 2 \end{bmatrix}. \text{ Using the}$$

calculation for the inverse from Problem 899 yields the following solution:

$$\begin{bmatrix} x \\ y \end{bmatrix} = \begin{bmatrix} -3 & 2 \\ 2 & -1 \end{bmatrix}\begin{bmatrix} 4 \\ 2 \end{bmatrix} = \begin{bmatrix} -8 \\ 6 \end{bmatrix}$$

So, the solution of the system is $x = -8$, $y = 6$.

916. b. The solution to the matrix equation

$$\begin{bmatrix} a & b \\ c & d \end{bmatrix}\begin{bmatrix} x \\ y \end{bmatrix} = \begin{bmatrix} e \\ f \end{bmatrix}, \text{ where } a, b, c, d,$$

e, and f are real numbers, is given by

$$\begin{bmatrix} x \\ y \end{bmatrix} = \begin{bmatrix} a & b \\ c & d \end{bmatrix}^{-1}\begin{bmatrix} e \\ f \end{bmatrix}, \text{ provided that}$$

the inverse matrix on the right side exists. From Problem 884, the given system can be written as the equivalent matrix equation

$$\begin{bmatrix} 2 & 3 \\ 1 & 1 \end{bmatrix}\begin{bmatrix} x \\ y \end{bmatrix} = \begin{bmatrix} 1 \\ -2 \end{bmatrix}. \text{ The solution}$$

is therefore given by

$$\begin{bmatrix} x \\ y \end{bmatrix} = \begin{bmatrix} 2 & 3 \\ 1 & 1 \end{bmatrix}^{-1}\begin{bmatrix} 1 \\ -2 \end{bmatrix}. \text{ Using the}$$

calculation for the inverse from Problem 900 yields the following solution:

$$\begin{bmatrix} x \\ y \end{bmatrix} = \begin{bmatrix} -1 & 3 \\ 1 & -2 \end{bmatrix}\begin{bmatrix} 1 \\ -2 \end{bmatrix} = \begin{bmatrix} -7 \\ 5 \end{bmatrix}$$

So, the solution of the system is $x = -7$, $y = 5$.

917. d. The solution to the matrix equation

$$\begin{bmatrix} a & b \\ c & d \end{bmatrix} \begin{bmatrix} x \\ y \end{bmatrix} = \begin{bmatrix} e \\ f \end{bmatrix}, \text{ where } a, b, c, d,$$

e, and f are real numbers, is given by

$$\begin{bmatrix} x \\ y \end{bmatrix} = \begin{bmatrix} a & b \\ c & d \end{bmatrix}^{-1} \begin{bmatrix} e \\ f \end{bmatrix}, \text{ provided that}$$

the inverse matrix on the right side exists. From Problem 885, the given system can be written as the equivalent matrix equation

$$\begin{bmatrix} -1 & 2 \\ 2 & -4 \end{bmatrix} \begin{bmatrix} x \\ y \end{bmatrix} = \begin{bmatrix} 3 \\ -6 \end{bmatrix}. \text{ Note that}$$

since $\det \begin{bmatrix} -1 & 2 \\ 2 & -4 \end{bmatrix} = 0$, it follows that

$$\begin{bmatrix} -1 & 2 \\ 2 & -4 \end{bmatrix}^{-1} \text{ does not exist, so we cannot}$$

apply this approach. Rather, we must inspect the system to determine whether there is no solution (which happens if the two lines are parallel) or if there are infinitely many solutions (which happens if the two lines are identical). The second equation in the system is obtained by multiplying both sides of the first equation by –2. So, the two lines are identical, and thus, the system has infinitely many solutions.

918. d. The solution to the matrix equation

$$\begin{bmatrix} a & b \\ c & d \end{bmatrix} \begin{bmatrix} x \\ y \end{bmatrix} = \begin{bmatrix} e \\ f \end{bmatrix}, \text{ where } a, b, c, d,$$

e, and f are real numbers, is given by

$$\begin{bmatrix} x \\ y \end{bmatrix} = \begin{bmatrix} a & b \\ c & d \end{bmatrix}^{-1} \begin{bmatrix} e \\ f \end{bmatrix}, \text{ provided that}$$

the inverse matrix on the right side exists. From Problem 886, the given system can be written as the equivalent matrix equation

$$\begin{bmatrix} 6 & 3 \\ 2 & 1 \end{bmatrix} \begin{bmatrix} x \\ y \end{bmatrix} = \begin{bmatrix} 8 \\ 3 \end{bmatrix}. \text{ Note that}$$

since $\det \begin{bmatrix} 6 & 3 \\ 2 & 1 \end{bmatrix} = 0$, it follows that

$$\begin{bmatrix} 6 & 3 \\ 2 & 1 \end{bmatrix}^{-1} \text{ does not exist, so we cannot}$$

apply the above approach. Rather, we must inspect the system to determine whether there is no solution (which happens if the two lines are parallel) or if there are infinitely many solutions (which happens if the two lines are identical). Multiplying both sides of the second equation by 3 yields the equivalent equation $6x + 3y = 9$. Subtracting this from the first equation yields the false statement $0 = -1$. From this, we conclude that the two lines must be parallel (which can also be checked by graphing them). Hence, the system has no solution.

919. a. The solution to the matrix equation

$$\begin{bmatrix} a & b \\ c & d \end{bmatrix}\begin{bmatrix} x \\ y \end{bmatrix} = \begin{bmatrix} e \\ f \end{bmatrix}$$, where a, b, c, d, e, and f are real numbers, is given by

$$\begin{bmatrix} x \\ y \end{bmatrix} = \begin{bmatrix} a & b \\ c & d \end{bmatrix}^{-1}\begin{bmatrix} e \\ f \end{bmatrix}$$, provided that the inverse matrix on the right side exists. From Problem 887, the given system can be written as the equivalent matrix equation

$$\begin{bmatrix} -3 & 4 \\ 4 & 2 \end{bmatrix}\begin{bmatrix} x \\ y \end{bmatrix} = \begin{bmatrix} 1 \\ -3 \end{bmatrix}$$. The solution is therefore given by

$$\begin{bmatrix} x \\ y \end{bmatrix} = \begin{bmatrix} -3 & 4 \\ 4 & 2 \end{bmatrix}^{-1} = \begin{bmatrix} 1 \\ -3 \end{bmatrix}$$. Using the calculation for the inverse from Problem 903 yields the following solution:

$$\begin{bmatrix} x \\ y \end{bmatrix} = \begin{bmatrix} -\frac{1}{11} & \frac{2}{11} \\ \frac{2}{11} & \frac{3}{22} \end{bmatrix}\begin{bmatrix} 1 \\ -3 \end{bmatrix} = \begin{bmatrix} -\frac{7}{11} \\ -\frac{5}{22} \end{bmatrix}$$

So, the solution of the system is $x = -\frac{7}{11}$, $y = -\frac{5}{22}$.

920. a. The solution to the matrix equation

$$\begin{bmatrix} a & b \\ c & d \end{bmatrix}\begin{bmatrix} x \\ y \end{bmatrix} = \begin{bmatrix} e \\ f \end{bmatrix}$$, where a, b, c, d, e, and f are real numbers, is given by

$$\begin{bmatrix} x \\ y \end{bmatrix} = \begin{bmatrix} a & b \\ c & d \end{bmatrix}^{-1}\begin{bmatrix} e \\ f \end{bmatrix}$$, provided that the inverse matrix on the right side exists. From Problem 888, the given system can be written as the equivalent matrix equation

$$\begin{bmatrix} 1 & -4 \\ 0 & 25 \end{bmatrix}\begin{bmatrix} x \\ y \end{bmatrix} = \begin{bmatrix} -2 \\ 1 \end{bmatrix}$$. The solution is therefore given by

$$\begin{bmatrix} x \\ y \end{bmatrix} = \begin{bmatrix} 1 & -4 \\ 0 & 25 \end{bmatrix}^{-1} = \begin{bmatrix} -2 \\ 1 \end{bmatrix}$$. Using the calculation for the inverse from Problem 904 yields the following solution:

$$\begin{bmatrix} x \\ y \end{bmatrix} = \begin{bmatrix} 1 & \frac{4}{25} \\ 0 & \frac{1}{25} \end{bmatrix}\begin{bmatrix} -2 \\ 1 \end{bmatrix} = \begin{bmatrix} -\frac{46}{25} \\ \frac{1}{25} \end{bmatrix}$$

So, the solution of the system is $x = -\frac{46}{25}$, $y = \frac{1}{25}$.

921. b. The solution to the matrix equation

$$\begin{bmatrix} a & b \\ c & d \end{bmatrix}\begin{bmatrix} x \\ y \end{bmatrix} = \begin{bmatrix} e \\ f \end{bmatrix}$$, where $a, b, c, d,$

e, and f are real numbers, is given by

$$\begin{bmatrix} x \\ y \end{bmatrix} = \begin{bmatrix} a & b \\ c & d \end{bmatrix}^{-1}\begin{bmatrix} e \\ f \end{bmatrix}$$, provided that

the inverse matrix on the right side exists.
From Problem 889, the given system can be
written as the equivalent matrix equation

$$\begin{bmatrix} -3 & 1 \\ 1 & -2 \end{bmatrix}\begin{bmatrix} x \\ y \end{bmatrix} = \begin{bmatrix} 5 \\ -9 \end{bmatrix}$$. The

solution is therefore given by

$$\begin{bmatrix} x \\ y \end{bmatrix} = \begin{bmatrix} -3 & 1 \\ 1 & -2 \end{bmatrix}^{-1} = \begin{bmatrix} 5 \\ -9 \end{bmatrix}$$. Using

the calculation for the inverse from Problem
905 yields the following solution:

$$\begin{bmatrix} x \\ y \end{bmatrix} = \begin{bmatrix} -\frac{1}{5} & -\frac{1}{5} \\ -\frac{1}{5} & \frac{3}{5} \end{bmatrix}\begin{bmatrix} 5 \\ -9 \end{bmatrix} = \begin{bmatrix} -\frac{1}{5} \\ \frac{22}{5} \end{bmatrix}$$

So, the solution of the system is $x = -\frac{19}{5}$,
$y = \frac{22}{5}$.

922. c. The solution to the matrix equation

$$\begin{bmatrix} a & b \\ c & d \end{bmatrix}\begin{bmatrix} x \\ y \end{bmatrix} = \begin{bmatrix} e \\ f \end{bmatrix}$$, where $a, b, c, d,$

e, and f are real numbers, is given by

$$\begin{bmatrix} x \\ y \end{bmatrix} = \begin{bmatrix} a & b \\ c & d \end{bmatrix}^{-1}\begin{bmatrix} e \\ f \end{bmatrix}$$, provided that

the inverse matrix on the right side exists.
From Problem 890, the given system can be
written as the equivalent matrix equation

$$\begin{bmatrix} 2 & 0 \\ 12 & -3 \end{bmatrix}\begin{bmatrix} x \\ y \end{bmatrix} = \begin{bmatrix} -2 \\ 4 \end{bmatrix}$$. The

solution is therefore given by

$$\begin{bmatrix} x \\ y \end{bmatrix} = \begin{bmatrix} 2 & 0 \\ 12 & -3 \end{bmatrix}^{-1} = \begin{bmatrix} -2 \\ 4 \end{bmatrix}$$. Using

the calculation for the inverse from Problem
906 yields the following solution:

$$\begin{bmatrix} x \\ y \end{bmatrix} = \begin{bmatrix} -\frac{1}{2} & 0 \\ 2 & -\frac{1}{3} \end{bmatrix}\begin{bmatrix} -2 \\ 4 \end{bmatrix} = \begin{bmatrix} -1 \\ -\frac{16}{3} \end{bmatrix}$$

So, the solution of the system is $x = -1$,
$y = -\frac{16}{3}$.

923. c. The solution to the matrix equation

$$\begin{bmatrix} a & b \\ c & d \end{bmatrix} \begin{bmatrix} x \\ y \end{bmatrix} = \begin{bmatrix} e \\ f \end{bmatrix}, \text{ where } a, b, c, d,$$

e, and *f* are real numbers, is given by

$$\begin{bmatrix} x \\ y \end{bmatrix} = \begin{bmatrix} a & b \\ c & d \end{bmatrix}^{-1} \begin{bmatrix} e \\ f \end{bmatrix}, \text{ provided that}$$

the inverse matrix on the right side exists. From Problem 891, the given system can be written as the equivalent matrix equation

$$\begin{bmatrix} 0 & 1 \\ -2 & -1 \end{bmatrix} \begin{bmatrix} x \\ y \end{bmatrix} = \begin{bmatrix} -4 \\ 0 \end{bmatrix}. \text{ The}$$

solution is therefore given by

$$\begin{bmatrix} x \\ y \end{bmatrix} = \begin{bmatrix} 0 & 1 \\ -2 & -1 \end{bmatrix}^{-1} = \begin{bmatrix} -4 \\ 0 \end{bmatrix}. \text{ Using}$$

the calculation for the inverse from Problem 907 yields the following solution:

$$\begin{bmatrix} x \\ y \end{bmatrix} = \begin{bmatrix} -\frac{1}{2} & -\frac{1}{2} \\ 1 & 0 \end{bmatrix} \begin{bmatrix} -4 \\ 0 \end{bmatrix} = \begin{bmatrix} 2 \\ -4 \end{bmatrix}$$

So, the solution of the system is $x = 2$, $y = -4$.

924. b. The solution to the matrix equation

$$\begin{bmatrix} a & b \\ c & d \end{bmatrix} \begin{bmatrix} x \\ y \end{bmatrix} = \begin{bmatrix} e \\ f \end{bmatrix}, \text{ where } a, b, c, d,$$

e, and *f* are real numbers is given by

$$\begin{bmatrix} x \\ y \end{bmatrix} = \begin{bmatrix} a & b \\ c & d \end{bmatrix}^{-1} \begin{bmatrix} e \\ f \end{bmatrix}, \text{ provided that}$$

the inverse matrix on the right side exists. The solution is, therefore, given by

$$\begin{bmatrix} x \\ y \end{bmatrix} = \begin{bmatrix} -1 & 0 \\ 2 & -1 \end{bmatrix}^{-1} = \begin{bmatrix} -2 \\ 1 \end{bmatrix}. \text{ Using}$$

the calculation for the inverse from Problem 908 yields the following solution:

$$\begin{bmatrix} x \\ y \end{bmatrix} = \begin{bmatrix} -1 & 0 \\ -2 & -1 \end{bmatrix} \begin{bmatrix} -2 \\ 1 \end{bmatrix} = \begin{bmatrix} 2 \\ 3 \end{bmatrix}$$

So, the solution of the system is $x = 2$, $y = 3$.

925. d. The solution to the matrix equation

$$\begin{bmatrix} a & b \\ c & d \end{bmatrix} \begin{bmatrix} x \\ y \end{bmatrix} = \begin{bmatrix} e \\ f \end{bmatrix}, \text{ where } a, b, c, d,$$

e, and *f* are real numbers is given by

$$\begin{bmatrix} x \\ y \end{bmatrix} = \begin{bmatrix} a & b \\ c & d \end{bmatrix}^{-1} \begin{bmatrix} e \\ f \end{bmatrix}, \text{ provided that}$$

the inverse matrix on the right side exists.

Note that since $\det \begin{bmatrix} 3 & 2 \\ 3 & 2 \end{bmatrix} = 0$, it follows

that $\begin{bmatrix} 3 & 2 \\ 3 & 2 \end{bmatrix}^{-1}$ does not exist. Therefore,

we cannot apply the approach. Rather, we must inspect the system to determine whether there is no solution (which happens if the two lines are parallel) or if there are infinitely many solutions (which happens if the two lines are identical). Subtracting the second equation from the first equation yields the false statement $0 = -3$. From this, we conclude that the two lines must be parallel (which can also be checked by graphing them). Hence, the system has no solution.

926. d. The solution to the matrix equation

$$\begin{bmatrix} a & b \\ c & d \end{bmatrix}\begin{bmatrix} x \\ y \end{bmatrix} = \begin{bmatrix} e \\ f \end{bmatrix},$$ where a, b, c, d,

e, and f are real numbers is given by

$$\begin{bmatrix} x \\ y \end{bmatrix} = \begin{bmatrix} a & b \\ c & d \end{bmatrix}^{-1}\begin{bmatrix} e \\ f \end{bmatrix},$$ provided that

the inverse matrix on the right side exists.

Note that since $\det\begin{bmatrix} 3 & -2 \\ 9 & -6 \end{bmatrix}$, it follows

that $\begin{bmatrix} 3 & -2 \\ 9 & -6 \end{bmatrix}^{-1}$ does not exist. Rather, we

must inspect the system to determine whether there is no solution (which happens if the two lines are parallel) or if there are infinitely many solutions (which happens if the two lines are identical). The second equation in the system is obtained by multiplying both sides of the first equation by 3. Therefore, the two lines are identical, so the system has infinitely many solutions.

927. c. The solution to the matrix equation

$$\begin{bmatrix} a & b \\ c & d \end{bmatrix}\begin{bmatrix} x \\ y \end{bmatrix} = \begin{bmatrix} e \\ f \end{bmatrix},$$ where a, b, c, d,

e, and f are real numbers is given by

$$\begin{bmatrix} x \\ y \end{bmatrix} = \begin{bmatrix} a & b \\ c & d \end{bmatrix}^{-1}\begin{bmatrix} e \\ f \end{bmatrix},$$ provided that

the inverse matrix on the right side exists. The solution is, therefore, given by

$$\begin{bmatrix} x \\ y \end{bmatrix} = \begin{bmatrix} -1 & -1 \\ -1 & 0 \end{bmatrix}^{-1}\begin{bmatrix} -1 \\ 1 \end{bmatrix}.$$ Using the

calculation for the inverse from Problem 911 yields the following solution:

$$\begin{bmatrix} x \\ y \end{bmatrix} = \begin{bmatrix} 0 & -1 \\ -1 & 1 \end{bmatrix}\begin{bmatrix} -1 \\ 1 \end{bmatrix} = \begin{bmatrix} -1 \\ 2 \end{bmatrix}$$

So, the solution of the system is $x = -1$, $y = 2$.

928. c. The solution to the matrix equation

$$\begin{bmatrix} a & b \\ c & d \end{bmatrix}\begin{bmatrix} x \\ y \end{bmatrix} = \begin{bmatrix} e \\ f \end{bmatrix},$$ where a, b, c, d,

e, and f are real numbers is given by

$$\begin{bmatrix} x \\ y \end{bmatrix} = \begin{bmatrix} a & b \\ c & d \end{bmatrix}^{-1}\begin{bmatrix} e \\ f \end{bmatrix},$$ provided that

the inverse matrix on the right side exists. The solution is therefore given by

$$\begin{bmatrix} x \\ y \end{bmatrix} = \begin{bmatrix} 0 & 2 \\ 4 & 0 \end{bmatrix}^{-1}\begin{bmatrix} 14 \\ -20 \end{bmatrix}.$$ Using the

calculation for the inverse from Problem 912 yields the following solution:

$$\begin{bmatrix} x \\ y \end{bmatrix} = \begin{bmatrix} 0 & \frac{1}{4} \\ \frac{1}{2} & 0 \end{bmatrix}\begin{bmatrix} 14 \\ -20 \end{bmatrix} = \begin{bmatrix} -5 \\ 7 \end{bmatrix}$$

So, the solution of the system is $x = -5$, $y = 7$.

Set 59 (Page 140)

929. c. First, rewrite the system as the following equivalent matrix equation as in Problem 913:

$$\begin{bmatrix} -3 & 7 \\ 1 & 5 \end{bmatrix} \begin{bmatrix} x \\ y \end{bmatrix} = \begin{bmatrix} 2 \\ 8 \end{bmatrix}.$$

Next, identify the following determinants to be used in the application of Cramer's rule:

$$D = \begin{vmatrix} -3 & 7 \\ 1 & 5 \end{vmatrix} = (-3)(5) - (1)(7) = -22$$

$$D_x = \begin{vmatrix} 2 & 7 \\ 8 & 5 \end{vmatrix} = (2)(5) - (8)(7) = -46$$

$$D_y = \begin{vmatrix} -3 & 2 \\ 1 & 8 \end{vmatrix} = (-3)(8) - (1)(2) = -26$$

So, from Cramer's rule, we have:

$$x = \frac{D_x}{D} = \frac{-46}{-22} = \frac{23}{11}$$

$$y = \frac{D_y}{D} = \frac{-26}{-22} = \frac{13}{11}$$

Thus, the solution is $x = \frac{23}{11}, y = \frac{13}{11}$.

930. b. First, rewrite the system as the following equivalent matrix equation as in Problem 882:

$$\begin{bmatrix} 1 & 0 \\ 0 & 1 \end{bmatrix} \begin{bmatrix} x \\ y \end{bmatrix} = \begin{bmatrix} a \\ b \end{bmatrix}.$$

Next, identify the following determinants to be used in the application of Cramer's rule:

$$D = \begin{vmatrix} 1 & 0 \\ 0 & 1 \end{vmatrix} = (1)(1) - (0)(0) = 1$$

$$D_x = \begin{vmatrix} a & 0 \\ b & 1 \end{vmatrix} = (a)(1) - (b)(0) = a$$

$$D_y = \begin{vmatrix} 1 & a \\ 0 & b \end{vmatrix} = (1)(b) - (0)(a) = b$$

So, from Cramer's rule, we have:

$$x = \frac{D_x}{D} = \frac{a}{1} = a$$

$$y = \frac{D_y}{D} = \frac{b}{1} = b$$

Thus, the solution is $x = a, y = b$.

931. b. First, rewrite the system as the following equivalent matrix equation as in Problem 915:

$$\begin{bmatrix} 1 & 2 \\ 2 & 3 \end{bmatrix} \begin{bmatrix} x \\ y \end{bmatrix} = \begin{bmatrix} 4 \\ 2 \end{bmatrix}.$$

Next, identify the following determinants to be used in the application of Cramer's rule:

$$D = \begin{vmatrix} 1 & 2 \\ 2 & 3 \end{vmatrix} = (1)(3) - (2)(2) = -1$$

$$D_x = \begin{vmatrix} 4 & 2 \\ 2 & 3 \end{vmatrix} = (4)(3) - (2)(2) = 8$$

$$D_y = \begin{vmatrix} 1 & 4 \\ 2 & 2 \end{vmatrix} = (1)(2) - (2)(4) = -6$$

So, from Cramer's rule, we have:

$$x = \frac{D_x}{D} = \frac{8}{-1} = -8$$

$$y = \frac{D_y}{D} = \frac{-6}{-1} = 6$$

Thus, the solution is $x = -8, y = 6$.

932. a. First, rewrite the system as the following equivalent matrix equation as in Problem 916:

$$\begin{bmatrix} 2 & 3 \\ 1 & 1 \end{bmatrix} \begin{bmatrix} x \\ y \end{bmatrix} = \begin{bmatrix} 1 \\ -2 \end{bmatrix}.$$

Next, identify the following determinants to be used in the application of Cramer's rule:

$$D = \begin{vmatrix} 2 & 3 \\ 1 & 1 \end{vmatrix} = (2)(1) - (1)(3) = -1$$

$$D_x = \begin{vmatrix} 1 & 3 \\ -2 & 1 \end{vmatrix} = (1)(1) - (-2)(3) = 7$$

$$D_y = \begin{vmatrix} 2 & 1 \\ 1 & -2 \end{vmatrix} = (2)(-2) - (1)(1) = -5$$

So, from Cramer's rule, we have:

$$x = \frac{D_x}{D} = \frac{7}{-1} = -7$$

$$y = \frac{D_y}{D} = \frac{-5}{-1} = 5$$

Thus, the solution is $x = -7, y = 5$.

933. d. First, rewrite the system as the following equivalent matrix equation as in Problem 917:

$$\begin{bmatrix} -1 & 2 \\ 2 & -4 \end{bmatrix} \begin{bmatrix} x \\ y \end{bmatrix} = \begin{bmatrix} 3 \\ -6 \end{bmatrix}.$$

Next, identify the following determinants to be used in the application of Cramer's rule:

$$D = \begin{vmatrix} -1 & 2 \\ 2 & -4 \end{vmatrix} = (-1)(-4) - (2)(2) = 0$$

$$D_x = \begin{vmatrix} 3 & 2 \\ -6 & -4 \end{vmatrix} = (3)(-4) - (-6)(2) = -24$$

$$D_y = \begin{vmatrix} -1 & 3 \\ 2 & -6 \end{vmatrix} = (-1)(-6) - (2)(3) = 0$$

Since applying Cramer's rule requires that we divide by D in order to determine x and y, we can conclude only that the system either has zero or infinitely many solutions. We must consider the equations directly and manipulate them to determine which is the case. To this end, as in Problem 917, we note that the second equation in the system is obtained by multiplying both sides of the first equation by -2. Therefore, the two lines are identical, so the system has infinitely many solutions.

934. d. First, rewrite the system as the following equivalent matrix equation as in Problem 918:

$$\begin{bmatrix} 6 & 3 \\ 2 & 1 \end{bmatrix} \begin{bmatrix} x \\ y \end{bmatrix} = \begin{bmatrix} 8 \\ 3 \end{bmatrix}.$$

Next, identify the following determinants to be used in the application of Cramer's rule:

$$D = \begin{vmatrix} 6 & 3 \\ 2 & 1 \end{vmatrix} = (6)(1) - (2)(3) = 0$$

$$D_x = \begin{vmatrix} 8 & 3 \\ 3 & 1 \end{vmatrix} = (8)(1) - (3)(3) = -1$$

$$D_y = \begin{vmatrix} 6 & 8 \\ 2 & 3 \end{vmatrix} = (6)(3) - (2)(8) = 2$$

Since applying Cramer's rule requires that we divide by D in order to determine x and y, we can only conclude that either the system has zero or infinitely many solutions. We must consider the equations directly and manipulate them to determine which is the case. To this end, as in Problem 918, note that multiplying both sides of the second equation by 3 yields the equivalent equation $6x + 3y = 9$. Subtracting this from the first equation yields the false statement $0 = -1$. From this, we conclude that the two lines must be parallel (which can also be checked by graphing them). Hence, the system has no solution.

935. a. First, rewrite the system as the following equivalent matrix equation as in Problem 919:

$$\begin{bmatrix} -3 & 4 \\ 4 & 2 \end{bmatrix} \begin{bmatrix} x \\ y \end{bmatrix} = \begin{bmatrix} 1 \\ -3 \end{bmatrix}.$$

Next, identify the following determinants to be used in the application of Cramer's rule:

$$D = \begin{vmatrix} -3 & 4 \\ 4 & 2 \end{vmatrix} = (-3)(2) - (4)(4) = -22$$

$$D_x = \begin{vmatrix} 1 & 4 \\ -3 & 2 \end{vmatrix} = (1)(2) - (-3)(4) = 14$$

$$D_y = \begin{vmatrix} -3 & 1 \\ 4 & -3 \end{vmatrix} = (-3)(-3) - (4)(1) = -5$$

So, from Cramer's rule, we have:

$$x = \frac{D_x}{D} = \frac{14}{-22} = -\frac{7}{11}$$

$$y = \frac{D_y}{D} = \frac{5}{-22} = -\frac{5}{22}$$

Thus, the solution is $x = -\frac{7}{11}, y = -\frac{5}{22}$.

936. c. First, rewrite the system as the following equivalent matrix equation as in Problem 920:

$$\begin{bmatrix} 1 & -4 \\ 0 & 25 \end{bmatrix} \begin{bmatrix} x \\ y \end{bmatrix} = \begin{bmatrix} -2 \\ 1 \end{bmatrix}.$$

Next, identify the following determinants to be used in the application of Cramer's rule:

$$D = \begin{vmatrix} 1 & -4 \\ 0 & 25 \end{vmatrix} = (1)(25) - (0)(-4) = 25$$

$$D_x = \begin{vmatrix} -2 & -4 \\ 1 & 25 \end{vmatrix} = (-2)(25) - (1)(-4) = -46$$

$$D_y = \begin{vmatrix} 1 & -2 \\ 0 & 1 \end{vmatrix} = (1)(1) - (0)(-2) = 1$$

So, from Cramer's rule, we have:

$$x = \frac{D_x}{D} = \frac{-46}{25}$$

$$y = \frac{D_y}{D} = \frac{1}{25}$$

Thus, the solution is $x = -\frac{46}{25}, y = \frac{1}{25}.$

937. c. First, rewrite the system as the following equivalent matrix equation as in Problem 921:

$$\begin{bmatrix} -3 & 1 \\ 1 & -2 \end{bmatrix} \begin{bmatrix} x \\ y \end{bmatrix} = \begin{bmatrix} 5 \\ -9 \end{bmatrix}.$$

Next, identify the following determinants to be used in the application of Cramer's rule:

$$D = \begin{vmatrix} -3 & 1 \\ 1 & -2 \end{vmatrix} = (-3)(-2) - (1)(1) = 5$$

$$D_x = \begin{vmatrix} 5 & 1 \\ -9 & -2 \end{vmatrix} = (5)(-2) - (-9)(1) = -1$$

$$D_y = \begin{vmatrix} -3 & 5 \\ 1 & -9 \end{vmatrix} = (-3)(-9) - (1)(5) = 22$$

So, from Cramer's rule, we have:

$$x = \frac{D_x}{D} = \frac{1}{25}$$

$$y = \frac{D_y}{D} = \frac{22}{5}$$

Thus, the solution is $x = -\frac{1}{5}, y = \frac{22}{5}.$

938. b. First, rewrite the system as the following equivalent matrix equation as in Problem 922:

$$\begin{bmatrix} 2 & 0 \\ 12 & -3 \end{bmatrix} \begin{bmatrix} x \\ y \end{bmatrix} = \begin{bmatrix} -2 \\ 4 \end{bmatrix}.$$

Next, identify the following determinants to be used in the application of Cramer's rule:

$$D = \begin{vmatrix} 2 & 0 \\ 12 & -3 \end{vmatrix} = (2)(-3) - (12)(0) = -6$$

$$D_x = \begin{vmatrix} -2 & 0 \\ 4 & -3 \end{vmatrix} = (-2)(-3) - (4)(0) = 6$$

$$D_y = \begin{vmatrix} 2 & -2 \\ 12 & 4 \end{vmatrix} = (2)(4) - (12)(-2) = 32$$

So, from Cramer's rule, we have:

$$x = \frac{D_x}{D} = \frac{6}{-6} = -1$$

$$y = \frac{D_y}{D} = \frac{32}{-6} = -\frac{16}{3}$$

Thus, the solution is $x = -1, y = -\frac{16}{3}.$

939. a. First, rewrite the system as the following equivalent matrix equation as in Problem 923:

$$\begin{bmatrix} 0 & 1 \\ -2 & -1 \end{bmatrix} \begin{bmatrix} x \\ y \end{bmatrix} = \begin{bmatrix} -4 \\ 0 \end{bmatrix}.$$

Next, identify the following determinants to be used in the application of Cramer's rule:

$$D = \begin{vmatrix} 0 & 1 \\ -2 & -1 \end{vmatrix} = (0)(-1) - (-2)(1) = 2$$

$$D_x = \begin{vmatrix} -4 & 1 \\ 0 & -1 \end{vmatrix} = (-4)(-1) - (0)(1) = 4$$

$$D_y = \begin{vmatrix} 2 & -4 \\ -2 & 0 \end{vmatrix} = (0)(0) - (-2)(-4) = -8$$

So, from Cramer's rule, we have:

$$x = \frac{D_x}{D} = \frac{4}{2} = 2$$

$$y = \frac{D_y}{D} = \frac{-8}{2} = -4$$

Thus, the solution is $x = 2, y = -4.$

940. b. Identify the following determinants to be used in the application of Cramer's rule for the matrix equation

$$\begin{bmatrix} -1 & 0 \\ 2 & -1 \end{bmatrix} \begin{bmatrix} x \\ y \end{bmatrix} = \begin{bmatrix} -2 \\ 1 \end{bmatrix}:$$

$$D = \begin{vmatrix} -1 & 0 \\ 2 & -1 \end{vmatrix} = (-1)(-1) - (2)(0) = 1$$

$$D_x = \begin{vmatrix} -2 & 0 \\ 1 & -1 \end{vmatrix} = (-2)(-1) - (1)(0) = 2$$

$$D_y = \begin{vmatrix} -1 & -2 \\ 2 & 1 \end{vmatrix} = (-1)(1) - (2)(-2) = 3$$

So, from Cramer's rule, we have:

$$x = \frac{D_x}{D} = \frac{2}{1} = 2$$

$$y = \frac{D_y}{D} = \frac{3}{1} = 3$$

Thus, the solution is $x = 2$, $y = 3$.

941. d. Identify the following determinants to be used in the application of Cramer's rule for the matrix equation

$$\begin{bmatrix} 3 & 2 \\ 3 & 2 \end{bmatrix} \begin{bmatrix} x \\ y \end{bmatrix} = \begin{bmatrix} -2 \\ 1 \end{bmatrix}:$$

$$D = \begin{vmatrix} 3 & 2 \\ 3 & 2 \end{vmatrix} = (3)(2) - (2)(3) = 0$$

$$D_x = \begin{vmatrix} -2 & 2 \\ 1 & 2 \end{vmatrix} = (-2)(2) - (1)(2) = -6$$

$$D_y = \begin{vmatrix} 3 & -2 \\ 3 & 1 \end{vmatrix} = (3)(1) - (3)(-2) = 9$$

Since applying Cramer's rule requires that we divide by D in order to determine x and y, we can only conclude that the system has either zero or infinitely many solutions. We must consider the equations directly and manipulate them to determine which is the case. To this end, as in Problem 925, subtracting the second equation from the

first equation yields the false statement $0 = -3$. From this, we conclude that the two lines must be parallel (which can also be checked by graphing them). Hence, the system has no solution.

942. d. Identify the following determinants to be used in the application of Cramer's rule for the matrix equation

$$\begin{bmatrix} 3 & -2 \\ 9 & -6 \end{bmatrix} \begin{bmatrix} x \\ y \end{bmatrix} = \begin{bmatrix} 4 \\ 12 \end{bmatrix}:$$

$$D = \begin{vmatrix} 3 & -2 \\ 9 & -6 \end{vmatrix} = (3)(-6) - (9)(-2) = 0$$

$$D_x = \begin{vmatrix} 4 & -2 \\ 12 & -6 \end{vmatrix} = (4)(-6) - (12)(-2) = 0$$

$$D_y = \begin{vmatrix} 3 & 4 \\ 9 & 12 \end{vmatrix} = (3)(12) - (9)(4) = 0$$

Since applying Cramer's rule requires that we divide by D in order to determine x and y, we can conclude only that the system has either zero or infinitely many solutions. We must consider the equations directly and manipulate them to determine which is the case. To this end, as in Problem 926, the second equation in the system is obtained by multiplying both sides of the first equation by 3. The two lines are identical, so the system has infinitely many solutions.

943. b. Identify the following determinants to be used in the application of Cramer's rule for the matrix equation

$$\begin{bmatrix} -1 & -1 \\ -1 & 0 \end{bmatrix} \begin{bmatrix} x \\ y \end{bmatrix} = \begin{bmatrix} -1 \\ 1 \end{bmatrix}:$$

$$D = \begin{vmatrix} -1 & -1 \\ -1 & 0 \end{vmatrix} = (-1)(0) - (-1)(-1) = -1$$

$$D_x = \begin{vmatrix} -1 & -1 \\ 1 & 0 \end{vmatrix} = (-1)(0) - (1)(-1) = 1$$

$$D_y = \begin{vmatrix} -1 & -1 \\ -1 & 1 \end{vmatrix} = (-1)(1) - (-1)(-1) = -2$$

So, from Cramer's rule, we have:

$$x = \frac{D_x}{D} = \frac{1}{-1} = -1$$

$$y = \frac{D_y}{D} = \frac{-2}{-1} = 2$$

Thus, the solution is $x = -1$, $y = 2$.

944. b. Identify the following determinants to be used in the application of Cramer's rule for the matrix equation

$$\begin{bmatrix} 0 & 2 \\ 4 & 0 \end{bmatrix} \begin{bmatrix} x \\ y \end{bmatrix} = \begin{bmatrix} 14 \\ -20 \end{bmatrix}:$$

$$D = \begin{vmatrix} 0 & 2 \\ 4 & 0 \end{vmatrix} = (0)(0) - (4)(2) = -8$$

$$D_x = \begin{vmatrix} 14 & 2 \\ -20 & 0 \end{vmatrix} = (14)(0) - (-20)(2) = 40$$

$$D_y = \begin{vmatrix} 0 & 14 \\ 4 & -20 \end{vmatrix} = (0)(-20) - (4)(14) = -56$$

So, from Cramer's rule, we have:

$$x = \frac{D_x}{D} = \frac{40}{-8} = -5$$

$$y = \frac{D_y}{D} = \frac{-56}{-8} = 7$$

Thus, the solution is $x = -5$, $y = 7$.

Section 8—
Common Algebra Errors

Set 61 (Page 144)

945. b. The answer should be $\frac{1}{9}$ because $(-3)^{-2} = \frac{1}{(-3) \cdot (-3)} = \frac{1}{9}$.

946. c. There is no error. Any nonzero quantity raised to the zero power is 1.

947. a. The statement should be $0.00013 = 1.3 \times 10^{-4}$ because the decimal point must move to the left four places in order to yield 0.00013.

948. c. There is no error. The power doesn't apply to the −1 in front of the 4. In order to square the entire −4, one must write $(-4)^2$.

949. a. This is incorrect because you cannot cancel terms in a sum; you can cancel only factors that are common to the numerator and denominator.

950. b. You must first get a common denominator before you add two fractions. The correct computation is: $\frac{3}{4} + \frac{a}{2} = \frac{3}{4} + \frac{2a}{4} = \frac{3+2a}{4}$

951. b. The placement of the quantities is incorrect. A correct statement would be "200% of 4 is 8."

952. c. There is no error. In order to compute 0.50% of 10, you multiply 10 by 0.0050 to get 0.05.

953. b. The sum $\sqrt{3} + \sqrt{6}$ cannot be simplified further because the radicands are different.

954. b. The first equality is incorrect; the radicals cannot be combined because their indices are different.

955. a. The third equality is incorrect because the binomial was not squared correctly. The correct denominator should be $2^2 + 2\sqrt{3} + (\sqrt{3})^2 = 7 + 2\sqrt{3}$.

956. a. The exponents should be multiplied, not added, so the correct answer should be x^{10}.

957. c. There is no error.

958. a. The first equality is wrong because you must multiply the numerator by the *reciprocal* of the denominator.

959. b. The correct answer should be x^{15} because $\frac{x^{12}}{x^{-3}} = x^{12}x^3 = x^{12+3} = x^{15}$.

960. a. The correct answer should be e^{8x} because $(e^{4x})^2 = e^{4x \cdot 2} = e^{8x}$.

Set 62 (Page 146)

961. a. The inequality sign must be switched when multiplying both sides by a negative real number. The correct solution set should be $(-\infty, -4)$.

962. a. There are two solutions of this equation, namely $x = -1$ and $x = 3$.

963. c. There is no error.

964. b. The value $x = -7$ cannot be the solution because it makes the terms in the original equation undefined—you cannot divide by zero. Therefore, this equation has no solution.

965. b. While $x = 1$ satisfies the original equation, $x = -1$ cannot because negative inputs into a logarithm are not allowed.

966. a. The denominator in the quadratic formula is $2a$, which in this case is 2, not 1. The complex solutions should be $x = \pm i\sqrt{5}$.

967. b. The signs used to define the binomials on the right side should be switched. The correct factorization is $(x - 7)(x + 3)$.

968. a. You must move all terms to one side of the inequality, factor (if possible), determine the values that make the factored expression equal to zero, and construct a sign chart to solve such an inequality. The correct solution set should be $[-2, 2]$.

969. a. The left side must be expanded by FOILing. The correct statement should be $(x - y)^2 = x^2 - 2xy + y^2$.

970. b. This equation has no real solutions because the output of a square root must be nonnegative.

971. c. There is no error.

972. b. The left side is not a difference of squares. It cannot be factored further.

973. b. You cannot cancel terms of a sum in the numerator and denominator. You can cancel only factors common to both. The complex fraction must first be simplified before any cancellation can occur. The correct statement is:

$$\frac{2x^{-1} - y^{-1}}{x^{-1} + 4y^{-1}} = \frac{\frac{2}{x} - \frac{1}{y}}{\frac{1}{x} + \frac{4}{y}} = \frac{\frac{2y}{xy} - \frac{x}{xy}}{\frac{y}{xy} + \frac{4x}{xy}} = \frac{\frac{2y - x}{xy}}{\frac{y + 4x}{xy}} =$$

$$\frac{2y - x}{xy} \cdot \frac{xy}{y + 4x} = \frac{2y - x}{y + 4x}$$

974. b. The first equality is incorrect because the natural logarithm of a sum is not the sum of the natural logarithms. In fact, the expression on the extreme left side of the string of equalities cannot be simplified further.

975. a. The first equality is incorrect because $2 \log_5 (5x) = \log_5(5x)^2 = \log_5(25x^2)$. The other equalities are correct as written.

976. c. There is no error.

Set 63 (Page 148)

977. b. The line $y = 0$ is the horizontal asymptote for f.

978. a. The line is vertical, so its slope is undefined.

979. a. The point is actually in Quadrant II.

980. b. The domain of f must be restricted to $[0, \infty)$ in order for f to have an inverse. In such case, the given function $f^{-1}(x) = \sqrt{x}$ is indeed its inverse.

981. c. There is no error.

982. c. There is no error.

983. c. There is no error.

984. a. The coordinates of the point that is known to lie on the graph of $y = f(x)$ are reversed; they should be $(2,5)$.

985. c. There is no error.

986. b. -2 is not in the domain of g, so that the composition is not defined at -2.

987. a. The point $(0,1)$ is the y-intercept, not the x-intercept, of f.

988. a. The graph of g is actually decreasing as x moves from left to right through the domain.

989. a. The graph of $y = f(x + 3)$ is actually obtained by shifting the graph of $y = f(x)$ to the left 3 units.

990. c. There is no error.

991. a. You cannot distribute a function across parts of a single input. As such, the correct statement should be $f(x - h) = (x - h)^4$.

992. a. The graph of $y = 5$ passes the vertical line test, so it represents a function. It is, however, not invertible.

Set 64 (Page 150)

993. a. Since adding the two equations results in the false statement $0 = 8$, there can be no solution of this system.

994. a. Since multiplying the first equation by -2 and then adding the two equations results in the true statement $0 = 0$, there are infinitely many solutions of this system.

995. c. There is no error.

996. a. The two matrices on the left side of the equality do not have the same dimension, so their sum is undefined.

997. b. The inner dimensions of the two matrices on the left side are not the same. Therefore, they cannot be multiplied.

998. b. The difference is computed in the wrong order. The correct statement should be

$$\det \begin{bmatrix} 4 & 2 \\ 1 & -1 \end{bmatrix} = (4)(-1) - (2)(1) = -6.$$

999. c. There is no error.

1000. a. You cannot add a 2×2 matrix and a real number because their dimensions are different. The sum is not well-defined.

1001. c. There is no error.

GLOSSARY

absolute value the absolute value of a is the distance between a and 0

addend a quantity that is added to another quantity. In the equation $x + 3 = 5$, x and 3 are addends

additive inverse the negative of a quantity

algebra the representation of quantities and relationships using symbols

algebraic equation an algebraic expression equal to a number or another algebraic expression, such as $x + 4 = -1$

algebraic expression one or more terms, at least one of which contains a variable. An algebraic expression may or may not contain an operation (such as addition or multiplication), but does not contain an equal sign

algebraic inequality an algebraic expression being compared to another algebraic expression, using $\neq$, $<$, $>$, $\leq$, or $\geq$, such as $x + 2 > 8$

base a number or variable that is used as a building block within an expression. In the term $3x$, x is the base. In the term 2^4, 2 is the base

binomial an algebraic expression that contains two terms, such as $(2x + 1)$

coefficient the numerical multiplier, or factor, of an algebraic term. In the term $3x$, 3 is the coefficient

composite number a number that has at least one other positive factor besides itself and 1, such as 4 or 10

constant a term, such as 3, that never changes value

coordinate plane a plane partitioned by an x-axis and a y-axis

cubic equation an equation in which the highest degree is 3. The equation $y = x^3 + x$ is cubic

decreasing function a function whose graph falls vertically from left to right

degree   The degree of a variable is its exponent. The degree of a polynomial is the highest degree of its terms. The degree of x^5 is 5, and the degree of $x^3 + x^2 + 9$ is 3.

distributive law   a term outside a set of parentheses that contains two terms should be multiplied by each term inside the parentheses: $a(b + c) = ab + ac$

dividend   the number being divided in a division problem (the numerator of a fraction). In the number sentence $6 \div 2 = 3$, 6 is the dividend

divisor   the number by which the dividend is divided in a division problem (the denominator of a fraction). In the number sentence $6 \div 2 = 3$, 2 is the divisor

domain   the set of all values that can be substituted for x in a function

equation   two expressions separated by an equal sign, such as $3 + 6 = 9$

exponent   a constant or variable that states the number of times a base must be multiplied by itself. In the term $3x^2$, 2 is the exponent

factor   If two or more whole numbers multiplied together yield a product, those numbers are factors of that product. Since $2 \times 4 = 8$, 2 and 4 are factors of 8.

factoring   breaking down a product into its factors

FOIL   an acronym that stands for First, Outer, Inner, Last, which are the pairs of terms that must be multiplied in order to find the terms of the product of two binomials. $(a + b)(c + d) = ac + ad + bc + bd$

function   an equation that associates a unique y-value to every x-value in its domain

greatest common factor   (GCF) the largest expression that can be factored out of every term in an expression

imaginary number   a number whose square is less than zero, such as the square root of -9, which can be written as $3i$

increasing function   a function whose graph rises vertically from left to right

inequality   two expressions that are compared using the symbol $\neq, <, >, \leq$, or $\geq$

integer   a whole number, the negative of a whole number, or zero. Examples of integers are 2 and -2

inverse functions   Two functions are inverses of each other if and only if each composed with the other yields the identity function ($y = x$). The graphs of inverse functions are reflections of each other over the line $y = x$.

like terms   two or more terms that have the same variable bases raised to the same exponents, but may have different coefficients, such as $3x^2$ and $10x^2$ or $7xy$ and $10xy$

linear equation   an equation that can contain constants and variables, and the exponents of the variables are 1; the equation $y = 3x + 8$ is linear

matrix   an array of real numbers composed of m rows and n columns

monomial   an expression that consists of products of powers of variables, such as $3x^2$

ordered pair   an x-value and a y-value, in parentheses separated by a comma, such as (4,2)

parallel lines   lines that have the same slope. Parallel lines never intersect

percent   p% = p/100. The expression 36% is equal to 36 out of 100.

perpendicular lines   lines that intersect at right angles; the product of the slopes of two perpendicular lines is -1

polynomial   an expression that is the sum of one or more terms, such as $x^2 + 2x + 1$, each with whole numbered exponents

prime factorization   the writing of a number as a product comprised of only prime numbers

prime number   a number whose only positive factors are 1 and itself, such as 3 or 7

product   the result of multiplication. In the number sentence $2 \times 4 = 8$, 8 is the product

proportion   an equation that shows two equal ratios, such as $\frac{16}{12} = \frac{4}{3}$

quadratic equation   an equation in which the highest degree is 2; the equation $y = x^2 + 1$ is a quadratic equation

radical   a root of a quantity

radicand   the quantity under a radical symbol; in $\sqrt[3]{8}$, 8 is the radicand

range   the set of all y-values that can be generated from x-values in an equation or function

ratio   a relationship between two or more quantities, such as 3:2

rational expression   the quotient of two polynomials

root   a value of x in the domain of a function for which $f(x)$ is 0

slope   a measurement of steepness of a line computed as follows: the change in the y-values between two points on a line divided by the change in the x-values of those points

slope-intercept form $y = mx + b$, where m is the slope of the line and b is the y-intercept

system of equations   a group of two or more equations for which the common variables in each equation have the same values

quotient   the result of division; in the number sentence $6 \div 2 = 3$, 3 is the quotient

term   a variable, constant, or product of both, with or without an exponent, that is usually separated from another term by addition, subtraction, or an equal sign, such as $2x$ or 5 in the expression $(2x + 5)$

trinomial   an expression comprised of three terms, such as $(6x^2 + 11x + 4)$

unknown   a quantity whose value is not given, usually represented by a letter

unlike terms   two or more terms that have different variable bases, or two or more terms with identical variables raised to different exponents, such as $3x^2$ and $4x^4$

variable   a symbol, such as x, that represents a number

vertical line test   the drawing of a vertical line through the graph of an equation to determine if the equation is a function; if a vertical line can be drawn anywhere through the graph of an equation such that it crosses the graph more than once, then the equation is not a function

x-axis   the horizontal line on a coordinate plane along which $y = 0$

x-intercept   the x-value of a point where a curve crosses the x-axis

y-axis   the vertical line on a coordinate plane along which $x = 0$

y-intercept   the y-value of the point where a curve crosses the y-axis

Using the codes below, you'll be able to log in and access additional online practice materials!

Your free online practice access codes are:

FVE41EA41WI2FMQJBN2B

FVE2D3GV64135BOLRQ5R

FVE7APTS01QOQ5LC4Q2Q

FVE860EQQSO161IJ6V6E

Follow these simple steps to redeem your codes:

- Go to **www.learningexpresshub.com/affiliate** and have your access code handy.

If you're a new user:

- Click the **New user? Register here** button and complete the registration form to create your account and access your products.
- Be sure to enter your unique access code only once. If you have multiple access codes, you can enter them all—just use a comma to separate each code.
- The next time you visit, simply click the **Returning user? Sign in** button and enter your username and password.
- Do not re-enter previously redeemed access codes. Any products you previously accessed are saved in the **My Account** section on the site. Entering a previously redeemed access code will result in an error message.

If you're a returning user:

- Click the **Returning user? Sign in** button, enter your username and password, and click **Sign In**.
- You will automatically be brought to the **My Account** page to access your products.
- Do not re-enter previously redeemed access codes. Any products you previously accessed are saved in the **My Account** section on the site. Entering a previously redeemed access code will result in an error message.

If you're a returning user with new access codes:

- Click the **Returning user? Sign in** button, enter your username, password, and new access code, and click **Sign In**.
- If you have multiple access codes, you can enter them all—just use a comma to separate each code.
- Do not re-enter previously redeemed access codes. Any products you previously accessed are saved in the **My Account** section on the site. Entering a previously redeemed access code will result in an error message.

If you have any questions, please contact LearningExpress Customer Support at LXHub@LearningExpressHub.com. All inquiries will be responded to within a 24-hour period during normal business hours: 9:00 A.M.–5:00 P.M. Eastern Time. Thank you!